NEUROMETHODS

Series Editor
Wolfgang Walz
University of Saskatchewan
Saskatoon, SK, Canada

For further volumes:
http://www.springer.com/series/7657

Neuromethods publishes cutting-edge methods and protocols in all areas of neuroscience as well as translational neurological and mental research. Each volume in the series offers tested laboratory protocols, step by step methods for reproducible lab experiments and addresses methodological controversies and pitfalls in order to aid neuroscientists in experimentation. *Neuromethods* focuses on traditional and emerging topics with wide ranging implications to brain function, such as electrophysiology, neuroimaging, behavioral analysis, genomics, neurodegeneration, translational research and clinical trials. *Neuromethods* provides investigators and trainees with highly useful compendiums of key strategies and approaches for successful research in animal and human brain function including translational "bench to bedside" approaches to mental and neurological diseases.

Nanotherapy for Brain Tumor Drug Delivery

Edited by

Vivek Agrahari

CONRAD / Eastern Virginia Medical School, Arlington, VA, USA

Anthony Kim

Department of Neurosurgery and Pharmacology, University of Maryland School of Medicine, Baltimore, MD, USA

Vibhuti Agrahari

Department of Pharmaceutical Sciences, College of Pharmacy, University of Oklahoma Health Science Center, Oklahoma City, OK, USA

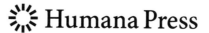 Humana Press

Editors
Vivek Agrahari
CONRAD / Eastern Virginia Medical
School
Arlington, VA, USA

Anthony Kim
Department of Neurosurgery and Pharmacology
University of Maryland School of Medicine
Baltimore, MD, USA

Vibhuti Agrahari
Department of Pharmaceutical
Sciences, College of Pharmacy
University of Oklahoma Health Science Center
Oklahoma City, OK, USA

ISSN 0893-2336   ISSN 1940-6045   (electronic)
Neuromethods
ISBN 978-1-0716-1054-1   ISBN 978-1-0716-1052-7   (eBook)
https://doi.org/10.1007/978-1-0716-1052-7

This Humana imprint is published by the registered company Springer Science+Business Media, LLC, part of Springer Nature.
The registered company address is: 1 New York Plaza, New York, NY 10004, U.S.A.

Preface to the Series

life sciences have two basic foundations: concepts and tools. The *Neuromethods* series focuses on the tools and techniques unique to the investigation of the nervous system and excitable cells. It will not, however, shortchange the concept side of things as care has been taken to integrate these tools within the context of the concepts and questions under investigation. In this way, the series is unique in that it not only collects protocols but also includes theoretical background information and critiques which led to the methods and their development. Thus it gives the reader a better understanding of the origin of the techniques and their potential future development. The *Neuromethods* publishing program strikes a balance between recent and exciting developments like those concerning new animal models of disease, imaging, in vivo methods, and more established techniques, including, for example, immunocytochemistry and electrophysiological technologies. New trainees in neurosciences still need a sound footing in these older methods in order to apply a critical approach to their results.

Under the guidance of its founders, Alan Boulton and Glen Baker, the *Neuromethods* series has been a success since its first volume published through Humana Press in 1985. The series continues to flourish through many changes over the years. It is now published under the umbrella of Springer Protocols. While methods involving brain research have changed a lot since the series started, the publishing environment and technology have changed even more radically. *Neuromethods* has the distinct layout and style of the Springer Protocols program, designed specifically for readability and ease of reference in a laboratory setting.

The careful application of methods is potentially the most important step in the process of scientific inquiry. In the past, new methodologies led the way in developing new disciplines in the biological and medical sciences. For example, Physiology emerged out of Anatomy in the nineteenth century by harnessing new methods based on the newly discovered phenomenon of electricity. Nowadays, the relationships between disciplines and methods are more complex. Methods are now widely shared between disciplines and research areas. New developments in electronic publishing make it possible for scientists that encounter new methods to quickly find sources of information electronically. The design of individual volumes and chapters in this series takes this new access technology into account. Springer Protocols makes it possible to download single protocols separately. In addition, Springer makes its print-on-demand technology available globally. A print copy can therefore be acquired quickly and for a competitive price anywhere in the world.

Saskatoon, SK, Canada   *Wolfgang Walz*

Preface

Brain cancer is the leading cause of cancer-related deaths in patients younger than 35 years [1]. Glioblastoma also known as glioblastoma multiforme (GBM) accounts for ~40% of primary brain tumors, taking more than 15,000 lives in the USA every year [1]. GBM tumors are characterized by extensive vascularization, a high mitotic index, cellular pleomorphism, genomic instability, and tissue necrosis. GBM is not curable with surgery alone because tumor cells invade into the surrounding brain, rendering complete resection unfeasible without major brain injury. The current standard of care for patients with GBM consists of surgery for maximal resection or biopsy followed by radiation and chemotherapy [2, 3]. Without treatment, most patients live fewer than 6 months. With the most aggressive combination therapies, the mean survival is still less than 18 months, often with devastating neurological consequences. Importantly, the survival statistics for patients with GBM have improved only incrementally during the last 30 years [4]. The limited progress and few recent advances strongly underscore the importance of developing new innovative therapeutic approaches and testing them in advanced animal models.

This *Neuromethods* book series entitled *Nanotherapy for Brain Tumor Drug Delivery* provides a guide on nanoformulations and other drug delivery approaches to both academic and industry scientists. The book includes a total of 12 chapters from the leading experts in the field, covering diverse research topics, methods, and protocols, in nanoparticle drug delivery, gene therapy, neurosurgical brain implant, exosomes, MRI-guided focused ultrasound (MRgFUS), and advanced preclinical GBM animal models.

The book begins with the chapter from **Dr. Giovanni Tosi,** who describes the state-of-the art and innovative nanomedicines for glioblastoma. **Dr. Ibrahim M. El-Sherbiny** reports passive and active targeting approaches for brain tumor treatment. This includes the methods of preparation of carrier systems, their characterizations, and in vitro/in vivo testing. **Dr. Paula Schiapparelli** develops strategies to modulate the blood–brain barrier for directed brain tumor targeting and provides an overview of different intended strategies to circumvent the blood–brain barrier. **Dr. Vibhuti Agrahari** describes the current and future prospects of nanobiotechnology strategies and briefly discusses the associated imaging techniques for brain tumor targeted therapy. **Dr. Aniket Wadajkar** and **Dr. Anthony Kim** report advanced surface-modified nanoparticle drug carriers for brain cancer treatment and discuss some of the cell surface molecules employed for targeting nano-drug carriers to GBM. **Dr. Ajeet Kaushik** reports inorganic nanostructures for brain tumor treatment, the advanced characterization techniques, and the related challenges and future prospects. **Dr. Jung Soo Suk** reports different strategies to enhance the distribution of nanotherapeutics in the brain and the approaches to maximize the convection-enhanced delivery (CED)-mediated therapeutic and/or nanoparticle delivery to the brain. **Dr. Graeme Woodworth** describes focused ultrasound (FUS)-mediated blood–brain barrier disruption for enhanced drug delivery to brain tumors. They also provide a summary of recent advances in strategies for detecting, controlling, and mapping FUS activity in the brain. **Dr. Betty Tyler** describes a neurosurgical implant-based strategy for brain cancer therapy and provides the perspective on the preclinical safety, efficacy, clinical trials, and subsequent FDA approval of this novel delivery system. **Dr. Jiangbing Zhou** uses novel gene therapy delivery approaches to treat brain cancer and confirms the potential of liposome-template hydrogel nanoparticles (LHNPs) as a powerful CRISPR/Cas9 delivery vehicle. **Dr. Venkat Nadithe** uses exosomes as a new approach to treat the CNS tumors and focuses on describing exosome isolation, characterization, and drug loading methods for treating glioblastoma. **Dr. Carla Vitorino** describes the protocol development for establishing the xenograft glioblastoma models for use in preclinical development.

In summary, this book brings forth the mix of basic principles and research advancements of brain tumor targeted drug delivery approaches, making an integrated scientific approach to benefit the scientific community working in this important therapeutic area, one of the most serious threat to human life. We hope that this *Neuromethods* book series will help accelerate the clinical translation of revolutionary nanotechnologies for treating brain tumors, to solve some of our most difficult clinical problems.

Arlington, VA, USA   *Vivek Agrahari*
Baltimore, MD, USA   *Anthony Kim*
Oklahoma City, OK, USA   *Vibhuti Agrahari*

References

1. Wen PY, Kesari S. Malignant gliomas in adults. N Engl J Med. 2008; 359(5):492–507

2. Stupp R, Mason WP, van den Bent MJ, Weller M, Fisher B, Taphoorn MJ, et al. Radiotherapy plus concomitant and adjuvant temozolomide for glioblastoma. N Engl J Med. 2005;352(10):987–96

3. Nabors LB, Portnow J, Ammirati M, Brem H, Brown P, Butowski N, et al. Central nervous system cancers, version 2.2014. Featured updates to the NCCN Guidelines. J Natl Compr Canc Netw. 2014;12(11):1517–23. PMCID: 4337873

4. Woodworth GF, Dunn GP, Nance EA, Hanes J, Brem H. Emerging insights into barriers to effective brain tumor therapeutics. Front Oncol. 2014;4:126. PMCID: 4104487

Contents

Contributors

VIBHUTI AGRAHARI • *Department of Pharmaceutical Sciences, College of Pharmacy, University of Oklahoma Health Science Center, Oklahoma City, OK, USA*

AWAIS AHMED ABRAR AHMED • *Emergency Medical Services Department, Faculty of Applied Medical Sciences, Jazan University, Jazan, Kingdom of Saudi Arabia*

RAWAN AL-KHARBOOSH • *Department of Neurosurgery, Mayo Clinic, Jacksonville, FL, USA; Mayo Clinic College of Medicine, Mayo Clinic Graduate School, Neuroscience, Rochester, MN, USA*

PAVLOS ANASTASIADIS • *Marlene and Stewart Greenebaum Comprehensive Cancer Center, University of Maryland School of Medicine, Baltimore, MD, USA; Department of Neurosurgery, University of Maryland School of Medicine, Baltimore, MD, USA*

JOÃO BASSO • *Faculty of Pharmacy, University of Coimbra, Pólo das Ciências da Saúde, Azinhaga de Santa Comba, Coimbra, Portugal; Centre for Neurosciences and Cell Biology (CNC), University of Coimbra, Faculty of Medicine, Coimbra, Portugal*

MARK BELL • *Bernard J. Dunn School of Pharmacy, Shenandoah University, Winchester, VA, USA*

CHRISTINE P. CARNEY • *Marlene and Stewart Greenebaum Comprehensive Cancer Center, University of Maryland School of Medicine, Baltimore, MD, USA; Department of Neurosurgery, University of Maryland School of Medicine, Baltimore, MD, USA*

JOSHUA CASAOS • *Department of Neurosurgery, Johns Hopkins University, Baltimore, MD, USA*

MIGUEL CASTELO-BRANCO • *Coimbra Institute for Biomedical Imaging and Translational Research (CIBIT), Edifício do ICNAS, Polo 3, Azinhaga de Santa Comba, Coimbra, Portugal; Institute of Nuclear Science Applied to Health (ICNAS), University of Coimbra, Polo 3, Azinhaga de Santa Comba, Coimbra, Portugal*

ANN T. CHEN • *Department of Neurosurgery, Yale University, New Haven, CT, USA*

ZEMING CHEN • *Department of Neurosurgery, Yale University, New Haven, CT, USA*

NINA P. CONNOLLY • *Marlene and Stewart Greenebaum Comprehensive Cancer Center, University of Maryland School of Medicine, Baltimore, MD, USA; Department of Neurosurgery, University of Maryland School of Medicine, Baltimore, MD, USA*

HONGGANG CUI • *Department of Chemical and Biomolecular Engineering and Institute for Nanobiotechnology, The Johns Hopkins University, Baltimore, MD, USA*

FEDERICA DA ROS • *Nanotech Lab, Te.Far.T.I., Life Sciences Department, University of Modena and Reggio Emilia, Modena, Italy*

JASON T. DUSKEY • *Hospital Pharmacy Unit, Azienda USL di Modena, Azienda Ospedaliera—Universitaria di Modena, Modena, Italy*

IBRAHIM M. EL-SHERBINY • *Center for Materials Science, University of Science and Technology (UST), Zewail City of Science and Technology, Giza, Egypt*

FLAVIO FORNI • *Hospital Pharmacy Unit, Azienda USL di Modena, Azienda Ospedaliera—Universitaria di Modena, Modena, Italy*

ANA FORTUNA • *Faculty of Pharmacy, University of Coimbra, Pólo das Ciências da Saúde, Azinhaga de Santa Comba, Coimbra, Portugal; Centre for Neurosciences and Cell Biology (CNC), University of Coimbra, Faculty of Medicine, Coimbra, Portugal; Coimbra*

Institute for Biomedical Imaging and Translational Research (CIBIT), Edifício do ICNAS, Polo 3, Azinhaga de Santa Comba, Coimbra, Portugal

NOAH GORELICK • *Department of Neurosurgery, Johns Hopkins University, Baltimore, MD, USA*

AMR HEFNAWY • *Center for Materials Science, University of Science and Technology (UST), Zewail City of Science and Technology, Giza, Egypt*

MOHD IMRAN • *Department of Chemical Engineering, Faculty of Engineering, Jazan University, Jazan, Kingdom of Saudi Arabia*

PRANJALI P. KANVINDE • *Marlene and Stewart Greenebaum Comprehensive Cancer Center, University of Maryland School of Medicine, Baltimore, MD, USA; Department of Neurosurgery, University of Maryland School of Medicine, Baltimore, MD, USA*

BABAK KATEB • *National Center for NanoBioElectronics, West Hollywood, CA, USA; California Neurosurgical Institute, Los Angeles, CA, USA; Brain Mapping Foundation, West Hollywood, CA, USA; Society for Brain Mapping and Therapeutics, West Hollywood, CA, USA*

SANDEEP KAUR • *Manchester University, Fort Wayne, IN, USA*

AJEET KAUSHIK • *Center for Personalized Nanomedicine, Institute of NeuroImmune Pharmacology, Department of Immunology and NanoMedicine, Herbert Wertheim College of Medicine, Florida International University, Miami, FL, USA; NanoBioTech Laboratory, Department of Natural Sciences, Division of Sciences, Art, & Mathematics, Florida Polytechnic University, Lakeland, FL, USA*

NAMIR KHALASAWI • *Department of Chemical & Biomolecular Engineering, Johns Hopkins University, Baltimore, MD, USA*

ANTHONY J. KIM • *Marlene and Stewart Greenebaum Comprehensive Cancer Center, University of Maryland School of Medicine, Baltimore, MD, USA; Department of Neurosurgery, University of Maryland School of Medicine, Baltimore, MD, USA; Department of Pharmacology, University of Maryland School of Medicine, Baltimore, MD, USA; Department of Pharmaceutical Sciences, University of Maryland School of Pharmacy, Baltimore, MD, USA*

MONTSERRAT LARA-VELAZQUEZ • *Department of Neurosurgery, Mayo Clinic, Jacksonville, FL, USA; School of Medicine, PECEM from the National Autonomous University of Mexico, Mexico City, Mexico*

VAIBHAV MUNDRA • *Manchester University, Fort Wayne, IN, USA*

VENKATAREDDY NADITHE • *St. Louis College of Pharmacy, St. Louis, MO, USA*

KARINA NEGRON • *Center for Nanomedicine, Wilmer Eye Institute, Johns Hopkins University School of Medicine, Baltimore, MD, USA; Department of Pharmacology & Molecular Sciences, Johns Hopkins University School of Medicine, Baltimore, MD, USA; Brain Trauma Neuroprotection and Neurorestoration Branch, Center for Military Psychiatry and Neuroscience, Walter Reed Army Institute of Research, Silver Spring, MD, USA*

YEN NGUYEN • *St. Louis College of Pharmacy, St. Louis, MO, USA*

NATALIA ODDONE • *Hospital Pharmacy Unit, Azienda USL di Modena, Azienda Ospedaliera—Universitaria di Modena, Modena, Italy*

ILARIA OTTONELLI • *Hospital Pharmacy Unit, Azienda USL di Modena, Azienda Ospedaliera—Universitaria di Modena, Modena, Italy*

ALFREDO QUINONES-HINOJOSA • *Department of Neurosurgery, Mayo Clinic, Jacksonville, FL, USA*

CHRISTINE PUJOL ROOKS • *Bernard J. Dunn School of Pharmacy, Shenandoah University, Winchester, VA, USA*

BARBARA RUOZI • *Hospital Pharmacy Unit, Azienda USL di Modena, Azienda Ospedaliera—Universitaria di Modena, Modena, Italy*

PAULA SCHIAPPARELLI • *Department of Neurosurgery, Mayo Clinic, Jacksonville, FL, USA*

JOSÉ SERENO • *Coimbra Institute for Biomedical Imaging and Translational Research (CIBIT), Edifício do ICNAS, Polo 3, Azinhaga de Santa Comba, Coimbra, Portugal; Institute of Nuclear Science Applied to Health (ICNAS), University of Coimbra, Polo 3, Azinhaga de Santa Comba, Coimbra, Portugal*

VALERIA SGARBI • *Hospital Pharmacy Unit, Azienda USL di Modena, Azienda Ospedaliera—Universitaria di Modena, Modena, Italy*

HAE SHIM • *St. Louis College of Pharmacy, St. Louis, MO, USA*

HAO SU • *Department of Chemical and Biomolecular Engineering and Institute for Nanobiotechnology, The Johns Hopkins University, Baltimore, MD, USA*

JUNG SOO SUK • *Center for Nanomedicine, Wilmer Eye Institute, Johns Hopkins University School of Medicine, Baltimore, MD, USA; Department of Chemical & Biomolecular Engineering, Johns Hopkins University, Baltimore, MD, USA; Department of Ophthalmology, Johns Hopkins University School of Medicine, Baltimore, MD, USA*

GIOVANNI TOSI • *Hospital Pharmacy Unit, Azienda USL di Modena, Azienda Ospedaliera—Universitaria di Modena, Modena, Italy; Department of Life Sciences, University of Modena and Reggio Emilia, Modena, Italy*

BETTY TYLER • *Department of Neurosurgery, Johns Hopkins University, Baltimore, MD, USA*

MARIA ANGELA VANDELLI • *Hospital Pharmacy Unit, Azienda USL di Modena, Azienda Ospedaliera—Universitaria di Modena, Modena, Italy*

CARLA VITORINO • *Faculty of Pharmacy, University of Coimbra, Pólo das Ciências da Saúde, Azinhaga de Santa Comba, Coimbra, Portugal; Centre for Neurosciences and Cell Biology (CNC), University of Coimbra, Faculty of Medicine, Coimbra, Portugal; Coimbra Chemistry Centre, Department of Chemistry, University of Coimbra, Coimbra, Portugal*

ANIKET S. WADAJKAR • *Marlene and Stewart Greenebaum Comprehensive Cancer Center, University of Maryland School of Medicine, Baltimore, MD, USA; Department of Neurosurgery, University of Maryland School of Medicine, Baltimore, MD, USA*

JEFFREY A. WINKLES • *Marlene and Stewart Greenebaum Comprehensive Cancer Center, University of Maryland School of Medicine, Baltimore, MD, USA; Department of Surgery, University of Maryland School of Medicine, Baltimore, MD, USA; Center for Vascular and Inflammatory Diseases, University of Maryland School of Medicine, Baltimore, MD, USA*

GRAEME F. WOODWORTH • *Marlene and Stewart Greenebaum Comprehensive Cancer Center, University of Maryland School of Medicine, Baltimore, MD, USA; Department of Neurosurgery, University of Maryland School of Medicine, Baltimore, MD, USA*

JIANGBING ZHOU • *Department of Neurosurgery, Yale University, New Haven, CT, USA*

Chapter 1

Glioblastoma: State of the Art of Treatments and Applications of Polymeric and Lipidic Nanomedicines

Valeria Sgarbi, Jason T. Duskey, Ilaria Ottonelli, Federica Da Ros, Natalia Oddone, Maria Angela Vandelli, Flavio Forni, Giovanni Tosi, and Barbara Ruozi

Abstract

Glioblastoma multiforme (GBM) is one of the most devastating tumors affecting more than 5 in 100,000 people. Unfortunately, its diagnosis is often discovered in late stages and is normally deadly, having a life expectancy of 12–15 months and a mere 3% of the affected patients living 3 years or more independent of race, sex, and age. Sadly, current treatments (i.e., chemotherapy, radiation, surgery) are extremely aggressive and extend the patient's life by little more than a year on average. Even when treatment appears successful, relapse is often experienced. These extreme treatments, combined with their lack of long-term success, call for new innovations. Among them, nanomedicine becomes one of the most promising approaches regarding possible applications in advancing or ameliorating GBM management. In this chapter, we will therefore analyze the state of the art and the most novel and outstanding innovation in terms of diagnosis and treatment options.

Key words Glioblastoma, Cancer, Brain, Treatment, Diagnosis, Nanomedicine, Innovation

1 Brain Tumors

Brain tumors are among the most aggressive forms of cancer ever known. Despite the progress in the current clinical approaches, the outcome of these diseases is often unfortunately fatal. The modern literature deeply investigated on the mechanism of tumor cell development and their features. Together with identification of novel targets or drugs, more effective than the current ones, a scenario to new therapies could be open. Cancer development follows a multistep process, which involves all genetic alteration that leads the normal cell to become a cancerous cell [1]. According to Hanahan and Weinberg, six physiological aspects are identified to be able to induce cancer [2]:

Vivek Agrahari et al. (eds.), *Nanotherapy for Brain Tumor Drug Delivery*, Neuromethods, vol. 163, https://doi.org/10.1007/978-1-0716-1052-7_1, © Springer Science+Business Media, LLC, part of Springer Nature 2021

1. Self-sufficiency in producing cell growth signals.
2. Insensitivity to cell proliferation inhibitors.
3. Evasion from programmed cell death mechanisms.
4. Uncontrolled replication.
5. Ability to induce angiogenesis (growth of new vascular tissue capable of nourishing the tumor).
6. Ability to invade tissues.

To which two other characteristics have been added:

1. Altered cellular metabolism that can support the growth of the cancerous cell.
2. Ability to escape the immune response.

Understanding the mechanisms of carcinogenesis allows preventing the development of tumors which is connected to an uncontrolled cellular replication due to the **increase of proto-oncogenic factors** (promoting cell proliferation) or by the **decrease of tumor suppressor genes** (factors inhibiting cell proliferation).

Central nervous system (CNS) tumors often have significant histopathological heterogeneity: for this reason, to harmonize the identification and finally establish a universal classification, the WHO firstly attempts to propose a classification of tumors in 1957 [3]. Now WHO classifies brain tumors into **four types (ICD-O, International Classification of Diseases for Oncology)**, depending on the degree of malignancy, which is based on the recognition of anaplasia by histological analysis (nuclear atypia, cellular pleomorphism, mitotic activity, endothelial hyperplasia and angiogenesis, necrosis, clinical features, and response to therapies).

1. **Grade I or pilocytic astrocytoma**, characterized by slow growth, absence of malignancy, and associated with long survival. Possibility of healing by surgery.
2. **Grade II or low-grade astrocytoma**, characterized by relatively slow growth, either malignant or nonmalignant; it may relapse to higher-grade tumors (e.g., low-grade astrocytoma that can turn into an anaplastic astrocytoma).
3. **Grade III or anaplastic astrocytoma**, malignant; it relapsed to tumors of higher grade. They present nuclear atypia and very fast mitotic activity. They are usually treated with radiotherapy and chemotherapy.
4. **Grade IV or glioblastoma multiforme** (65% of gliomas), rapid growth, very aggressive malignancy, poor survival expectancy. It presents extensive necrosis and evolutionary capacity both pre· and post-intervention.

2 Glioblastoma

2.1 Incidence

Up to today, 240,000 cases of brain and CNS tumors are estimated worldwide each year. In many European countries, 2–3 cases of glioblastoma (GBM) per 100,000 people per year are diagnosed [4]. Based on data provided by the Central Brain Tumor Registry of the United States (CBTRUS), more than 3 new cases of primary malignant brain tumor per 100,000 people are diagnosed each year in the United States, and 80% of these are malignant gliomas. The average age of diagnosis is 64 years. This index is the highest among all CNS malignancies, followed by second-degree astrocytomas (0.56/100,000) and other nonspecific malignant gliomas (0.46/ 100,000) [5]:

1. **Age**: it is mainly diagnosed in people of average age of 64 years. It is uncommon in children (3%) and the risk increases in elderly people (there are peaks between 75 and 84 years).

2. **Gender**: the risk of developing GBM is 1.6 times greater in men than in women. Furthermore, the development of primary GBM is more common in men, while the risk of developing secondary GBM is higher in women.

3. **Ethnic group:** the highest incidence is found in the white race, followed by the black race, the Asians and the inhabitants of the Pacific islands, and finally by the Native Americans. From 2006 to 2010, it was found that whites have a risk index two times greater than blacks to contract GBM.

Life expectancy for those affected by GBM does not exceed 12–15 months. If not treated, expectations do not exceed 3 months of life. The incidence in the diagnosis has increased compared to the last 20 years, thanks to the progress of diagnosis by radiological means [3].

GBM is the most aggressive and common form of malignant gliomas of the astrocytic line [5]: it corresponds to Grade IV brain tumors according to WHO classification, and it is incurable in almost cases [6]. The most common symptoms include increasingly aggressive headaches accompanied by vomiting and nausea (due to increased intracranial pressure), neurological deficits, confusion, memory loss, visual impairment, motor dysfunction and paralysis, changing personality, and seizures [1]. The percentage of those who can survive for about 5 years is 33.3%.

2.2 Etiology

Etiologically, some causes of tumor emergence were recognized [7] such as primary exposure to radiological and x-ray therapy, pesticides, smoke, oil refining products, or direct exposure to synthetic rubber production.

In a small subgroup of people with gliomas (5%), the tumor is associated with hereditary syndromes such as neurofibromatosis, tuberous sclerosis, and Li-Fraumeni syndrome. Genomic studies (genome-wide association studies [GWAS]) have identified the presence of 8 SNPs (DNA single-nucleotide polymorphism) related to the increasing in the risk of developing glioma on 7 genes [8]. The others represent sporadic cases.

Glioblastoma was one of the most studied cancers by the Cancer Genome Atlas Research Network, which identified recurrent alterations in three major cell-signaling pathways associated with carcinogenesis [5]:

1. **MGMT:** O^6-methylguanine-DNA-methyltransferase **(MGMT)** *is a DNA repairing protein that removes alkyl groups from the O^6 of guanine and makes the cells resistant to the alkylation of DNA by temozolomide (TMZ).* When methylated, the promoter for MGMT is silenced, leading to increased sensitivity of the tumor cell to TMZ. The MGMT promoter is methylated in 50% of new GBM cases, and it is associated with tumors that carry IDH1 mutations, too. This is the reason why this marker is more common in secondary GBM than in primary ones.

2. **IDH1 (isocitrate dehydrogenase) mutation**: IDH mutation is more common in second and third degree astrocytomas and oligodendromas, and 90% of cases are IDH1. It is very common in young adults. The gliomas carrying the IDH1 mutation derive from low-grade gliomas, generally having a frontal location. They are typical markers of second degree gliomas and are associated with a more favorable prognosis.

3. **EGFR (epidermal growth factor receptor)**: it is a transmembrane tyrossine-kinase encoded by chromosome 7p12 and whose signaling pathway modulates many cellular functions including growth, migration, and survival. *In GBM, it promotes invasiveness, tumor growth, and resistance to radiotherapy and chemotherapy.* The amplification of the EGFR and the increase of its gene expression lead to an increase in its activity. Approximately 60–70% of GBM carry EGFR mutations. Of the primary GBM, 40% leads to EGFR amplification and is associated with an unfavorable prognosis.

4. **Mutations of TP53 (tumor protein 53)**: it affects 60–70% of secondary GBM and 25–30% of primary GBMs. It is more common in young patients.

5. **ATRX (ATP-dependent helicase II or x-linked helicase) mutation**: it is specific to tumors that carry mutations of IDH1 and TP53. It is present in grade 2 and 3 astrocytomas (71%), oligoastrocytomas (86%), and secondary GBMs (57%).

6. **TERT (telomerase reverse transcriptase)**: it is involved in telomere maintenance, essential for cell growth. Its mutation is particularly common in primary GBMs and is related to EGFR amplification, but inversely related to changes in IDH1 and PT53. It is associated with a lower life expectancy.

7. **Loss of chromosome 10**: it affects 80–90% of GBM. PTEN (phosphatase and Tensin homolog, gene coding for an onco-suppressor) is located here, changed in 20–40% of primary GBM.

8. **Loss of chromosome 1p and 19q**: it is one of the most studied mutations and is associated with oligodendroglial morphology and better life expectancy. They are associated with mutations of IDH1.

9. **G-CIMP (cytosine-phospate-guanine island methylator phenotype)**: it is present in 10% of GBM and associated mainly with proneural GBM and IDH1 mutations. It is rare in primary GBMs.

2.3 Prognosis

Numerous attempts were made to improve the prognosis of this tumor, and in some surviving patients, magnetic resonances have been conducted to obtain more precise images of these tumor cells. Extensive necrosis and presence of multifocal lesions and edema seem to be negatively correlated with the survival of patients. The tumor mass is located above all in the temporal and frontal lobe, but it can also extend to the cerebellum, to the encephalic trunk, and to the spinal cord [1].

Surgical operations, radiotherapy, and concomitant chemotherapy with temozolomide and carmustine are the standard interventions for treating patients under the age of 70. However, the resistance to therapy and the speed at which these cancer cells change need to find methods that provide means that can better penetrate the blood-brain barrier (BBB), making therapy more effective and thus increasing tumor regression [7].

GBM has a very poor **prognosis**: only a few patients reach 2 and half years of survival, and less than 5% of patients achieve 5 years post-diagnosis survival. The hope of survival in the following year from the diagnosis is 35%. A study of the population affected by GBM has shown that the first quarter of the second year after diagnosis represents the peak of mortality rates [5].

Several variables influence the prognosis of GBM patients:

1. **Location of the tumor**: the GBM of the frontal lobe has a greater life expectancy compared to GBM in other sites.

2. **Gender**: the life expectancy in men and women affected by GBM is about 1 year (36.7% and 32.8%, respectively). It declines precipitously in the second year (13.7% for both sexes), reaching a rate of 4.7% and 4.6% for the few who survive 5 years.

3. **Race**: there are no significant differences between the white and black races, while the inhabitants of the Pacific Islands show a greater life expectancy.

4. **Age**: the increase in age is associated with lower survival. Patients between the ages of 70 and 74 and those over the age of 74 have a significant risk of death compared to patients aged 60–65.

2.4 Clinical Subtypes GBM is often described in two clinical forms: primary GBM and secondary GBM. The primary is the most common (95% of cases) and develops ex novo in about 3–6 months in older patients. The secondary develops from a previous astrocytoma (in about 10–15 years) in younger people [7].

They respond in the same way to the currently known treatments, and both have the same unlucky final outcome in almost all cases.

1. **Primary GBM** presents a mutated and amplified form of EGFR (epidermal receptor for growth factor), an amplification of MDM2 (gene suppressing pro-apoptotic protein p53), mutation of PTEN gene, and CDKN2A deletions (encoding for p16 and p14 proteins, both onco-suppressors) [3]. The tumor cells are small, circular, or tightly stretched; they are densely packed and have a hyperchromatic nucleus. Sometimes it resembles an anaplastic oligodendroglioma, with strictly atypical cellular forms [3].

2. **Secondary GBM** expresses more TGF-β (protein that regulates cell proliferation), leading to an autocrine stimulation that promotes cell growth. It also presents an overexpression of PDGFR (platelet-derived growth factor, which promotes mitosis), mutations in the p53 protein (tumor protein, tumor suppressor), and IDH1 (gene coding for the homonymous enzyme responsible for detoxification by oxidative stress) [7].

Based on the molecular heterogeneity of GBM and the alterations described above, the **Cancer Genome Atlas (TCGA)** divides the tumor into molecular subclasses [9]:

1. **Classic** shows aberrations that lead to EGFR amplification, loss of chromosome 10, amplification of chromosome 7, and gene expression of markers typical of astrocytes (21% of GBM).

2. **Mesenchymal** shows mutations of NF1 (neurofibromin1) and PTEN, a gene expression of mesenchymal markers (CD44, extracellular matrix adhesion protein, and VEGF) and a lower amplification of EGFR compared to other gliomas.

3. **Proneural** shows focal amplifications of PDGFRA and mutations of IDH1 and TP53, a gene expression typical of oligodendrocytes. It rises at a young age (31% of GBM). Sometimes it may present necrosis, which is associated with a poor prognosis [3].

4. **Neural** presents a gene expression typical of normal brain tissue, but markers are typical of the classic and proneural subtype, with amplification of chromosome 7 and loss of chromosome 10, and mutations of p53 (16% of GBM).

The secondary and primary GBM are therefore histologically indistinguishable, but they change in the genetic and epigenetic profile. Most GBMs with a change in IDH1 are secondary (80%), while only 5% of primary GBMs show this mutation. These molecular differences suggest how to act with personalized therapies according to the cases, to obtain an efficacy in the treatment [3].

2.5 Pathogenesis

Since the limits in diagnosis and subsequently the lack of efficacious treatments of this type of cancer are clear, several researches have been pursued, in particular focusing on the identification of a specific type of **brain tumor stem cells** that have the ability to begin the process of carcinogenesis and which could also represent a new therapeutic target [1]. The fact that certain types of cells can proliferate and differentiate in the CNS suggested the idea that it may exist a specific CNS cell having characteristics of a normal stem cell, but behaving like a tumor cell.

The brain, like other organs, made up of well-defined cellular hierarchies in development and homeostasis. It gives birth to cancer by **ontogeny**, a process that leads embryonic cells (undifferentiated stem cells) to the development of an organism with cells that have well-defined and different characteristics from tissue to tissue. Therefore, GBM precursors are certain types of stem cells [10]. These brain stem cells (CSCs) soon proved themselves resistant to chemotherapy and radiotherapy, emphasizing their central role in progressive and invasive capacity, as well as the ability to relapse of brain tumors [11]. Although the term "stem cell" is used, this does not mean that all cancer cells originate from modified stem cells. There are functional criteria that define whether a brain cell is a CSC, as cellular regeneration, continuous proliferation, ability to generate a tumor following a second intracranial transplant, expression of particular cell markers, and ability to generate progeny of cells belonging to different cell lines.

Recognizing these CSCs, however, remains very difficult, since many transcription factors and membrane proteins identified for CSCs are also common to NSPCs. Using the cytometry technique, several membrane proteins associated with CSCs have been identified (CD133, CD15, CD44, L1CAM, and A2B5) as shown in Table 1. These proteins mediate the interaction of the cell with the surrounding environment. For example, **CD133**, the first marker identified by binding to a specific antibody, is a glycoprotein expressed in normal neuronal stem cells, which becomes overexpressed in cells that have the greatest ability to regenerate and proliferate [11]. Furthermore, **the high expression of this protein is associated with a higher glioma malignancy.**

Table 1
Stem cell markers in GBM, adapted from [9]

Stem cell markers	Type	Cellular regulation
CD133	Surface glycoprotein	Positively associated with more aggressive tumors
L1CAM	Adhesion molecule	Neuronal adhesion molecule required for the maintenance of growth and survival of CD133 positive glioma cells with staminal-like properties
CD44	Cell surface marker	Positively associated with more aggressive tumors, localized together with IDH1 on the endothelial cells of the niche in which the stem cells are located
A2B5	Surface glycoside	Association with more aggressive tumors
IDH1	Transcriptional regulator	It promotes autonomous cell regeneration
CD15 (aka-SSEA-1 or LeX)	Cell surface protein	Stem cell markers in CD133 negative tumors
Integrin α6	Trans-membrane receptor	It regulates autonomous cell regeneration, proliferation and tumor formation interacting with the extracellular matrix

In addition to CD133, the overexpression of **L1CAM**, another molecular adhesion surface protein, is also associated with increased glioma aggressiveness. It helps to maintain the growth and survival of CD133 positive stem cells [9].

Also the other membrane markers, **CD15** and **CD44**, associated with particular subgroups of GBM, are not 100% correlated to the tumor development, because these proteins are also expressed by stem cells of other tissues, and even here they can be false positives [11].

3   Diagnosis and Treatment Guidelines

3.1   Diagnosis

AROME is a medical organization aimed at increasing collaboration between oncologists and other professional health figures involved in cancer research in the Mediterranean area. In 2010, the first Guidelines, structured as minimum requirements that must be proposed in the diagnosis and treatment of GBM and essential actions that any oncologist should perform to provide an acceptable cancer treatment were published. On the other hand, they try to rationalize the therapy and create a better use of the means available to treat a greater number of patients in a more efficient manner, with a better quality-price ratio [13]. The proposal to create Clinical Guidelines aims to improve the treatment and diagnosis of cancer and assist clinical decisions by rationalizing the use of accessible resources and prioritizing research. The heterogeneity of the Guidelines is often viewed negatively because it is considered a step backwards from the attempt to standardize methods of approach to GBM, but the AROME Group believes that this heterogeneity is necessary if we also consider countries with limited resources. In 2010 they published the first minimum and standard requirements that did not include CNS tumors, and in 2013 they created a neuro-oncology group made up of teachers and experts in the field. The diagnosis of GBM is based only on a pathological examination. Clinical symptoms and characteristic images of the lesions can help to anticipate their recognition [13].

3.1.1 Clinical Characteristics

Depending on the location of the tumor, clinical symptoms of glioblastoma consist of increased intracranial pressure, resulting in migraines, diplopia (double vision), and vertigo accompanied by nausea and vomiting, convulsions, focal neurological deficits, and cognitive disorders. The examination of the fundus of the eye may highlight papilledema supporting the diagnosis of increased intracranial pressure and the presence of presumed intracranial lesions.

In addition to the patient's medical history, the neurological examinations performed, and the received treatments, the following factors are also considered: (1) age, (2) height, (3) weight, (4) epileptic history, (5) results of biological tests, and (6) clinical autonomy of the patient according to the Karnofsky scale (health assessment scale, calculated based on the patient's quality of life).

3.1.2 Image Instruments

Imaging techniques are advantageous as they allow noninvasive examination of brain tissue by probing microscopic water movements to evaluate cell density and tissue architecture indirectly. Applied to brain tumors, they allow evaluating the degree of aggression and prognosis. The main parameter to consider is the *ADC (apparent diffusion coefficient)*, which is inversely related to tumor cell density, so its value is lower in high-grade tumors. A decrease in ADC was also seen in patients who received radiation and chemotherapy, and this is an indication of tumor progression [14]. The study considers a region of interest called ROI, within which evaluations are carried out. The examined parameters are:

1. *Cerebral blood flow (CBF)*: entity of the blood volume that transits in the unit of time in the ROI.
2. *Cerebral blood volume (CBV)*: absolute blood volume present at a given moment in the ROI.
3. *Mean transit time (MTT)*: average time taken by the blood to transit in ROI.

4. *Time to peak (TP)*: time taken to reach maximum concentration in ROI.

To access to good imaging results, the most used imaging techniques are PWI (magnetic resonance of perfusion), DWI (magnetic resonance of diffusion), DSC-MRI (dynamic susceptibility contrast MRI), DCE-MRI (dynamic contrast enhanced-MRI), arterial spin label MR perfusion (ASL), MRS (magnetic resonance spectroscopy), PET (positron emission tomography), and amino acid PET.

Besides, we reported the different levels of recommendation for the management of imaging in GBM patients [15].

Level I of recommendation: whenever possible, MRI with gadolinium should be used as a contrast agent, since it allows to identify the differentiation of GBM from other tumors.

Level II of recommendation: computed tomography with the addition of contrast agent can contribute to the identification of GBM from other tumors.

Level III of recommendation: the addition of PET to standard diagnostic methods helps to provide details that can improve the accuracy of brain injury detection, including those related to brain tumors.

3.1.3   Tumor Samples

Level I of recommendation: the diagnosis of a malignant glioma should be based on histopathological analysis of brain tissue.

For the regression of neurological symptoms, surgical treatment is recommended; however for those patients for whom surgery is not feasible, taking a sample of tumor tissue by biopsy is essential to obtain a valid diagnosis before performing any treatment. In exceptional cases, even biopsy may be contraindicated, for example, when the tumor's location is too deep or affects the dialectical brain areas or for patients at high risk of bleeding [13].

Level II of recommendation: frozen tissues and cytopathological evaluations are recommended for the intraoperative diagnosis of malignant glioma. Consultation with a neuro-pathologist specialized in the diagnosis of brain tumors is recommended for problematic cases.

In some cases, antitumor treatment can be initiated without pathological evidence, but a multidisciplinary council meeting is required first, and both the patient and the family must be informed of the potential risk of diagnosis and treatment being misdiagnosed [13, 15].

Level III of recommendation: the incorporation of clinical and radiographic information with the final diagnosis is recommended. The WHO brain tumor classification criteria are internationally recognized and can be used to establish the diagnosis of malignant glioma [15].

Table 2
Molecular markers that characterize the primary GBM and the secondary GBM adapted from [12]

Primary GBM	Secondary GBM
• 90–95% of GBM • they generally do not show changes in IDH • they have an accumulation of the following mutated molecules: RTK (receptor tyrosine kinase), RB1 (retinoblastoma), and p53 which is associated with mutations/deletions of CDKN2A/ARF (49%), MDM2 amplifications (14%), and mutations and deletions of TP53 (35%) • EGFR amplification (40–50%) • deletion of PTEN (80%) • mutations of PI3K (15%) • amplification of PDGFRA (18%)	• presents mutations of IDH1 (85%), TP53 (62%) together with inactivation of the ATRX • amplification of PDGFRA much more evident than the primary GBM • methylation of the MGMT

3.1.4 Molecular Markers   The diagnosis of GBM is mainly based on morphological characteristics according to the WHO classification. In oncology, two main
categories of biomarkers are recognized. *Prognostic biomarkers* give
information on the probable outcome of the disease regardless of
the treatment received, while the *predictive biomarkers* provide
information on the expected results with the application of specific
interventions, so they can help to select the most suitable between
multiple therapies and are particularly useful in targeted therapy [6]
(*see* Table 2).

The most evident molecular characteristics in gliomas are:

1. **IDH-1 mutation:** the IDH-1 mutation is recognized as one of
   the factors that best distinguishes gliomas. IDH-2 mutation is
   less frequent, while IDH-1 mutation is found in 70–80% of
   grade II brain tumors, grade III gliomas, and secondary GBM.
   It is rare in other neoplasms of the CNS such as ependymomas,
   pilocytic astrocytomas, and gangliogliomas; therefore **finding
   a mutation of the IDH-1 leads already to suggest that it is
   an invasive glioma**. The most frequent mutation is R132H, at
   codon 132, in which a histidine amino acid replaces one of
   histidine. Tumors with this mutation progress more slowly
   than those that do not show it [16]. Among all IDH-mutant
   gliomas, 30–40% shows a co-deletion of the 1p/19q chromosome, typical of oligodendrogliomas [17]. It has a higher life
   expectancy and is a predictive biomarker for chemosensitivity
   [18]. Moreover, 60% shows mutations of the common TP53
   especially in II and III degree astrocytomas and secondary
   GBM (an 80% of 939 tumors) and loss of ATRX
   (ATP-dependent helicase, gene encoding for a regulator of
   the packaging degree of chromatin), which was observed in
   57% of secondary GBMs [19].

2. **Mutation of the MGMT promoter**: the gene for MGMT is located on chromosome 10q26 and encodes for the ubiquitous expression of DNA repairing enzymes that remove alkyl residues from the O6 position of guanine [20]. The action of the MGMT (O6-methylguanine-DNA methyltransferase) is to eliminate these alkylations to avoid the development of errors within the nucleotide sequence that could induce the appearance of carcinogenesis [6]. The standard therapy for GBM involves the use of radiation and chemotherapy with temozolomide, which acts by making cross-links in the DNA of the tumor cell in several places using the oxygen of guanine. This cross-linking is reversible due to the action of the MGMT, so it is expected that low levels of this enzyme will induce a greater sensitivity of the tumor cell to these alkylating agents. The expression of MGMT levels is largely determined by the methylation of its promoter: if it is methylated, then MGMT is silenced (40–50% of GBM) [12]. Regardless of other clinical or therapeutic factors, this methylation is prognostic for greater patient survival, as demonstrated by a phase III trial conducted by the German Neuro-Oncology Working Group (NOA), which compared the efficacy of radiotherapy alone with the efficacy of treatment with temozolomide [6].

3. **EGFR mutation:** is more common in primary GBM than in secondary GBM (50% of primary GBM). EGFR is a transmembrane glycoprotein. Its mutation leads to an increase in cell proliferation and inhibition of apoptosis, cell migration, and radio· and chemoresistance. It does not give favorable prognosis, neither long life span for the patients who carry it. The mutated form of EGFR is called EGFR-VIII and is not normally expressed in normal cell; therefore it is an excellent tumor marker identified by immunostaining techniques [6].

3.2 Therapy

Also for therapy, AROME draws up some guidelines and establishes the "Standard of Care" for patients in good clinical condition.

3.2.1 Neurosurgery

Level I of recommendation: There is insufficient evidence to support a level I of recommendation.

Level II of recommendation: Based on accessible data and the general consensus of retrospective data, it is recommended that a "maximal safe resection" be undertaken in adults with a recent diagnosis of supratentorial malignant glioma (the maximum cytoreductive procedure must ensure that the postoperative neurological deficit is minimized).

Level III of recommendation: It is recommended that biopsy, partial or total resection, be considered in the initial treatment of the malignant glioma, depending on the condition of the patient and the size and location of the tumor [15].

Maximal and safe surgery is a decisive point in the diagnosis and treatment of GBM. Neurosurgical procedures should be performed with a minimum delay of 2 weeks (minimum requirement) after clinical and radiological diagnosis. When it is feasible, the best strategy is an optimal intervention, which improves the outcome of other therapies and the patient's quality of life. The quality and extent of the resection can be assessed using brain-imaging techniques with or without contrast medium, within 48 h of the procedure. CT scan with contrast medium is a minimum requirement for postoperative imaging procedures [13].

3.2.2 Radiotherapy

Level I of recommendation: Radiation therapy is recommended for the treatment of recent diagnosis of malignant glioma in adults. The treatment scheme should include dosages above 60 Gy provided in 2 Gy fractions per day on the clinical volume to be treated (CTV). A scheme that provides hypofractionated radiation should be used for patients with poor prognosis and limited survival, without compromising the response. Hyperfractionation and accelerated fractionation have not been shown to be superior to conventional fractionation and are not recommended [15].

Level II of recommendation: It is recommended that radiotherapy planning should include 1–2 cm of margin around the radiographically defined tumor area or MR-weighted abnormality [15].

The clinical volume to be treated (CTV) includes the area of intervention in the case of total surgical resection, or the residual area of the tumor plus 2 cm more, depending on the volume of the tumor mass (below or above 250 cm^3). Instead, to plan the volume to be treated later (PTV), add 0.5 cm to the CTV [13]. The impact of surgical-radiotherapy intervention delay on the outcome of therapy, compared with other prognostic parameters such as patient characteristics (age, weight, etc.), tumor aggressiveness, and molecular biology is unknown. Lawrence et al. report a deleterious impact on the survival of patients treated less than 2 weeks after diagnosis and better results in patients treated with radiotherapy more than 5 weeks after surgery [21].

3.2.3 Cytotoxic Chemotherapy

Level I of recommendation: Treatment with concurrent temozolomide and post-irradiation is recommended in patients between 18 and 70 years of age with adequate systemic health. This recommendation is supported on evidence from a single first-class study by EORTC [15].

In 2004 the European Organization for Research and Treatment of Cancer (EORTC) presented phase III studies demonstrating how GBM treatment with radiotherapy and temozolomide increased life expectancy from 12.1 to 14.6 months [22]. In combination with radiotherapy, temozolomide (TMZ) should be administered in a dose of 75 mg/m^2 and may be supplemented with a dosage of 150–200 mg/m^2. In the absence of tumor progression, the minimum number of cycles is 6 (minimum requirement). The number of maximum cycles is yet to be evaluated [13].

Level II of recommendation: Biodegradable polymers impregnated with carmustine are recommended in patients for whom craniotomy is indicated. This recommendation is based on the evidence of two class II studies [22].

Level III of recommendation: The addition of temozolomide to radioactive therapy is an option for patients with a recent diagnosis of glioblastoma who are over 65 years with a performance status of Karnofsky scale over 50 [15]. *Class III studies suggest that for patients 70 years of age or older with a recent diagnosis of glioblastoma, temozolomide alone is a tolerated alternative to radiation and its benefits may be comparable to those obtained with radiotherapy alone"* [23]. *Radiation therapy, followed by treatment with one of the nitrosoureas, is recommended for patients not suitable for temozolomide* [15].

Recent data on radiotherapy treatment for GBM with methylations of the MGMT treated with temozolomide revealed that delays greater than 6 weeks between intervention and radiotherapy did not affect the prognosis and survival of the patient. Therefore, the council recommends a delay of 4–6 weeks as the minimum requirement. It is important to test periodically all tumors for methylation of the MGMT, in order to make decisions on the adoption in therapy of adjuvants in the treatment [13].

4   Critical Issue in Current Managements

4.1   Surgery

The surgical operation is part of the "Standard of Care" for the treatment of GBM. In general, it is the first operation that is carried out in an attempt to reduce the tumor mass and simultaneously obtain a sample of tissue that can be analyzed to obtain a more precise diagnosis and observe certain molecular characteristics of the tumor to identify the right therapy and increase its success. The effectiveness of surgical treatment is generally determined by patient survival [24]. The property of GBM tumor lesions is to be able to infiltrate functional brain regions, so almost all surgical interventions have a high risk of leading the patient to neurological deficits (speech or motor) that reduce the quality of life of the patient, delay treatment success, and accelerate death [25]. Consequently, total resection is not planned in cases in which dialectical brain regions are involved. The planning of a surgical operation is therefore based on the benefit/risk ratio between a maximum safe resection and the risk of inducing neurological deficits [24].

Surgery, therefore, remains one of the central treatments in the therapy against GBM and helps eliminate most of the symptoms due to the presence of the tumor mass in the brain. However, this approach would consider a number of risks: it is not known how long the beneficial effects of the intervention can be maintained and does not guarantee the total recovery from the disease. The cancer often has relapses that make it even more aggressive than it was already. Therefore, significant improvements are needed in the treatment of GBM, which are able to significantly extend the life expectancy of this currently fatal pathology [25].

4.2 Radiotherapy

The surgery alone, as already mentioned, does not allow the total removal of the tumor mass (99%), and the remaining 1% of tumor cells that are not removed proliferate again, leading to the development of relapses. Therefore, chemotherapy and radiotherapy are always associated with surgery. Radiologists to try to remedy the problem generally add 2 cm more margin to the apparent margins of the tumor to be treated with radiotherapy, to identify a CTV that is able to include the remaining infiltrative cells. This CTV, however, also includes areas of healthy brain tissue highly sensitive to radiation; therefore the total dose needs to be reduced in order not to exceed the limits of tolerance of healthy brain cells. The reduced dosage, however, is not sufficient to completely inhibit the tumor cells, causing relapses in the treated area. In addition, these cells are more infiltrative than proliferative, and therefore radiotherapy and chemotherapy, which destroy high-proliferative cells, are ineffective in this case [26].

A tumor such as GBM is constantly evolving, and the high heterogeneity of tumor cells would require adapting the dose of radiation for each type of patient depending on the type of tumor, rather than outlining a homogeneous dose (currently 60 Gy) that is administered independently of the characteristics of tumor cells [27]. Not all cells, in fact, have the same degree of invasiveness and depending on these characteristic groups of more or less invasive cells should be treated with different doses of radiation. The difficulty, however, consists precisely in being able to identify how aggressive the tumor is: in the absence of this possibility, all GBM are inevitably treated in the same way. Less invasive tumors will therefore respond better to therapy [26].

4.3 Drug-Delivery Limits

Difficulties in drug delivery: Many conventional methods for drug delivery, such as the oral or intravenous administration, do not allow the achievement of therapeutic concentrations of the drug within intracranial tumors but, on the contrary, may reach toxic systemic concentrations. In fact, in order to allow the entry of the drug into the brain, higher doses are clinically administered, easily reaching the toxic dosage limit [28]. Currently, five potential approaches to enable the achievement of high intra-tumor drug concentrations without systemic toxicity are recognized, which involve local delivery of chemotherapies directly into the brain. All of these treatments display quite level of invasiveness, as they try to direct the chemotherapy directly where the tumor is located:

1. *Increase the permeability of BEE to drugs by chemical or structural alteration of the pharmacological molecule.* Many chemotherapeutics are large, positively charged and hydrophilic. Lomustine and semustine are two more lipophilic variants of carmustine, but clinical trials have not shown an improvement in treatment efficacy compared to carmustine.

2. *Interstitial drug delivery through catheters.* This approach involves bypassing the BEE by infusing the drug via catheter directly into the site where the tumor is located. The graft developed on the basis of this idea is the Ommaya reservoir, a device with a fluid reservoir implanted under the scalp, which has a catheter that fits into the ventricle. This allows the drug to enter directly into the cerebrospinal fluid and then into the brain. Many chemotherapeutic agents such as carmustine, methotrexate, bleomycin, and cisplatin, biological agents (IL-2), and interferon (INF-Y) were introduced directly into the tumor through this method [29]. Although the success of these therapies is widely documented in individual cases, the overall survival benefit has not been tested on large-scale clinical trials [30].

3. *Temporary destruction of BBB using hyperosmotic agents such as mannitol,* which recalls the water outside the endothelial cells, reducing its size and thus opening the spaces between the endothelial cells. Other molecules such as bradykinin, on the other hand, induce a real destruction of BEE. In both cases, the increase in drug penetration in the brain parenchyma is confirmed. However, this is not due to an increase in efficacy. Therefore, although the osmotic destruction of BEE increases the passage of hydrophilic substances in the brain parenchyma, it does not improve their transfer into the tumor [30].

4. *Drug delivery enhanced by CNS drug convention,* which could represent an effective drug-delivery method especially in areas such as the brain stem, where the tumor removal surgery is not always applicable. This method involves the use of a catheter, too. The convention is the result of a simple pressure gradient and is independent of molecular weight, unlike diffusion. It can be used for drug delivery of high drug concentrations in large brain regions without structural or functional impairment. The disadvantage consists in the fact that the tissue surrounding the catheter receives a quantity of drug sufficient to give the action, while more distant regions receive increasingly smaller concentrations. The drug spreads in the capillaries or is eliminated by efflux transporters, and its concentration in the brain decreases considerably. It is important to point out that there is no study showing positive effects on the use of this technique on patients. In addition, the CED implant is implanted surgically and subjects the patient to a high risk of infections [31].

5. *Use of polymers and microchips for direct therapy*, first described by Langer and Folkman in 1976, reports the sustained and predictable release of macromolecules from an ethylene-vinyl acetate co-polymer (EVAc). This polymer is inert and therefore not biodegradable. It releases its agents by diffusion through the micropores of its matrix. The degree of diffusion depends on the chemical properties of the drug, including water solubility, electrical charge, and molecular mass. The FDA has not given approval for its use in the brain, because once released the drug matrix, which is not biodegradable, remains as a permanent foreign body, and this is the main limit, but is used in many other fields like glaucoma, diabetes, asthma, and contraceptive therapy [30]. The next step was to create biodegradable polymers, such as PCPP-SA (polyanhydride-poly [bis (*p*-carboxyphenoxy) propane-sebacic acid]), which by a spontaneous reaction with water breaks into dicarboxylic acid. The hydrophobicity of PCPP-SA protects the drug from the interaction with the surrounding aqueous environment, and the erosion process is limited to the surface of the polymer, ensuring a constant drug release. The advantage lies in the absence of nonbiodegradable residues, and the decomposing compound that forms is neither toxic nor mutagenic. The problem with this system is that it does not allow the release of many hydrophilic compounds, and the release of carboplatin is not constant [32]. Today the therapy for GBM involves the use of carmustine polymers (BCNU).

The use of more innovative local delivery systems, such as catheters, CEDs, and biodegradable wafers, does not guarantee the total safety of the treatment: in fact, these grafts release a quantity of drugs that declines as they move away from the implant, and therefore the tumor is not treated in all its parts with the same concentration of drug. This can lead to the development of treatment resistance for those cancer cells that do not receive a toxic dosage. Furthermore, the implantation of these systems involves a surgical operation that subjects the patient to a high risk of infections.

The **lack of selective targeting for GBM does not prevent the distribution of the drug even in nontarget tissues** (heart, kidney, lungs, liver, spleen, bone marrow). The accumulation of chemotherapy in nontarget organs is the main cause of side effects, even serious, during therapy which can significantly worsen the quality of life of the patient and his state of health, sometimes leading to death.

4.4 Life Expectancy and Therapeutic Patterns

Life expectancy, following the treatment guidelines that involve surgical resection, followed by radiotherapy and chemotherapy, is however very poor (a few months of extra life from the diagnosis, usually no more than 6). The risk/benefit ratio is not favorable and significantly worsens the patient's living conditions due to the appearance of serious side effects due to therapy.

There is no precise treatment scheme for relapses, and the therapies used in this case prove to be ineffective in almost all cases.

Following these considerations, it is evident that the means currently used to combat GBM are not sufficient to guarantee the recovery from the disease. It is necessary to look towards innovative therapies, with less risk of application and greater effectiveness, which are able to provide patients with a more concrete hope of recovery, or in any case that allow them to live with the disease in the best possible way, and providing a longer life expectancy.

5   The Blood-Brain Barrier

The blood-brain barrier (BBB) is a dynamic interface between blood and central nervous system (CNS) providing for the cerebral homeostasis, and it protects the brain from toxic and pathogenic agents [33]. This barrier is strongly needed as CNS is the most critical biological system of the entire human body and correct neuronal functions need a highly controlled extracellular space [34]. Principal functions of the BBB are (a) to preserve cerebral homeostasis regulating ionic balance, (b) to defend brain from extracellular contamination, (c) to support CNS with nutrients through specific transport systems, and (d) to direct inflammatory cells to act in response to the changes of the surrounding space [8].

5.1   Structure

In the brain, capillaries are differentiated and develop in BBB, which exists first as a *selective diffusion barrier*, featured by a cerebrovascular endothelium characterized by the presence of tight junctions and absence of fenestrations [29].

BBB is composed of [33]:

1. **Basement membrane**: surrounding the endothelial cells of the cerebral microcirculation (BMVEC). It consists of two overlapping layers, composed of different classes of molecules belonging to the extracellular matrix [35].

2. **Neurons:** able to regulate vessels' function in response to metabolic needs by inducing the expression of unique enzymes for BMVECs.

3. **Microglia cells:** playing a fundamental role in the CNS immune response, they change their phenotype in response to changes in brain homeostasis.

4. **Pericytes:** covering 22–32% of the blood capillaries and are able to synthesize many components of the basement membrane. They play a central role in the differentiation of BBB. They communicate with vascular cells through a series of *gap*

junctions and adhesion molecules. They are involved in the transport through the BBB of substances. They have contractile capabilities and regulate blood flow at the level of the brain capillaries [34].

5. **Astrocytes:** they are glial cells whose cytoplasmic expansions form lamellar extroflexions that approach the external surface of the BBB and the basement membrane. They interpose between the basal membrane and neurons, mediating both the metabolic and physical relationships between blood and nerve cells.

6. **Endothelial cells** (**EC**): representing the *anatomical unit of BBB,* they *regulate the transport and metabolism of substances from the blood to the brain and vice versa,* by exploiting different actions:

   (a) They prevent the entry of toxic substances into the brain and provide for efficacious communication with adjacent cells, allowing them to maintain brain homeostasis.

   (b) They differ from other endothelial cells for the selective ability in transporting substances.

   (c) They are 50/100 times more adherent than the endothelial cells of the other blood capillaries (this prevents any diffusion activity of molecules through their membrane).

   (d) They have *no fenestrations,* but not all brain endothelial cells have this characteristic [36].

   (e) They have low pinocytotic activity (ability to form vesicles for the transport of substances).

   (f) They have a continuous basal membrane.

   (g) They have on the luminal surface a complex of various carbohydrates that has negative charges able to reject many negatively charged compounds [35].

   (h) They have a high number of mitochondria (8–11% of total volume), thus letting high energy for the active transport of substances inside and outside the brain.

   (i) On their surface, we can find many *adhesion molecules.*

   (j) Their cytoskeleton is made up of actin and myosin filaments, which can interact together to provide contractile activity to the cell.

   (k) They have a *polarized structure* that allows them to develop certain transport properties [35].

The endothelium, astrocytes, pericytes, neurons, and the matrix constitute a **neurovascular unit**, able to illustrate how the brain can respond to the development of pathologies [34].

5.2 Intercellular Junctions of BBB

Specific intercellular junctions establish the interaction between endothelial cells and BBB [33]:

1. **Tight junctions (TJ)**: main junctions responsible for BBB's properties; they are located in the apical region of endothelial cells and regulate the diffusion of substances between the apical area and the basolateral zone of the plasma membrane. They form a complex series of parallel and interconnected transmembrane protein strands, arranged as a series of multiple barriers and formed by transmembrane proteins that interact with transmembrane proteins of adjacent cells. Based on the number of times that cross the plasma membrane, we can distinguish them in:

   (a) *Claudins*: they are a type of TJ (24 family members identified) with a molecular weight between 20 and 27 kDa. They are responsible for the **reduced permeability of the BBB.** They interact by **homophilic bonds** (equal structures) between the extracellular loops. They pass the membrane four times [34].

   (b) *Occludins*: they are TJ of molecular weight of about 65 kDa, and they were the first to be discovered, as well as the best known. As claudins, they pass the membrane four times [34]. *High levels of occludins decrease paracellular permeability. They have calcium-dependent adhesiveness.*

   (c) *Junctional adhesion molecules (JAM)*: proteins with a molecular weight of 40 kDa, belonging to the IgG superfamily. We can distinguish them in JAM1, JAM2, and JAM3. JAM1 are involved in cell adhesion by a **homophilic bond** [34] and together with the occludins and the claudins constitute the TJs. They regulate the transendothelial migration of leukocytes. *Their possible dysfunction could be associated with disorders in the transfer of leukocytes associated with CNS disorders.* They pass through the plasma membrane one time.

2. **Cytoplasmic proteins**: proteins that are used to recognize TJs; they are part of the MAGUK family (membrane-associated guanilate kinase). The *ZO proteins*, a sub-membrane protein associated with TJ, are part of this family. The ZO1 forms the central portion of the TJ and connects it to the cytoskeleton by binding to the carboxy-terminal end of the occludin. *Loss of ZO1 is related to an increase in membrane permeability.*

3. **Adherent junctions (AJ)**: they are located below the TJ, in the basal region of the lateral plasma membrane. They form the *adhesion belt* and allow the adhesion of the BMVECs to one another, the inhibition of cellular contact during the growth

and remodelling processes, and the regulation of the paracellular permeability. The constituents of the AJ allow cell-cell adhesion:

(a) *Cadherins* are calcium-dependent membrane glycoproteins. They are the principal constituents of the AJ. They are assembled around cellular junctions and mediate cell adhesion in a calcium-dependent manner; when overexpressed, they inhibit cell proliferation, decreasing membrane permeability and migration.

(b) *Catenins*, their main role is to anchor cadherins to the cytoskeleton. There are four types: α, β, located in the interendothelial junctions of the BBB capillaries, Y and δ. β-catenin is also involved in cell signaling mechanisms. The upregulation of this protein could contribute to the maintenance of TJ and barrier function.

The binding of an agonist (vasoregulatory agent, free radicals, NO, PGE2, or other inflammatory agents) with its respective receptor expressed on the endothelial surface initiates the signaling cascade called **protein kinase/RhoA/Rho-kinase (ROCK)** and **MAPK (mitogen-activated protein kinase)**, which causes the phosphorylation of TJ and adherent junctions. This causes the junction complex to dissociate itself from its anchorage with the cytoskeleton, and it leads to a weakening of cell-cell adhesion.

At the same time, the intracellular calcium concentration increases, activating the eNOS and the MLCK (myosin light-chain kinase): the myosin chain is phosphorylated, and its binding to actin is then activated. This promotes cytoskeletal contraction, and the consequent cell retraction, which contributes to the increase in the permeability of BBB [33].

5.3 Transport across the BBB

We can divide the transport routes through the BBB into two great categories, depending on the pathways:

1. **Paracellular**, between adjacent cells by opening the junctions. In this case, the molecules pass by diffusion, following their concentration gradient.

2. **Transcellular**, through the cell. In this case, the small lipid molecules (oxygen, carbon dioxide, ethanol) pass through the cell membrane by diffusion. Hydrophilic molecules, on the other hand, like amino acids and glucose, need specific transport mechanisms:

   To evaluate how the BBB modifies its functions according to physiological and pathological events that occur in our organism, different experimental systems in vivo, ex vivo, and in vitro have been used. The more the system in question is similar to the in vivo system, the higher its complexity [33]. Important parameters to be assessed for permeability studies are:

3. **TEER** (*trans-endothelial electrical resistance*): useful indicator of the paracellular flows of ions; TERR values could be of help in a better understanding of the impediment of the passage of small ionic molecules through the BBB. It is considered *one of the most accurate measures of membrane integrity and barrier functionality:* its decrease reflects an increase in membrane permeability and a loss of function.

4. **Permeability**: an increase in the permeability of BBB is observed in many CNS pathologies, inflammatory processes, infections, ischemia, epileptic manifestations, and trauma. As probes to measure membrane permeability, fluorescent dyes (fluorescein sodium or fluorescein isothiocyanate) or isotopes are used, which also help to quantify the molecules that cross the BBB. Sucrose, inulin, and mannitol are the main tracers of paracellular transport. Albumin, on the other hand, is a marker of endothelial permeability and has been observed in several vesicles. In the case of glioblastoma, a co-culture of bovine, murine, swine, or human endothelial cells with C6 murine glioma cells has been shown to increase the strength with which the junctions adhere to each other. Malignant glial cells, instead, like those of glioblastoma, have shown to increase the permeability of BEE by secreting angiogenic factors such as VEGFR [13].

5.4 Blood-Brain-Tumor Barrier

In comparison with BBB, the blood-brain-tumor barrier (BBTB) presents [37, 38] an endothelial layer with relevant breaks, compromised TJ, damaged basement membrane, and disorders in the interaction between endothelium and astrocytes [39]. This partial opening of the TJs and the *increase of the fenestrations inevitably lead to an increase* in permeability of the BBTB. High-grade tumors such as anaplastic astrocytomas (WHO grade III) and **glioblastomas (WHO grade IV)** show mild to moderate increasing in permeability [38]. In some subtypes of tumor cells, the ABC transporters are more expressed, responsible for the outflow of drugs from the cerebral parenchyma and therefore of chemoresistance of glioblastoma [39]. Although there are several histological types of brain tumors, many of these cause **cerebral edema**, which is one of the most frequent causes of death and morbidity. *The increase in permeability of BBTB is associated with TJ defects, and these abnormalities are related to an increase in tumor malignancy* [40]. The invasive potential of glioma causes a large proliferation of tumor cells outside the area that has a broken BBTB, and mainly in those brain areas in which the BBB is still intact [39]. Occludins, claudin-1, and claudin-5 undergo a downregulation in glioma and are instead absent in the TJ of the vessels of metastatic tumors. In addition to being downregulated, the occludin is also phosphorylated: this inhibits the interactions with ZO-1 and ZO-2 and increases the permeability of TJ [40]. The reason why the tumors show defects in these proteins is that they have a reduced number of astrocytes, which we know are essential for maintaining the structure and functionality of BBB and an excessive secretion of angiogenic factors [40]. In fact, many malignant tumors are so active that the nutritive support provided by the blood is no longer sufficient to compensate for their needs. This results in a hypoxia stage that probably favors the secretion of *angiogenic factors such as VEGFR* (in GBM it has been shown that hypoxia favors the secretion of this factor) [41].

In this view, insurmountable barriers, efflux systems, and resistances are connected to the presence of BBB, but also BBTB strategically prevents optimal therapy for brain tumors, since the TJs, posed on the endothelium of the brain capillaries, greatly prevent the entrance of molecules from the blood to the brain. TJs of the cerebral endothelium form a continuous lipid layer that allows access only to small, electrically neutral lipid molecules. Many of the chemotherapeutic agents do not possess these characteristics. Moreover, there is little pinocytotic activity, so the molecular transport that occurs by transcytosis is severely compromised. The presence of ABC transporters, active transport proteins that cause the outflow of the drug from the brain parenchyma, on the luminal section of the cerebral endothelium, significantly affects the transport of molecules through the BBB and prevents the access of many CNS chemotherapeutics [14].

The BBTB should also be considered. The tumor capillaries, as mentioned previously, tend to be abnormal, dilated, and tortuous, with greater permeability. The greater the permeability of these capillaries, the greater the intra-tumor interstitial pressure is. This results in a net flow of fluid from the center to the periphery of the tumor and surrounding tissue, called peri-tumoral edema. In this setting, many capillaries are partially or completely collapsed, and therefore the penetration of pharmacological agents is largely limited. The presence of these barriers has encouraged the development of new drug delivery methods for CNS tumors [14].

In conclusion, the inability of chemotherapeutics to cross the BBB and the BBTB and the fact that they cause the appearance of side effects is due to their chemical nature and their size: being in fact mainly charged and large molecules, the amount that can penetrate into the cerebral parenchyma is relatively poor. Easily they are distributed in other organs, causing toxicity.

1. If the drug is able to pass the BBB, it is often transferred again outside the cerebral parenchyma due to the presence of efflux pumps on the membrane of the endothelial cells.

2. The chemotherapy currently in use, for the dosage given and the nature of cancer cells, in the long terms cause the appearance of resistance, making the tumor more aggressive and more difficult to treat.

6   The Promise of Nanomedicine

Since 1970s many efforts have been done to improve efficacy of the treatments for brain tumors, since their complete elimination in most cases have never been achieved [42]. Many progresses have been made today with radiotherapy and chemotherapy, although it is clear that through these treatments it is not possible to have a remarkable impact on patient survival. Indeed, life expectancy for patients with GBM submitted to therapy generally does not go further the year and a half, at best. The principal limitation consists of poor pharmacokinetics and an inappropriate biodistribution of therapeutics, which causes insufficient penetration inside the tumor. Drugs are quickly removed from the systemic circulation and accumulate in many healthy organs, giving toxicity [42].

Among the number of approaches designed for GBM targeting, nanomedicine surely represent one of the most promising in terms of advantages in higher selectivity of treatments and decrease in side effect of the delivered drugs. Nanomedicine, namely, the application of nanotechnologies for the diagnosis, monitoring, prevention, treatment, and understanding of disease mechanisms to obtain clinical benefits, is designed to obtain more effective therapeutic agents against the disease and less harmful to the patient, compared with conventional therapies [43]. As said, this relatively new discipline is based on the use of nanomaterials ranging in size from 10 to 500 nm for various applications, including the treatment of brain tumors.

The use of nanomedicines (i.e., polymeric nanoparticles, liposomes, etc.) could lead to potential advantages:

1. Improved pharmacokinetics and biodistribution of loaded drugs.

2. Increased therapeutic efficacy allowing a greater accumulation of the drug in the tissue.

3. Reduced dose-dependent adverse drug reactions, minimizing the possibility of the drug to accumulate in non-target tissues.

4. Ability to combine therapy with multiple drugs in a single nanoformulate.

5. Ability to manipulate the surface of the nano-carrier with a range of molecules that increase the specificity to the target tissue and reduce plasma proteins' interaction, improving blood circulation.

Due to all of these advantages, as obvious, nanomedicine approach was investigated to overcome the major limitations in brain tumor treatments, as brain structural complexity, the heterogeneous and invasive nature of the tumor, the difficulty in establishing tumor margins, the insufficient accumulation of the therapeutic agent at the tumor site, and the possible acquisition of resistance to chemotherapy.

Besides all of these factors, one of the main obstacles is represented by the presence of three distinct barriers at CNS level, both to be overcome to let the drug being active in the site of action. The major obstacle is BBB which, as previously described, due to its complex structure, prevents the entry of many harmful molecules both exogenous (toxins) and endogenous, but also of chemotherapeutics against the brain tumor, thus compromising the success of the therapy. The second barrier that prevents the passage of many substances is the cerebrospinal fluid-blood barrier. Finally, the blood-brain-tumor barrier (BBTB), even if characterized by an increased permeability compared to BBB, is featured by a high interstitial pressure that prevents the penetration of drugs into the tumor site. Moreover, the difference of the interstitial spaces between the endothelial cells of the tumor capillaries does not guarantee a homogeneous distribution of the medicinal product, and this significantly compromises the effectiveness of the therapy [30]. Many chemotherapeutics used in cancer treatment can damage DNA, without being able to distinguish between healthy cells and cancer cells. This increases the risk of considerable side effects [42].

6.1 Nanomedicines for Drug Delivery

In the last years, different nanosystems for the delivery of therapeutic agents into the brain were investigated [42]:

1. **Polymeric nanoparticles (NPs) and lipid nanoparticles:** Nanocarriers (10–250 nm) made of natural or synthetic polymers (i.e., chitosan, albumin, polylactide-*co*-glycolic acid) where the drug is stably encapsulated within the matrix of the polymers or inside an aqueous inner, showing good protection of the loaded drugs against physiological or environmental conditions and triggering a controlled release of the loaded drugs. Moreover, due to the presence of chemical reactive groups featuring the polymeric chain, both the nanosystem and even the polymers could be modified with selective ligands which will be finally located onto the surface of the NPs, thus facilitating the passage across the BBB. Depending on the nature of the polymer and its time of degradation, they allow a prolonged release of the drug over time. When the material forming the NPs is lipid, two types of nanocarriers (10–1000 nm) exist depending on the formulation: *solid lipid nanoparticles (SLNs)* formed by a lipid matrix stabilized by a surfactant, and *lipid nanocapsules (LNCs)* formed by an oleic core surrounded by lipophilic and hydrophilic surfactants.

2. **Liposomes (LPs):** Representing the first nanoparticles systems approved by FDA in clinical use, LPs are biocompatible and biodegradable nontoxic lipid systems (50–300 nm) formed by

a double phospholipid layer with an inner aqueous reservoir that allow the encapsulation of a large number of hydrophilic drugs. In order to improve their stability, their surface can be modified with hydrophilic agents such as polyethylene glycol (PEG), which make them stealth to our immune system.

3. **Micelles:** Spherical nanostructures (20–200 nm) with an inner core of lipid polymers and a shell of hydrophilic polymers formed by an assembly of amphiphilic polymers in aqueous solution.

4. **Dendrimers:** Nanocarriers (2–15 nm) formed by branched polymeric molecules which carry ligands on the ending of ramifications.

6.2 Key Features of Nanomedicines

Each nanoparticle system displays the advantage of being individually created depending on size, shape, and chemical surface in order to meet the finalities of the proposed functions. All of these features will strongly impact on the final destiny of the drug delivery systems, as too large dimension will never be able to get the final target, and too small size will cross not selectively any physiological barrier. The same consideration could also been drawn if considering shape and surface charge, with different fates depending on the properties exposed by nanomedicines. Moreover, the surface exposition (surface chemistry) will also impact on the protein corona adsorbed onto nanosystems, which will drive strongly towards safety or toxicity aspects of the circulating nanosystems.

Besides these aspects, we should consider that nanomedicine has an extended surface area in relation to their volume; they could improve drug's solubility, prolonging plasma half-life of the loaded drugs and finally controlling their release. One of most attractive features is the ability of nanomedicines in encapsulating both hydrophobic and hydrophilic molecules, gene material, proteins, enzymes, and virtually any kind of drugs, just choosing the most suitable material composition and the right formulation procedure. Besides, to obtain a tailored release of the drug, a number of nanosystems were designed to respond to various target site stimuli, such as pH values and temperature variations, therefore accelerating the drug release within selected biological conditions. These unique features of nanomedicines would perfectly fit the need for novel therapeutic options, but we should underline different cargo molecules (i.e., imaging agents, fluorescent probes, contrast agents, or even quantum dots), thus switching the aim from therapy to imaging goal. As an example, nanoparticles made of *quantum dots* (semiconductor materials) have a narrow emission spectrum and excellent photo stability and can be used therefore as fluorescent probes for the diagnosis of brain tumors at the molecular level [44].

Besides all of these aspects, regarding brain tumors, one of the most critical points is the lack of selectivity. Thus, nanomedicine could be greatly of help in novel and targeted therapy selectively directed to the cancer cells.

6.3 Targeting Brain Tumors

Two types of tumor targeting are currently known [45]:

1. *Passive targeting*: is based on the passive diffusion of nanocarriers according to concentration gradient. In this kind of approach, nanocarrier size represents one of the main characteristics to achieve passive targeting and biodistribution within brain tumors. Nanocarriers of 10–100 nm may benefit from the neovascularization of the tumor tissue, the discontinuity of the capillaries, and the compromised lymphatic system, which alters tissue drainage, so NPs can fenestrate and accumulate easier within the tumor tissue than the healthy brain.

   Neo-vasculature events occurring in cancer development could therefore help in passive targeting, but also the EPR effect (enhanced permeability and retention) could impact on the success of this strategy. In fact, the accumulation of nanocarriers within the tumor site could be enhanced by EPR effect describing the condition in which tumor capillaries, poorly differentiated and with inter-endothelial spaces, allow the extravasation of the NPs. The increase in permeability is mainly associated with the reduced drainage capacity of the tumor, due to a compromised lymphatic system, whereby the nanoparticles are able to accumulate in the tumor tissue and to exert their chemotherapeutic action [43]. This phenomenon is not observed with other low molecular weight chemotherapies that rely on the diffusion to penetrate the tumor, which cannot discriminate between healthy tissue and diseased tissue. However, passive targeting shows major limitations, as it is widely known that when administered intravenously, if not having "stealth" properties and very small particle size, nanocarriers are easily *opsonized* and removed from the circulation by cells of the reticuloendothelial system (RES). Only a small fraction of nanoparticles can therefore reach the tumor. In the case of brain tumors, EPR effect is inefficient due to a dense brain matrix and a high interstitial fluid pressure that prevents the spread of NPs in the brain parenchyma [46].

2. *Active targeting*: specific surface engineering is required to increase the ability to target selective molecules and receptors expressed on target cells or structure. This approach has been widely investigated, with some relevant preclinical and phase 1 and 2 results with different types of nanocarriers.

   In the case of brain cancer, from a general point of view, the main targeting goals are surely represented by the BBB (which must be overcome if it is in a healthy state) and brain cancer cells. Therefore, a number of studies were conducted in order to explore potentiality in exploiting **receptor-mediated endocytosis mechanisms (transcytosis)**, which could allow the translocation across the BBB of nanosized particles inside the brain compartment. The most present approaches consist of exploiting the interaction of the surface ligand on the nanocarrier (transferrin, transferrin receptor antibody, lactoferrin, melanotransferrin, folic acid, and mannose) with specific receptors, followed by the formation of endocytic vesicles able to uptake the nanocarriers. The process continues with BBB transcytosis and endocytosis in the cerebral parenchyma. Although this mechanism is the most promising resource in terms of "transport route through the BBB," it shows some limits. Most of the receptor-mediated approaches are based on strong bonds among the receptor and the ligand conjugated on the surface of the nanocarrier, and if this bond turns out to be too strong, the nanocarriers will hardly be able to detach itself, resulting in a low degree of exocytosis [47]. Furthermore, a receptor saturation mechanism may occur due to a limited number of receptors and a high amount of NPs attempting to interact with them. This hinders the mechanism of endocytosis. Therefore, while in the past the binding of the receptor ligand and their affinity played a major role, today it is known that the ligand needs to have specificity for its receptor, but the binding affinity should not be so high as to prevent the detachment of the nanocarrier and the mechanism of exocytosis [47].

3. *Long circulating strategy:* another example of a strategy to overcome BEE and achieve brain targeting is **pegylation**, or the use of hydrophilic polymers, such as polyethylene glycol (PEG), to cover the surface of the nanoparticle and make it more resistant to opsonization by the RES. Pegylation increases the plasma half-life of the particle, allowing it to travel long distances and reach the tumor site, due to the discontinuity of endothelial cells in tumor capillaries, which is not always a constant feature in all types of tumors, and it is not the case of GBM or other brain tumors.

7 Nanoparticles for Brian Tumor Therapy

In this meta-analysis study, particular attention was been focused, on the use of polymeric/lipid nanocarriers for GBM therapy. To evaluate the efficacy of the nanosystems currently under study, and their potential use in clinical practice, it was decided to consider certain parameters/protocols/conditions, which are considered pivotal to fully describe the potential advantages/disadvantages of the proposed treatment, compared to conventional therapies.

The first protocol is connected to quality and the numbers of the in vivo studies conducted on nanoparticles, which are still relatively insufficient and often under development. As evident from literature review, we assist to a lack of standardized and unique well-established animal model of GBM. In fact, most of the preclinical research is performed on animal models (mice and rats) undergone to an intracranial implantation of different types of tumor cells obtained by surgical resection. The most common are U87-MG e C6 glioma, but these models were often debated since it was demonstrated that they do not fully reproduce the human GBM. Therefore, the most interesting innovation in preclinical studies was achieved by using 101/8 GBM line in an orthotopic animal model to establish malignant grade IV glioblastoma model, with a high capacity for brain growth and a low tendency to necrosis [48, 49]. This GBM model was considered as highly comparable, both from histological and morphological point of view to human glioblastoma [48].

Then, the experimental protocols are often not fully completed if considering number and types of tested samples, number of animals (generally 5–10 per group), and way of administration (i.v. or i.p.). To be fully complete in designing a protocol for in vivo testing of nanomedicine, which could really give reliable outputs in terms of efficacy in treatment, we consider as pivotal and minimum samples the following: (1) saline solution or placebo, (2) free drug solution, (3) empty unmodified nanocarrier, (4) unloaded targeted nanocarriers, and (5) GBM or BBB targeted nanocarriers loaded with the drug.

In order to give a critical aspect to the evaluation of the experiments conducted with nanocarriers against glioblastoma and to consider the validity of these in vivo studies, relevant parameters have been identified, such as formulation aspects, therapeutic efficacy, treatment safety, and anticancer drugs used.

1. **Formulation aspects:** A number of formulation aspects, which deeply impact on the efficiency and success of the proposed treatment, would be recognized in order to better describe the strategy and the approach used in the experiments.

   (a) *Composition*: the *type of polymer* used should be firstly evaluated to assure biocompatibility of the nanocarriers with the organism, to avoid any toxic reactions, as well as its hydrophilic or lipophilic nature, which would impact on the choice of protocols for the most suitable encapsulation efficiency with drugs of different chemical-physical features.

   (b) *Encapsulation efficiency*: this parameter reveals *the amount of drug that can be encapsulated inside the nanocarrier*, expressed as a weight/weight percentage. High encapsulation efficiency would be important to provide a sufficient quantity of active ingredient to be transported and released to the tumor site. This characteristic strongly depends on the nature of the drug and on the nature of the material of which the nanocarriers are made.

(c) *Specific surfaces and ligands*: presence or absence of a ligand on the surface of the nanoparticle is a pivotal feature for successful surface engineering. Therefore, after surface conjugation with proper ligand, the *binding specificity for structures on tumor cells is the critical point*. The presence of a surface ligand allows nanocarriers to perform an active targeting, which is important to increase selectivity towards specific structures overexpressed by the tumor cell (receptors, surface molecules) and therefore allows a greater accumulation of drug in the tumor mass, limiting in this way the accumulation in non-target tissues, too. Surface engineering could be also used to improve translocation across the BBB, and if also cancer targeting ligands are present, *dual-targeting* approach represents the approach. The greater the ligand is specific to the tumor, the greater is the selectivity of the treatment and the accumulation of the drug at the level of the tumor mass. This fact limits also the risk of side effects due to the toxicity of the chemotherapeutic agent on non-target organs.

2. **Therapeutic efficacy**: to assess whether nanosystem has possible future applications at clinical levels, its effectiveness must be verified. *Therapeutic efficacy* evaluation could be analyzed by taking into consideration the following parameters:

(a) *Average survival*: represents the effect in terms of life expectancy of animals used in the experiments of the treatment with nanocarriers loaded with active anticancer drugs compared to the placebo or free drug solution. The obtained results must demonstrate a significant increase in the life expectancy of the animal to be valid, expressed at least as the number of days of life following the last cycle of therapy.

(b) *Tumor mass*: is the tumor volume generally expressed in cm^3 or mm^3, calculated at the end of the treatment. A significant reduction of the tumor mass after treatment with the drug carried by nanocarrier, compared to the initial volume or compared with the placebo treatments or free drug solution, indicates a good therapeutic efficiency.

(c) *Tumor inhibition index*: is expressed as a percentage, and it indicates the effective capacity of the drug carried by nanocarrier to reduce tumor proliferation and growth. The data obtained are always compared with those provided by the placebo and the drug's solution.

3. **Treatment safety**: toxic effects that may occur following treatment with nanocarriers should be carefully investigated or at least considered, since these outcomes would negatively affect the overall evaluation of the proposed approach, together with the evaluation of therapeutic efficacy, in order to speed up the transfer from preclinical to clinical phase.

   In this context, the risk of toxicity could be extrapolated by designed experiments on:

   (a) *Biodistribution*: identifies as a critical aspect *where* nanocarriers tend to accumulate, their fate or their tropism. In fact, if a considerable amount of nanocarriers (and therefore of the loaded drug) reaches non-target tissues instead of the tumor, the risk of developing more or less serious side effects is considerably high and may lead to ineffective therapy. As the approach of nanomedicine is also thought to circumvent one of the main limits of current conventional therapies as the poor selectivity for cancer cells, having a biodistribution in favor of the tumor mass with none or limited accumulation in other organs (liver, spleen, lungs, kidney, heart) would guarantee the selectivity of treatment and a significant reduction in toxicity.

   (b) *Animal weight loss*: expressed as BMW (body mass weight), it is considered one of the most significant *toxicity indices*. A considerable weight loss normally means a relevant toxicity associated to the treatment. Normally, the weight of the animal is monitored for all the treatment, and results are compared with control treatments as placebo or free drug's solution, to highlight both advantages and disadvantages of the tested approach based on drug-loaded nanocarriers.

   (c) *Death due to toxicity*: any death occurring during the experimentation following hyper-dosage or manifestation of toxic effects should be reported, as it is useful in determining the risk/benefit ratio of the therapy, as well as the maximum dose of treatment.

   (d) *Integrity and status of BBB*: to assess an overall evaluation of the approach, the state of BBB and its integrity should be stated. It is important to consider a couple of aspect: the first is concerning the toxicity of any therapy (nanomedicine or free drug) to the BBB, which should not be influenced on its permeability. Secondly, we have to consider whether the BBB is healthy or not in function of the state and severity of the tumor, as it will deeply impact on the targeting strategy. This last consideration is mainly connected to the design of the NPs, which will be selectively targeting both BBB crossing and cancer cells or only cancer cells in the case the BBB is completely disrupted.

4. **Anticancer drugs**: considering the wide and broad spectrum of anticancer drugs or active molecules used in therapy of GBM, we decided to focus mainly on the most used drugs yet used in the treatment of these kinds of brain tumors. Therefore, the drugs taken into consideration are those of greater interest in the therapy for their effectiveness against cancer cells, such as paclitaxel, docetaxel, doxorubicin, temozolomide, and nitrosoureas (such as carmustine and lomustine). The aim of this specific aspect of meta-analysis is also to identify for each type of drug the most complete and most beneficial in terms of efficacy and toxicity, considering all the parameters mentioned above and comparing different studies, and therefore to identify the one which has greater chance of entering in clinical phase in the future.

7.1 Paclitaxel

Paclitaxel (PTX) is one of the most effective chemotherapeutic agents in antitumor therapy and is able to stabilize microtubule activity and block cell mitosis. It demonstrated antitumor activity against glioma cells in vitro and on animal models of brain tumors [50]. Because of its low therapeutic index and lipophilic nature, its clinical application in the treatment against GBM is very limited. In addition, phase II studies have shown evidence of resistance development and poor penetration across the BBB [51]. For these reasons, PTX was one of the most studied anticancer drugs to be delivered by means of nanomedicine-based approach against GBM. In fact, the studies examining paclitaxel as a chemotherapeutic agent (Table 3 and Fig. 1) were related to **polymeric nanoparticles** [51–58], **polymeric micelles** [59], and **solid lipid nanoparticles (LNCs)** [60].

The most widely used polymers are **PEG-PCL** (poly-ethylene glycol-*co*-poly-ε-caprolactone) [51, 55, 59, 61], **PEG-PLA** (poly-lactic acid) [52, 58, 59], and **PEG-PLGA** (poly (lactic-*co*-glycolic acid)) [53, 54, 57]. They are all biodegradable polymers with high hydrophobicity, able to encapsulate hydrophobic active ingredients such as PTX.

PEG (poly-ethylene glycol) conjugated on the surface of the nanocarriers increases stealth properties of the systems, thus increasing half-life of the drug preventing the opsonization by the RES and allowing intravenous administration. LNCs, on the other hand, have a core of triglycerides such as Labrafac and are used for an intrathecal administration route using CED [60].

For all of these formulations, the **encapsulation efficiency** is very high, namely, over 90% for LNCs [60] NPs made of PEG-PLGA [56], slightly below 90% for NPs made of PEG-PCL [55]. The only cases of low encapsulation efficiency close to 30% were recognized for NPs made of PEG-PLA [58, 59]. These results and any differences must be related to the type of polymer used and the formulation technique applied. In general we can assume that

Table 3
Summary of applications of nanomedicines for paclitaxel delivery to glioblastoma

Considered parameters	Angiopep-2 [51]	IL-13 e Angiopep-1 [54]
Survival days more than placebo	15	15
Difference between the tumor mass given by the placebo and that given by the nanoparticle system (in mm^3)	36 mm^3	16 mm^3
Tumor inhibition index	65.6%	73.4%
Encapsulation efficiency	86%	78%
Weight loss and other toxicities	Not relevant	Not relevant
Biodistribution	Greater at tumor level RES Liver	Greater at tumor level Spleen Liver

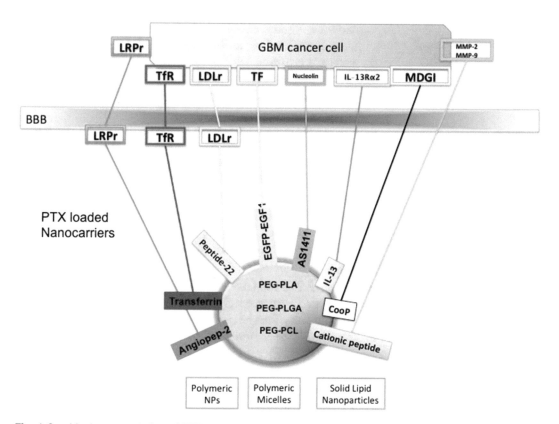

Fig. 1 Graphical representation of PTX-loaded nanocarriers used in GBM treatment

from a formulation point of view, if not considering the presence of PEG, the use of polyesters as co-polymers (PLGA/PCL) would favor a better stable encapsulation of the drug, probably due to solubility aspect.

Different types of ligands were utilized in these experiments aiming to BBB crossing or to target GBM directly:

1. **Angiopep-2** is a specific ligand of the **LRP receptor** (low-density lipoprotein receptor-related protein), a member of the low-density lipoprotein receptor family, highly expressed on BBB and mediating transcytosis of many ligands through the BBB such as lactoferrin and melanotransferrin. Luckily, LRP is also overexpressed on glioma cells; therefore this kind of receptor is a very valuable target for both BBB and glioma [51, 55, 62]. *Role in targeting: GBM targeting + BBB crossing.*

   · **Transferrin (Tf)**: Tf binds to its transferrin receptor (TfR) which is normally overexpressed at GBM cells level, but it would be useful also to cross the BBB as the receptor is also expressed at the BBB level and triggers endocytosis, enabling nanocarrier transfer across the BBB [58]. *Role in targeting: GBM targeting + BBB crossing.*

2. **Peptide-22 (Ac-[cMPRLRGC]-NH2)**: cyclic peptide showing high binding affinity for the LDL receptor, overexpressed at the tumor cell membrane level, but not on healthy brain cell membrane. While it is almost known that a number of peptides could display competition in LDL receptor with endogenous biological substrates of LDL-R, this peptide was demonstrated not to compete with circulating LDLs [57], thus assuring a favorable kinetics of interaction with the receptor also under physiological conditions. *Role in targeting: GBM targeting + BBB crossing.*

3. **EGFP-EGF1**: protein derived from coagulation factor VII that shows a specific binding affinity for the tissue factor (TF), overexpressed at the level of the tumor cells, without inducing coagulation events [58]. *Role in targeting: GBM targeting.*

4. **Aptamer AS1411 (Ap)**: oligonucleotide (single strand of DNA) folded into a specific three-dimensional structure that has high affinity for the nucleoline, a protein highly expressed in the plasma membrane of the cancer cells [53]. *Role in targeting: GBM targeting.*

5. **Interleukin-13 (IL-13)** interacts with its IL-13Rα2 receptor (an IL-13 receptor subunit) overexpressed on GBM cells [54]. *Role in targeting: GBM targeting.*

6. **Cationic peptide**: cationic peptide that manages to be easily internalized within cells. It consists of low molecular weight protamine (**LMWP** and **ALMWP**). It is a peptide linker of MMP (matrix metalloproteinases), which targets MMP-2 and MMP-9, overexpressed by GBM cells by which they characterize its high invasiveness [61]. *Role in targeting*: *GBM targeting*.

7. **CooP**: peptide that binds specifically to the mammary-derived growth inhibitor (MDGI), a member of a family of intracellular lipophilic proteins, including those that bind retinoic acid capable of binding fatty acids, which is overexpressed in certain subtypes of tumor cells including U87-MG [52]. *Role in targeting*: *GBM targeting*.

Among the selected studies on paclitaxel [10], only two approaches were describing the use of NPs [56] or LNCs [60] without any specific GBM ligands. In the case of NPs, authors reported a co-encapsulation within the same NPs with paclitaxel and temozolomide [56].

Interestingly, 40% of the selected studies utilized dual-targeting approach to assure both BBB crossing and GBM targeting. Among them, Zhang and colleagues [59] used micelles consisting of PEG-PCL and modified with **transferrin**, whose receptor is also expressed at the level of BBB and conjugation with **cyclo-[Arg-Gly-Asp-D-Phe-Lys] (c[RGDfK])**, a **peptide** that binds specifically to integrins overexpressed by glioma cells.

Similarly, in another study [57], targeted PEG-PLA NPs were produced using **peptide-22** to yet demonstrate to be able to interact with LDLR expressed on GBM as well as on endothelial cells, thus assuring with only one ligand a possible double targeting goal.

The same approach, meaning using the same ligand which could be used for both GBM and BBB endothelial cell targeting, was exploited by producing PEG-PLA NPs modified with **EGF-EGF1**, a specific ligand for tissue factor (TF) expressed at the level of BBB and GBM [58].

Finally, also Wang and colleagues analyzed PEG-PLGA NPs consisting of PEG-PGLA and modified with **IL-13** as ligand. It also includes the use of **angiopep-1**, a peptide that improves delivery through the BBB [55].

Regarding the therapeutic efficacy and results related to toxicity and biodistribution, the most relevant criticism is connected to lack of harmonization of the data, since some studies were limited to the analysis of the average survival, without considering the volume of the tumor mass or the tumor inhibition index or considering toxicity but not biodistribution.

The most complete researches could be therefore identified not only in relation to the presence of the most promising data in terms of therapeutic efficacy but also based on the completeness of the data provided, including biodistribution and toxicity studies.

The first study involves the use of NPs constituted by PEG-PCL and modified with **angiopep-2 (ANG-NPs)**. In this study, encapsulation efficiency of PTX-loaded NPs was close 86%. After treatment, comprehensive data are reported on therapeutic efficacy regarding average survival (37 days versus 22 days for placebo), tumor mass (19.35 mm^3 versus 55.72 mm^3 for placebo), and tumor inhibition index (65.6%). In addition, the authors provided information on toxicity by demonstrating that injection of 100 mg/kg per day of ANG-NP for 7 days does not result in significant weight loss or death. The blood parameters within 24 h of the last administration show no alterations; this is a sign that there is no evident renal or hepatic toxicity [55].

Biodistribution analysis were conducted by means of fluorescence, in comparison with control NPs, describing higher and significant signals related to ANG-NPs at the level of tumor mass at any time after injection, in a range between 2 and 24 h. The fluorescent signal given by ANG-NP is distributed throughout the brain, including the site of the tumor, indicating the fact that this type of particles accumulates in the brain not only for the EPR effect but also for the presence of the specific ligand. Due to accumulation of ANG-NPs (as well as control NPs) at the RES organs level by passive uptake and activation of the RES, only 2–10% of NPs can reach the target tissue.

The second study examined is the one conducted on polymeric micelles made up of PEG-PCL and modified with **transferrin**. In this study, Taxol, paclitaxel-loaded micelles (PM), transferrin-modified paclitaxel-loaded micelles (TPM), and TRPMs (transferrin-modified paclitaxel-conjugate-loaded micelles; micelles modified with Tf and loaded with PTX conjugated with the specific ligand for integrins) were tested. In this study, encapsulation efficiency of PTX-loaded NPs was close 83%. The average survival time registered after TRPMs treatment was 42.8 days versus 34.5 days of placebo. However, there are no data concerning the volume of the tumor mass and the tumor inhibition index [59]. Regarding safety, the toxicity was evaluated considering the body weight variation of mice (BMW), showing no weight variations with saline treatment contrarily to the remaining control groups (PTX solution). On the contrary, Taxol, PM,and TPM treatments lead to weight loss since the first administration and reach 80% of BMW on the thirtieth day of therapy. TRPMs did not show a relevant variation of BMW, thus confirming higher safety compared to other formulations, letting open possible increased frequency of administration, to maintain high drug concentration at tumor level, without relevant side effects [59]. Biodistribution studies showed significantly higher accumulation of the TRPMs in the tumor mass, in the peritumoral region and in normal brain tissue than other tested samples 1 h after administration. The higher level of tropism of the TRPM (0.3%/g tissue after 24 h) in the tumor and peritumoral region could be related to a higher expression of integrins, thus corroborating the rationale of using specific ligands for integrins.

The third significant study described the delivery of PTX by means of NPs made of PEG-PGLA **modified with IL-13**, to target IL-13 tumor receptor (encapsulation efficiency of PTX-loaded NPs was close 78%). In this study, the average survival is 32 days versus 17 days for placebo, the tumor mass at the end of the treatment is 6 mm^3 (22 mm^3 for placebo), and the tumor inhibition index is 73.4%. Regarding safety aspects, no significant weight loss and animal death were observed with Pep-NPs treatment. Liver, heart, spleen, lung, and kidney sections showed no signs of necrosis or degeneration compared with saline solution. The intravenous administration of 100 mg/kg of Pep-IL-13-NP-PTX for 6 days does not show signs of systemic toxicity. At every hour the concentration at the tumor mass follows the Pep-IL-13-NP-PTX>NP-PTX>Taxol order, while 4 h after the administration, the PTX concentration of Pep-IL-13-NP-PTX at the tumor level is 1.53 times greater than NP-PTX (28 ng/tissue versus 18 ng/issue) and 1.32 times greater than Taxol (15 ng/tissue). The peak is observed at 1 h after the administration, in which the Pep-IL-13-NP-PTX reaches 105 ng/tissue against the 60 ng/tissue of the NP-PTX and the 43 ng/tissue of the Taxol. The accumulation of NPs of both types is also observed in the liver and spleen, very low concentrations in the kidneys and lungs, but nothing in the heart [54].

7.2 Docetaxel

Docetaxel (DTX), like paclitaxel, although demonstrating an excellent antitumor efficacy against different types of tumor (breast, lung, ovarian cancer) and a good inhibition activity against glioma cells in vitro, shows poor applications in in vivo therapy, due to low selectivity on GBM and poor penetration capacity of BBB [62].

Despite the possible application of nanomedicine aiming to overcome this limitation, poor literature (Table 4 and Fig. 2) is present describing DTX loaded NPs, two of which involve the use of PEG-PCL-based NPs [38–62], and one involving the use of hybrid NPs of PEG-PLGA and phospholipids [63]. The choice to apply a phospholipidic monolayer external to the nanoparticle surface was investigated aiming to reduce the risk of toxicity and increase the biocompatibility as well as to improve the encapsulation of the drug, pharmacokinetics, and protection from external agents, such as pH and temperature, which could degrade the drug. The phospholipid layer also allows conjugation with GBM-specific ligands [63]. The encapsulation efficiency is indicated only in one of the three articles, and since DTX is a lipophilic drug, considering the type of polymers used, it is obviously very high (78%) [63].

Table 4
Summary of applications of nanomedicines for docetaxel delivery to glioblastoma

Considered parameters	IL-13 [38]	Angiopep-2 [62]	RGD [63]
Survival days more than placebo	18	13	41
Difference between the tumor mass given by the placebo and that given by the nanoparticle system (in mm^3)	Not indicated	Not indicated	Not indicated
Tumor inhibition index	Not indicated	Not indicated	73.4%
Encapsulation efficiency	Not indicated	Not indicated	78%
Weight loss and other toxicities	Not indicated	Not indicated	Not indicated
Biodistribution		Greater at tumor level	Greater at tumor level at 4 h after injection

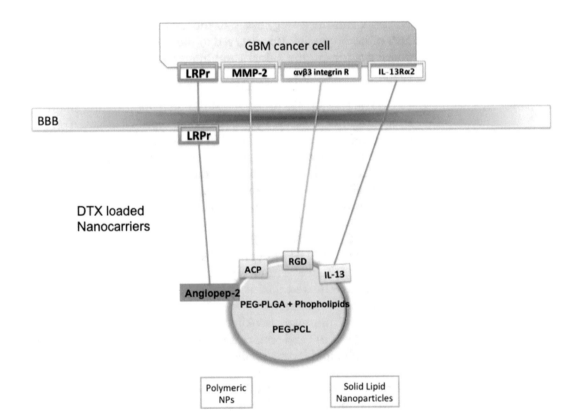

Fig. 2 Graphical representation of DTX-loaded nanocarriers used in GBM treatment

In these studies, the ligands use for surface engineering and targeting aim were mainly devoted to specifically target GBM cells, but as shown before, some BBB crossing extent was applied.

1. **Angiopep-2**: already analyzed for paclitaxel. *Role in targeting: GBM targeting + BBB crossing.*

2. **ctivatable-cell-penetrating peptide (ACP)**: peptide able to target MMP-2 (metalloprotease-2) overexpressed on the surface of tumor cells [62]. *Role in targeting: GBM targeting + BBB crossing.*

3. **RGD**: arginine-glycine-aspartic acid peptide that targets αvβ3 integrin receptors, overexpressed in 90% of GBM cells [38–63]. *Role in targeting: GBM targeting + BBB crossing.*

4. **Interleukin-13 (IL-13)** which, as already mentioned for paclitaxel, interacts with its IL-13Rα2 receptor (an IL-13 receptor subunit) overexpressed on GBM cells [11]. *Role in targeting: GBM targeting.*

The two studies conducted on the PEG-PCL NPs show only the average survival data as result on therapeutic efficacy, as reported for 32 days (versus 19 days for placebo) [62] and 35 days (17 days for placebo) [38]. The possible toxic effects were not examined in any of the three studies.

Considering the biodistribution, only one of the two works [62] analyzed the higher or lower presence of NPs in the brain compared to the other organs. After 2 h from the injection, the ANG-ACP-NP (angiopep-2 and ACP dual-modified NPs) and ANG-NP (angiopep-2 modified NPs) showed a high accumulation within the brain within 24 h (ANG-ACP-NP> ANG-NP> ACP-NP). The control NPs (NPs loaded with DTX) showed, as expected, a poor ability to penetrate the BBB and reach the tumor site.

The study conducted on hybrid NPs is the only one with information on the encapsulation efficiency, as well as information on the average survival, equal to 57 days (16 days for placebo). Also in this case, no information concerning any toxicity is indicated, while for biodistribution only a semi-quantitative analysis by fluorescence is reported, with data indicating that at 4 h from the injection the intensity of the fluorescent signal in the brain is very more powerful for the RGD-LP than the other formulations. Saline solution does not produce any signal [63].

7.3 Doxorubicin

Doxorubicin (DOX) is a chemotherapeutic agent belonging to the anthracycline family, which binds to cellular DNA by inhibiting the synthesis of nucleic acids and the mitotic process. Its antitumor efficacy against GBM was demonstrated by intrathecal administration, with remarkable increase in survival index, but if injected systemically (intravenously), DOX is poorly able to penetrate

Table 5
Summary of applications of nanomedicines for doxorubicin delivery to glioblastoma

Considered parameters	Polysorbate-80 [48, 64, 67]	Polysorbate-80 [49]	IL-13 [65]
Survival days more than placebo	35% long term with remission>180 days	Not indicated	100 days
Difference between the tumor mass given by the placebo and that given by the nanoparticle system (in mm^3)	1.15 major	31.34 mm^2	11 mm^3
Tumor inhibition index	Not indicated	Not indicated	Not indicated
Encapsulation efficiency	70%	78%	90%
Weight loss and other toxicities	Maximum tolerated dose 7.5 mg/kg weight	Not indicated	Not relevant
Biodistribution	1% of brain uptake 1 h after administration Accumulation at heart level 17 times less than free DOX	Not indicated	Not indicated

BBB, as it is a substrate of glycoprotein-P, the main efflux pump that determines the escape of the chemotherapeutic agent from the cerebral parenchyma, pouring it back into the vascular lumen. This prevents the achievement of cytotoxic concentrations at the tumor level [64].

The studies conducted on nanoparticles systems (Table 5 and Fig. 3), in which doxorubicin is encapsulated, analyze the use of non-pegylated nanoparticles made of biodegradable polymers as PBCA (poly(butyl-cyanoacrylate)), PLGA [48] and PICHA (poly (isoexyl-cyanoacrylate)) [49], and liposomes based on dipalmitoyl-phosphatidylcholine (DPPtdCho), cholesterol, PEG-distearoyl-phosphatidylethanolamine (PEG-DSPE), and DSPE-PEG-malei-mide in a molar ratio of 10:5:0.5:0.25 [65].

Due to its chemical-physical characteristics (amphiphilicity), DOX could be stably encapsulated into polymeric NPs but could also form stable complexes with these types of carriers by Van der Waals interactions [49]. Encapsulation efficiency is high in all the cases, as over 90% for liposomes [65], close to 80% for PICHA NPs [49] or close to 70% for PBCA-based NPs [48].

To better reach the targets, also in these researches several ligands were exploited, with the evidence that only one study involves the use of a specific ligand to GBM, while in the others the main focus was for BBB crossing aim and mainly on P-Gp to increase the transport of the drug in the tumor.

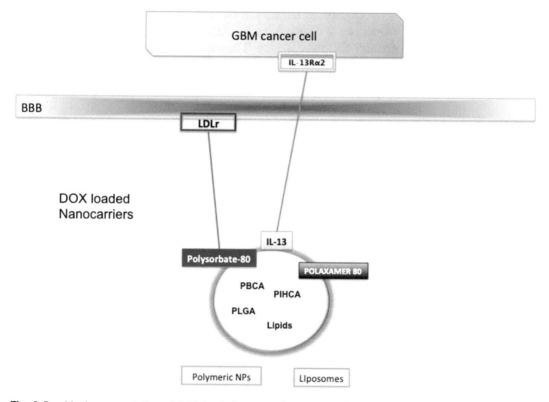

Fig. 3 Graphical representation of DOX-loaded nanocarriers used in GBM treatment

1. **Polysorbate-80**: non-ionic surfactant able mimics interaction of APO-E proteins with LDL receptors at BBB level [49].

2. **Poloxamer-188**: non-ionic triblock copolymers able to fluidize the lipid cell membrane of endothelial cells and increase the permeability [64] and that it lends itself to both NP-PLGA and NP-PBCA [66].

3. **Interleukin-13 (IL-13)**: already mentioned for paclitaxel, interacts with its IL-13Rα2 receptor (an IL-13 receptor subunit) overexpressed on GBM cells [65].

The study conducted on **NP-PBCA** [64] represents the most complete one, concerning data on therapeutic efficacy and toxicity. The most promising result was obtained in the group treated with 3 cycles of 1.5 mg/kg of DOX-NP + Ps80 (polysorbate-80-modified doxorubicin-loaded NPs), with evidence of a significant increase in survival, and more than 20% of the animals showed long-term remission (>180 days). The average survival is prolonged in the 3 × 2.5 mg/kg group; therefore there is a dose-dependent antitumor effect. However, animals with a long-term survival index died before 180 days, probably due to high-dose toxicity [64].

Concerning the size of the tumor mass, the untreated animals developed a tumor mass of maximum size with low necrotic areas around the twelfth day. Therefore, this temporal time point has been chosen as a comparison and histological evaluation time. After treatment with doxorubicin (delivered or free), the animals show a tendency to increased tumor growth accompanied by a slight inflammatory infiltration in the areas surrounding the tumor. Untreated animals, on the other hand, have a more invasive and widespread tumor, with high inflammation and many necrotic areas. The cells have a high rate of apoptosis and proliferation. The untreated group has a 4.3 times higher tumor mass, whereas for the group treated with NP-DOX + Ps the tumor mass is only 1.15 times greater compared to the dimensions that were recorded at the beginning of the treatment [64]. In the blood, quite low concentration differences were recorded between all four formulations up to 4 h. In the heart, classically site of accumulation of doxorubicin where it causes toxicity, the concentrations of both nanoparticulate formulations remained very low, while for free doxorubicin the concentrations were 17 times greater than those of NPs.

This is an important consideration given that the limited use of doxorubicin as an anticancer agent is due to its cardiotoxicity. It was observed that the highest concentration of polymer (polysorbate-80) in sick rats treated with NPs loaded with doxorubicin was about 1% of the dose after 1 h from the injection. In healthy rats, uptake of these NPs in the brain was lower, about 0.44% of the dose [48]. The toxicity of doxorubicin bound to NPs could be the same or relatively minor compared to the free one. The maximum tolerated dose is about 7.5 mg/kg body weight [67].

To assess full toxicity profile, body weight, blood biochemical parameters, hematologic parameters, and urine analysis were determined, showing that the toxicity for free DOX and DOX-NP + Ps is dose dependent, and the major side effects by administering the maximum tolerated dose (7.5 mg/kg) were partial or total mortality, considerable weight loss (16–25%), adynamia, tremors, and piloerection. From the third day diarrhea has occurred. No significant side effects were recorded with the dosage of 4.5 mg/kg. For DOX-NP (doxorubicin-loaded NPs) and DOX-NP + Ps, weight loss is significantly less and does not differ much from the control group. Mortality for a single administration at therapeutic dosages for DOX-NP + Ps is 12% [67].

Instead, the study that uses **IL-13 conjugated liposomes** shows an average survival greater than 100 days, with four of the seven treated mice that continued to survive more than 200 days after the last injection and generally characterized by a good physical state. The tumor was initially 5 mm^3: treatment with liposome control (without DOX) showed a tumor growth greater than 100% in 5 weeks, reaching 12 mm^3, while treatment with liposomes conjugated with IL-13 and loaded with DOX for a period of 6 weeks led to the reduction of the tumor up to 1 mm^3. Finally, the treatment with unconjugated liposomes and loaded with DOX has seen an absence of variation of the tumor mass which remains approximately the same (5 mm^3) after 5 weeks [65]. No signs of renal toxicity (plasma creatinine levels) or hepatic toxicity (plasma levels of GOT and ALT) were observed compared to the control group. In this study, no focus was recovered on biodistribution.

Regarding the study of DOX delivery by PICHA NPs, there are no data on average survival, but only on the tumor mass, equal to 3.96 mm^2 ± 5.16 mm^2 (compared with placebo, 35.3 mm^2 ± 24.0 mm^2) with a proliferation index of 35% (80% placebo). No data on biodistribution or toxicity were reported [49].

7.4 Temozolomide and Nitrosoureas Drugs

Temozolomide (TMZ) is a drug that is already used in therapy against GBM according to the Guidelines, but due to its poor selectivity (it also attacks the hematopoietic cells giving myelosuppression), its poor solubility in physiological conditions, and facility of hydrolyzing, it has many limitations in therapeutic use [27].

Lomustine and carmustine (BCNU) are two nitrosoureas widely investigated for cancer treatment, but showing a relevant number of drawbacks related to efficacy and bioavailability.

Lomustine is not currently used in therapy against GBM due to inability in administration since it displays a very high level of lipophilicity and very large molecular dimensions. Moreover, beside the limitation in administration, lomustine fails to exceed the BBB, distributing in other organs (in particular bone marrow) and giving serious toxicity (myelosuppression as for TMZ) [43]. Carmustine, instead, is currently present in the GBM therapy according to the Guidelines and is administered intracranially in the form of biodegradable wafers (Gliadel), since in aqueous solution the active principle is very unstable. This treatment, however, allows to increase the life expectancy for only 3 months and is associated in any case, due to its invasiveness, to the appearance of complications such as infections, epileptic seizures, cerebral edema, and abnormal healing of wounds. It also leads to the development of resistance; therefore it is often associated with O6-benzylguanine (BG), which blocks the activity of MGMT, a DNA repairing enzyme [69].

The studies conducted on nanoparticles systems based on TMZ and nitrosoureas (Fig. 4 and Table 6) involve the use of NPs based on PLA (polylactic acid) [70], chitosan [27] sometimes associated with PLGA (poly (lactic-*co*-glycolic acid)) [69], and QCPQ (*N*-palmitoyl-*N*-monomethyl-*N,N*-dimethyl-*N,N,N*-trimethyl-6-*O*-glycol-chitosan) [68]. All of the materials used in all the studies are considered as biodegradable polymers of synthetic origin, except chitosan, which is a natural cationic polysaccharide originating from chitin, a component of the exoskeleton of many

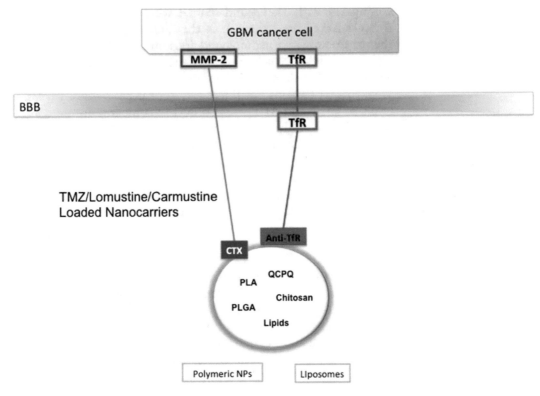

Fig. 4 Graphical representation of carmustine, lomustine, and TMZ-loaded nanocarriers used in GBM treatment

Table 6
Summary of applications of nanomedicines for TMZ and nitrosoureas delivery to glioblastoma

Considered parameters	[69]	IL-13 [72]
Survival days more than placebo	13 (death at day 16)	77 (death 45 days after the end of treatment)
Difference between the tumor mass given by the placebo and that given by the nanoparticle system (in mm³)	10.5 mm diameters	90 mm³
Tumor inhibition index	Not indicated	Not indicated
Encapsulation efficiency	Not indicated	46%
Weight loss and other toxicities	Not indicated	Not relevant
Biodistribution	Liver Spleen RES	Not indicated

crustaceans. It contains many amino groups that can be used to covalently link a large number of active ingredients and ligands. QCPQ is a polymer of an amphiphilic nature that also allows the oral intake of peptide or highly lipophilic drugs, and if injected intravenously it protects the drugs from degradation. It is also able to adhere to the luminal surface of endothelial cells, allowing transport of drugs through the BBB. It allows more than 40% of the drug to be loaded into the NPs [71].

Liposomal systems were less investigated compared to NPs, and one of the few papers described the use of DOTAP:DOPE liposomes to encapsulate TMZ [46].

Regarding encapsulation efficiency, even if in the case of lomustine and carmustine is not clearly reported, data concerning TMZ described a higher encapsulation efficiency when polymeric NPs are used (60%) compared to liposome (45%).

In these researches, only few studies exploited ligands for targeting devoted on both BBB crossing and GBM targeting, namely:

1. **Chlorotoxin (CTX)**: 36-amino acid peptide able to target selectively GBM cells in a specific way, since the target on which it acts is MMP-2 and allows the internalization of NPs at the intracellular level. It is also able to inhibit tumor invasion [27].

2. **Single-chain transferrin anti-receptor antibody (Anti-Tf)**: able to target the transferrin receptor, overexpressed at BBB and GBM cells. At the level of BBB, it actively regulates transcytosis through endothelial cells, so it can optimize the transport of the nanocarriers in the brain parenchyma [72].

The output coming out from these papers indicate a heterogeneity in protocols as, for example, biodistribution data, without considering therapeutic efficacy and toxicity in some cases [27, 70]. In the case of nitrosoureas-based NPS, biodistribution and survival are mentioned (lomustine) [68], while for BCNU it is also analyzed the volume of the tumor mass [69]. On the other side, the research developed on TMZ-based liposomes also considered toxicity data [72]. PLP-CS NPs loaded with carmustine (unmodified) were found to be highly accumulated in the liver and spleen, due to accumulation in the RES. No significant increases were recorded in the heart and kidney. The peak is recorded 1 h after administration. Considering efficacy in treatment, the animal experiments showed an average survival of 23 days (compared to 10 days of the placebo treatment), while the volume of the tumor mass calculated on the twelfth day was 2.4 ± 1.2 mm in diameter, against the data obtained after treatment with placebo (12.9 ± 2.2 mm) and with death of all rats on day 16th [69].

Interestingly, **liposomes modified with anti-Tf** were investigated by using two animal models obtained by intracranial implantation of two types of tumor cells. In the first animal model with U87-luc2 implantation, tumor growth was monitored until day 51. In this case, the experiment showed that both the free TMZ and the TMZ carried in liposomes are able to reduce the tumor mass, but considering anti-Tf engineered TMZ-Liposomes system, the level of inhibition was significantly higher. Furthermore, the experiments demonstrated that the tumor continued to grow, even if more slowly, following treatment with free TMZ, but when anti-Tf engineered TMZ-liposomes were administered, tumor growth was inhibited up to 3 weeks after the end of treatment. As confirmation on the efficacy, at day 19, for the untreated group the volume of the tumor mass was 100 mm^3, while for anti-Tf engineered TMZ-liposome decreased down to 10 mm^3. The average survival was 77 days for 37.5% of mice (with death 45 days from the end of treatment with liposomes). In the second animal model, obtained with an intracranial implant with T98G cells, the tumor mass measured at the 18th day after treatment with TMZ delivered by liposomes was considerably lower (55%) compared to the one treated with free TMZ, which was only 14%. Moreover, mice treated with anti-Tf engineered TMZ-liposomes showed constant BMW, while those treated with free TMZ show a significant weight loss on day 10, and then resume on day 18. From these results, we can conclude that the treatment with TMZ encapsulated in TfR targeted liposomes lead to a significant decrease in toxicity of TMZ and therefore a higher efficacy rate. In fact, at a dose of 75 $mg/m^2/$ day, free TMZ showed myelosuppression in 35–45% of cases, while when it is encapsulated in liposomes, there are no significant hematological effects. There are no creatine phosphokinase (CPK) increases, and cardiotoxicity markers were identified, whereas for free TMZ, CPK levels are 6 times higher than normal [72].

8   Conclusion and Perspectives

To facilitate a better view of the different applications reviewed, the classification and the comparison between nanocarriers on the basis of material, amount of encapsulated drug, chosen ligand, designed target, and encapsulation efficiency, together with average survival, tumor mass volume, and tumor inhibition index, representing the therapeutic efficacy of the different nanoparticulate systems examined, is reported in Table 7.

To finalize the analysis of the critical issues as lack of ligand, toxicity, and biodistribution, we described a comparison on the presence of biodistribution and toxicity studies, together with dual targeting mechanisms, while the absence of data related to these parameters represents a critical issue. From the comparison, it

Table 7

Summary of principal nanomedicines applied to glioblastoma treatments on the basis of material, drug, encapsulation efficiency, ligand and target, type of animal on which the experiments were conducted, the type of tumor cell used, and the route of administration, average survival, tumor mass volume, inhibition of tumor growth biodistribution, toxicity, and presence of dual-targeting (*PTX* paclitaxel, *DTX* docetaxel, *DOX* doxorubicin, *TMZ* temozolomide, *BCNU* carmustine, *LNC* liposomes, *CS* chitosan)

Types of carrier	Material	Drug	E.E.	Ligand	Target	In vivo experiments	Therapeutic efficacy	Added value	Critical issues	References
NPs	PEG-PCL	PTX	86%	Angiopep-2 (ANG)	LRP (GBM and BBB)	Animals: Mice IV administration Tumor cells: U87-MG glioma	*Average survival:* 37 day (NPs) vs. 22 day (placebo) *Tumor mass volume:* 19.35 mm³ (NPs) vs. 55.72 mm³ (placebo) *Tumor inhibition:* 65.6%	*Dual targeting:* YES *Biodistribution:* • Increased localization at tumor level and brain parenchyma at any time point • Accumulation in tumor tissue (2–10%). *Toxicity:* • Absence of weight loss or death • Unchanged blood parameters (24 h after admin) • NO renal or hepatic toxicity	Accumulation of ANG-NP in the liver, in the other organs and in the RES Slight accumulation of ANG-NP also in the kidneys and lungs	[51, 55]
Micelles	PEG-PCL	PTX	83%	Transferrin (Tf) c[RGDfK]	TfR (BBB and GBM) Integrin (GBM)	Animals: Mice IV administration Tumor cells: U87-MG glioma	*Average survival:* 42.8 day (NPs) vs. 34.5 day (placebo)	*Dual targeting:* YES *Biodistribution:* • Accumulation in tumor tissue (0.7%/g tissue) • Lower side accumulation *Toxicity:* • Unchanged BMW values	Lack of data on tumor mass volume and tumor inhibition index increased accumulation in lungs compared to other formulations	[59]

(continued)

Table 7
(continued)

Types of carrier	Material	Drug	E.E.	Ligand	Target	In vivo experiments	Therapeutic efficacy	Added value	Critical issues	References
NPs	PEG-PLA	PTX	30%	Peptide-22	LDLR (GBM)	Animals: Mice IV administration Tumor cells: C6-glioma	Average survival: 32 day (NPs) vs. 14 day (placebo)	Dual targeting: Biodistribution: • Tested NPs and control NPs show similar biodistribution. • Higher tumor accumulation (fluorescence signals 2.48-fold higher) after 24 h • High accumulation in the liver and spleen • Low accumulation in heart, lungs, and kidneys	Dual targeting: NO Low encapsulation efficiency Lack of data on tumor mass volume and tumor inhibition index Lack of toxicity studies	[57]
Ps	PEG-PLA	PTX	33%	EGF-EGFl	Tissue Factor (GBM and BBB)	Animals: Mice IV administration Tumor cells: C6-glioma	Average survival: 27 day (NPs) vs. 14 day (placebo)	Dual targeting: YES Biodistribution: • Accumulation in tumor tissue (fluorescence, 2.38-fold higher) after 6 h • Low accumulation in other organs compared to the control	Low encapsulation efficiency Lack of data on tumor mass volume and tumor inhibition index Lack of toxicity studies High hepatic accumulation	[58]

NPs	PEG-PLGA	PTX	45%	**AS1411** Nucleolin (GBM)	Animals: Mice IV administration Tumor cells: C6-glioma	Average survival: 31 day (NPs) vs. 18 day (placebo)	Dual targeting: YES Biodistribution: • Tumor accumulation 2.44 times greater than control	Lack of data on tumor mass volume and tumor inhibition index Lack of toxicity studies Higher concentration in the liver according to the order: Ap-PTX-NP> NP-PTX> taxol	[53]
LNC	Triglycerides (Labrafac)	PTX	93%	Not present Not present	Animals: Rats Intrathecal administration Tumor cells: glioma 9 L	Average survival: about 54 day (LNC) vs. 27 day (placebo) –1/10 long-term survival	Toxicity studies: • No signs of relevant toxicity • No significant weight loss • No sign of brain necrosis	Absence of ligands justified by route of administration Lack of data on tumor mass volume and tumor inhibition index Data on biodistribution not required due to route of administration	[60]
NPs	PEG-PLGA	PTX	78%	**IL-13 Angiopep-1** IL-13Rα2 (GBM) BBB	Animals: Mice IV administration Tumor cells: C6-glioma	Average survival: 32 day (NPs) vs. 17 day (placebo) Tumor mass volume: 6 mm³ (NPs) vs. 22 mm³ (placebo) Tumor inhibition: 73.4%	Dual targeting: YES Biodistribution: • Tumor Accumulation (Pep-IL-13-NP-PTX> NP-PTX> Taxol) at each time-point • High PTX tumor (1.53 times greater than	The accumulation of NPs of both types is also observed in the liver and spleen	[18]

(continued)

Table 7
(continued)

Types of carrier	Material	Drug	E.E.	Ligand	Target	In vivo experiments	Therapeutic efficacy	Added value	Critical issues	References
								control) • Low concentrations in kidneys, lungs, and heart *Toxicity studies:* • No death • Not significant weight loss • No sign of necrosis • No signs of systemic toxicity at high doses		
NPs	PEG-PCL	PTX	49%	LMWP ALMWP	*MMP-2 e MMP-9 (GBM)*	Animals: Mice IV administration Tumor cells: C6-glioma	*Average survival:* 48 day (NPs) vs. 20.5 day (placebo)	*Biodistribution:* • Higher tumor accumulation (2.86-fold more than control), 24 h after administration • Zero concentrations in kidneys and lungs	*Dual targeting:* NO Lack of data on tumor mass volume and tumor inhibition index Increased accumulation in the liver and spleen Lack of toxicity studies	[61]

Type	Polymer	Drug	Encapsulation	Ligand	Target	Model	Efficacy	Biodistribution/Toxicity	Limitations	Ref
NPs	PEG-PLA	PTX	44%	Coop	MDGI (GBM)	Animals: Mice IV administration Tumor cells: U87-MG glioma	Average survival: 47.5 day (NPs) vs. 20 day (placebo)	Biodistribution: • Increased tumor accumulation • Better penetration of the cerebral parenchyma • Similar concentrations in liver and spleen for both formulations • Less accumulation in the heart, lung, and kidney	Dual targeting: NO Lack of data on tumor mass volume and tumor inhibition index Lack of toxicity studies	[52]
NPs	PEG-PLGA	PTX +TMZ	90% 65%	Not present	Not present	Animals: Mice IV administration Tumor cells: U87-MG glioma	Tumor mass volume: 0.5 cm^3 (NPs) vs. 4 cm^3 (placebo) Tumor inhibition: 90%	Toxicity: • Significant weight loss only registered with free drugs	Lack of any ligand Lack of data on average survival Lack of biodistribution studies	[56]
NPs	PEG-PCL	DTX	Not indicated	Angiopep-2 ACP	LRP (GBM and BBB) MMP-2 (GBM)	Animals: Mice IV administration Tumor cells: C6-glioma	Average survival: 32 day (NPs) vs. 19 day (placebo)	Dual targeting: YES Biodistribution: • Higher brain accumulation (24 h after treatment) (ANG-ACP-NP> ANG-NP> ACP-NP)	Lack of encapsulation efficiency Lack of data on tumor mass volume and tumor inhibition index Lack of toxicity studies	[62]
Hybrid NPs	PEG-PLGA and phospholipids	DTX	78%	RGD	Integrin $\alpha_v\beta_3$ receptors (GBM)	Animals: Rats IV administration Tumor cells: C6-glioma	Average survival: 57 day (NPs) vs. 16 day (placebo)	Biodistribution: • Cerebral fluorescent signal intensity much more powerful for RGD-L-P than other formulations	Dual targeting: NO Lack of data on tumor mass volume and tumor inhibition index Lack of toxicity studies	[63]

(continued)

Table 7
(continued)

Types of carrier	Material	Drug	E.E.	Ligand	Target	In vivo experiments	Therapeutic efficacy	Added value	Critical issues	References
NPs	PEG-PCL	DTX	Not indicated	RGD IL-13	Integrin $\alpha_\nu\beta_3$ receptors (GBM IL-13α2 (GBM)	Animals: Mice IV administration Tumor cells: C6-glioma	*Average survival:* 35 day (NPs) vs. 17 day (placebo)	*Dual targeting:* YES	Lack of encapsulation efficiency Lack of data on tumor mass volume and tumor inhibition index Lack of toxicity studies Lack of biodistribution studies	[38]
NPs	PBCA	DOX	70%	Polysorbate-80	pG-P (BBB)	Animals: Rats IV administration Tumor cells: 101/8 GBM	*Average survival:* 35% long-term with remission >180 day *Tumor mass volume:* (after 12 day) Increase 1.15-fold greater (NPs) vs. 4.3 fold (placebo)	*Biodistribution:* • Brain drug concentration (6 μg/g) corresponding to 1.2% of the administered dose • Low levels of accumulation of NPs in the heart and testicles *Toxicity:* • Maximum tolerated dose close to 7.5 mg/kg • -Dose-dependent toxicity • Not significant weight loss • Significant reduction of cardiac and testicular toxicity	*Dual targeting:* NO Selective target for GBM missing Lack of data tumor inhibition index	[48]

Type	Composition	Target (barrier)	Drug	Encapsulation	Experimental model	Results	Toxicity	Comments	Ref
NPs	PLGA	Poloxamer-188 / BBB	DOX	/	Animals: Rats, IV administration, Tumor cells: 101/8 GBM	*Average survival:* 40% long term	/	*Dual targeting:* NO, Selective target for GBM missing, Lack of encapsulation efficiency, Lack of data on tumor mass volume and tumor inhibition index	[48]
LNC	DPPtdCho cholesterol PEG-DSPE DSPE-PEG-M	Il-13 ILR-13 (GBM)	DOX	90%	Animals: Rats, IV administration, Tumor cells: U87-MG glioma	*Average survival:* More than 100 day (4 out of 7 more than 200 day); *Tumor mass volume:* From 5 to 1 mm^3 in 6 weeks after LNC treatment	*Toxicity:* • Absence of renal of hepatic toxicity	*Dual targeting:* NO, Lack of data on tumor inhibition index, Lack of biodistribution studies	[65]
NPs	PICHA	Polysorbate-80 pG-P (BBB)	DOX	78%	Animals: Rats, IV administration, Tumor cells: 101/8 GBM	*Tumor mass volume:* 3.96 mm^2 ± 5.16 mm^2 (NPs) vs. 35.3 mm^2 ± 24.0 mm^2 (placebo); *Proliferation index:* 35% (NPs) vs. 80% (placebo)	/	*Dual targeting:* NO, Selective target for GBM missing, Lack of data on average survival, Lack of biodistribution studies, Lack of toxicity studies	[49]

(continued)

Table 7
(continued)

Types of carrier	Material	Drug	E.E.	Ligand	Target	In vivo experiments	Therapeutic efficacy	Added value	Critical issues	References
NPs	CS	TMZ	49%	CTX	MMP-2 (GBM)	Animals: Mice IV administration Tumor cells: None	/	*Biodistribution:* • BBB state evaluation (healthy state) *Toxicity:* • Specific targeting with CTX reduces the amount of TMZ to be administered • Potential reduced bone marrow toxicity • Increased half-life of TMZ	*Dual targeting:* NO Lack of studies related to pharmacological efficacy Lack of studies related to BMW	[27]
NPs	PLA	TMZ	61%	Not present	Not present	Animals: Mice IV administration Tumor cells: None	/	*Biodistribution:* • Accumulation of encapsulated TMZ in brain tissue 4 times greater than the free drug (80 μg/ml compared to 20 μg/ml of free TMZ) • Similar accumulation at the pulmonary level for encapsulated TMZ and free TMZ	Lack of any ligand Lack of studies related to pharmacological efficacy Lack of toxicity studies	[70]

							• Same amount of TMZ at the pulmonary level for PLA-NP-TMZ and free TMZ	
LNC	DOTAP: DOPE TMZ	46%	Anti-Tf	TfR (GBM and BBB)	Animals: Mice Administration IV Tumor cells: U87-luc2 e T98G	Average survival: 77 day (45 day of survival after the end of treatment) Tumor inhibition: 55% (inhibited up to 3 weeks from the end of treatment)	Dual targeting: YES • In vivo experiments conducted on two types of different tumor cells Toxicity: • BMW practically constant • No significant hematologic events, reduction of myelosuppression	Lack of data on tumor inhibition index Lack of biodistribution studies [46]
NPs	GCPQ Lomustine	/	Not present	Not present	Animals: Mice IV administration Tumor cells: U87-MG glioma	Average survival: 33, 17 day (NPs) vs. 17, 14 day (placebo)	Biodistribution: • Reduction of bone marrow and liver exposure to the drug about 25 and 38%, respectively • Concentration in the brain is 2 times greater than the lomustine solution Toxicity: • The highest dose of MET-lomustine (13 mg/kg) does not increase myelosuppression • Decreased platelet count for both formulations • Unchanged BMW	Lack of any ligand Lack of encapsulation efficiency Lack of data on tumor mass volume and tumor inhibition index [68]

(continued)

Table 7
(continued)

Types of carrier	Material	Drug	E.E.	Ligand	Target	In vivo experiments	Therapeutic efficacy	Added value	Critical issues	References
NPs	PLGA-CS	BCNU + BG	/	Not present	Not present	Animals: Rats IV administration Tumor cells: F98-glioma	*Average survival:* 23 day (NPs) vs. 10 day (placebo) *Tumor mass volume:* Calculated at day 12 2.4 ± 1.2 mm diameters (at day 4 no evidences of tumor)	*Biodistribution:* • Accumulation in liver and spleen (accumulation in the RES) • There are no significant increases in the heart and kidney • Peak 1 h after administration	Lack of any ligand Lack of encapsulation efficiency Lack of data on tumor inhibition index Lack of toxicity studies	[69]

is evident how the different studies, if well conducted and describing promising results in terms of efficacy, are not harmonized in terms of presence of full preclinical tests.

The absence of correct and coherent tests, with proper control samples and with standardized protocols represents the most important reason for the slow progression from preclinical to clinical testing and therefore strongly impacting on the translatability of nanomedicine into clinical practice.

Moreover, in order to be fully considered as promising in GBM treatment, the harmonization of the protocols of preclinical testing will help in replicating the results in independent laboratories, which is the first requisite for the assessment of the reliability of the obtained tests.

Concluding, the use of polymeric and lipid nanomedicine could really represent a relevant and promising approach for the treatment of glioblastoma, since preclinical data really demonstrated efficacy in treatment.

Even if the application of nanomedicine to treat brain cancer is still under investigations, due to the literature evidences, we can describe high potential in the management of both diagnosis and cure.

This hope for future amelioration of the current strategies of GBM therapy would only be possible if some critical issues will be solved and investigated in more depth:

1. *Design of nanomedicine*: the design of the nanomedicine will strongly influence the final outcome, in particular considering the stage of GBM and the pathological features which will drive the planning of the nanomedicines in terms of ligands and surface engineering, targeting efficiency, and finally bioavailability of the loaded drug. Moreover, the choice of the polymer will strongly impact on the final release of drugs as well as on the biodegradability and biocompatibility of the nanomedicines which will be therefore considered, along with the drug loaded inside, as a new "drug" as whole.

2. *Scale-up process*: if a nanomedicine would aim to be translated into clinical practice, above all the preclinical outcomes on efficacy and efficiency in targeting, the scale-up process will deeply govern its future application. In this view, the choice of the polymer used, the technology of nanoformulation, the use of organic solvents, the application of mild and "green" chemical reaction, and most of all the yield of production must be taken into high consideration. This issue is of particular importance, since at laboratory/academic level, generally, the researchers do not plan the production of the nanomedicines considering also these aspects, which are pivotal from an industrial point of view.

3. *Full characterization of the nanomedicine:* regulatory bodies (EMA, FDA, etc.) are now focusing not only on the efficacy of nanomedicines but also on the safety aspects. In particular, full characterization of nanomedicines in terms of average size, shape, morphology, drug loading, drug release kinetics, stability over time, and other key parameters must be detailed not only in aqueous or saline solutions but also in relevant environments (i.e., plasma) to be easily transferred and to furnish reliable data for future clinical application. In this view, a clear and defined standardization of the characterization procedures (which is now under development in a very high number of laboratories) will furnish future directions for creation of gold standard procedures and formulations. This aspect goes in the way of creating nanoproducts featured by a very high quality level, which will be ready for scale-up process and also for authorization for clinical use.

Acknowledgments

We gratefully thank UNIMORE for supporting activities in nanomedicine and FAR grant, Veronesi Grant (FUV), and the PhD School "Clinical and Experimental Medicine" of University of Modena and Reggio Emilia and thanks to MAECI Project 2019-2021 "Nanomedicine for BBB crossing in oncologic pathologies" and "Euronanomed III, Silk-fibroin interventional nano-trap for the treatment of glioblastoma".

References

1. Lima FR, Kahn SA, Soletti RC et al (2012) Glioblastoma: therapeutic challenges, what lies ahead. Biochem Biophys Acta 1826:338–349

2. Hanahan D, Weinberg RA (2011) Hallmarks of cancer: the next generation. Cell 144:646–674

3. Louis DN, Ohgaki H, Wiestler OD (2007) The 2007 WHO classification of Tumours of the central nervous system. Acta Neuropathol 114:97–109

4. Perkins A, Liu G (2016) Primary brain tumors in adults: diagnosis and treatment. Am Fam Physician 93:211–217

5. Thakkar JP, Dolecek T, Horbinski C et al (2014) Epidemiologic and molecular prognostic review of glioblastoma. Cancer Epid Biomark Prevent 23:1985–1996

6. Siegal T (2015) Clinical impact of molecular biomarkers in gliomas. J Clin Neurosci 22:437–444

7. Alifieris C, Trafalis DT (2015) Glioblastoma multiforme: pathogenesis and treatment. Pharmacol Therap 152:63–82

8. Ostrom QT, Bauchet L, Davis FG et al (2014) The epidemiology of glioma in adults: a "state of the science". Neuro-Oncology 16 (7):896–913

9. Jhanwar-Uniyal M, Labagnara M, Friedman M et al (2015) Glioblastoma: molecular pathways, stem cells and therapeutic targets. Cancers 7:538–555

10. Reya T, Morrison SJ, Clarke MF et al (2001) Stem cells, cancer, and cancer stem cells. Nature 414:105–111

11. Lathia JD, Mack SC, Mulkearns-Hubert EE et al (2015) Cancer stem cells in glioblastoma. Cold Spring Harbor Laboratory Press, New York, pp 1203–1217

12. Appin CL, Brat DJ (2015) Biomarker-driven diagnosis of diffuse gliomas. Mol Asp Med 45:87–96

13. French Panel; North Africa Panel (2016) Guidelines, "minimal requirements" and standard of care in glioblastoma around the Mediterranean area: a report from the AROME (Association of Radiotherapy and Oncology of the Mediterranean area) neuro-oncology working party. Crit Rev Oncol Hematol 98:189–199

14. Huang RY, Neagu MR, Reardon DA et al (2015) Pitfalls in the neuroimaging of glioblastoma in the era of antiangiogenic and immuno/targeted therapy—detecting illusive disease, defining response. Front Neurol 3:1–16

15. Olson JJ, Fadul CE, Brat DJ et al (2009) Management of newly diagnosed glioblastoma: guidelines development, value and application. J Neuro-Oncol 93:1–23

16. Appin CL, Brat DJ (2015) Molecular pathways in gliomagenesis and their relevance to neuropathologic diagnosis. Adv Anat Pathol 22:50–58

17. The Cancer Genome Atlas Research Network (2015) Comprehensive, integrative genomic analysis of diffuse lower grade gliomas. N Engl J Med 372:2481–2498

18. Wang J, Su HK, Zhao HF (2015) Progress in the application of molecular biomarkers in gliomas. Biochem Biophys Res Comm 465:1–4

19. Yan H, Parsons W, Jin G et al (2009) IDH1 and IDH2 mutations in gliomas. New Eng J Med 360:765–773

20. Pegg AE (2000) Repair of O(6)-alkylguanine by alkyltransferases. Mutat Res 462:83–100

21. Lawrence YR, Blumenthal DT, Matceyevsky D et al (2011) Delayed initiation of radiotherapy for glioblastoma: how important is it to push to the front (or the back) of the line? J Neuro-Oncol 105:1–7

22. Stupp R, Mason WP, van den Ben MJ et al (2005) Radiotherapy plus concomitant and adjuvant temozolomide for glioblastoma. New Engl J Med 352:987–996

23. Chinot OL, Barrie M, Frauger E et al (2004) Phase II study of temozolomide without radiotherapy in newly diagnosed glioblastoma multiforme in an elderly populations. Cancer 100:2208–2214

24. Bailey R, Han SJ, Cha S et al (2014) Impact of extent of resection for gliomas. Aus J Surg 1:1035

25. Gulati S, Jakola AS, Nerland US et al (2011) The risk of getting worse: surgically acquired deficits, perioperative complications, and functional outcomes after primary resection of glioblastoma. World Neurosurg 76:572–579

26. Price SJ, Gillard JH (2011) Imaging biomarkers of brain tumour margin and tumour invasion. British J Radiol 84:159–167

27. Fang C, Wang K, Stephen ZR et al (2015) Temozolomide nanoparticles for targeted glioblastoma therapy. ACS Appl Mater Interfaces 7:6674–6682

28. Parrish KE, Sarkaria JN, Elmquist WF (2015) Improving drug delivery to primary and metastatic brain tumors: strategies to overcome the blood–brain barrier. Clin Pharmacol Ther 97:336–346

29. Boiardi A, Eoli M, Pozzi A et al (1999) Locally delivered chemotherapy and repeated surgery can improve survival in glioblastoma patients. Ital J Neurol Sci 20:43–48

30. Lesniak MS, Upadhyay U, Goodwin R et al (2005) Local delivery of doxorubicin for the treatment of malignant brain tumors in rats. Anticancer Res 25:3825–3831

31. Laquintana V, Trapani A, Denora N et al (2009) New strategies to deliver anticancer drugs to brain tumors. Exp Opin Drug Deliv 6:1017–1032

32. Leong KW, Brott BC, Langer R (1985) Bioerodible polyanhydrides as drug-carrier matrices. I: characterization, degradation, and release characteristics. J Biomed Mater Res 19:941–955

33. Cardoso FL, Brites D, Brito MA (2010) Looking at the blood–brain barrier: molecular anatomy and possible investigation approaches. Brain Res Rev 64:328–363

34. Hawkins B, Davis TP (2005) The blood-brain barrier/neurovascular unit in health and disease. Pharmacol Rev 57:173–185

35. Daneman R (2012) The blood–brain barrier in health and disease. Ann Neurol 72:648–672

36. Antel PJ, Biernacki K, Wosik K et al (2001) Glial cell influence on the human blood-brain barrier. Glia 36:145–155

37. Deli MA, Abrahám CS, Kataoka Y et al (2005) Permeability studies on in vitro blood–brain barrier models: physiology, pathology, and pharmacology. Cell Mol Neurobiol 25:59–127

38. Gao H, Yang Z, Cao S et al (2014) Tumor cells and neovasculature dual targeting delivery for glioblastoma treatment. Biomaterials 35:2374–2382

39. Van Tellingen O, Yetkin-Arik B, de Gooijer MC et al (2015) Overcoming the blood–brain tumor barrier for effective glioblastoma treatment. Drug Resist Updat 19:1–12

40. Papadopoulos MC, Saadoun S, Woodrow CJ et al (2001) Occludin expression in microvessels of neoplastic and non-neoplastic human brain. Neuropath Appl Neurobiol 27:384–395

41. Shibata T, Giaccia AJ, Brown JM (2000) Development of a hypoxia responsive vector for tumor-specific gene therapy. Gene Ther 7:493–498

42. Saenz del Burgo L, Hernández RM, Orive G et al (2014) Nanotherapeutic approaches for brain cancer management. Nanomedicine: NBM 10:905–919

43. Qian er Meel R, Vehmeijer LJ, Kok RJ et al (2013) Ligand-targeted particulate nanomedicines undergoing clinical evaluation: current status. Adv Drug Del Rev 65:1284–1298

44. Gao X, Cui Y, Levenson RM et al (2004) In vivo cancer targeting and imaging with semiconductor quantum dots. Nat Biotechnol 22:969–976

45. Tosi G, Musumeci T, Ruozi B et al (2013) The "fate" of polymeric and lipid nanoparticles for brain delivery and targeting: strategies and mechanism of blood-brain-barrier crossing and trafficking into the central nervous system. J Drug Deliv Sci Technol 32:66–76

46. Kim S, Harford JB, Pirollo KF et al (2015) Effective treatment of glioblastoma requires crossing the blood brain barrier and targeting tumors including cancer stem cells: the promise of nanomedicine. Biochem Biophys Res Commun 468:1–5

47. Gabathuler R (2010) Approaches to transport therapeutic drugs across the blood-brain barrier to treat brain disease. Neurobiol Dis 37:48–57

48. Kreuter J, Gelperina S (2008) Use of nanoparticles for cerebral cancer. Tumori 94:271–277

49. Wohlfart S, Khalansky AS, Bernreuther C et al (2011) Treatment of glioblastoma with poly (isohexyl cyanoacrylate) nanoparticles. Int J Pharm 415:244–251

50. Régina A, Demeule M, Che C et al (2008) Antitumor activity of ANG1005, a conjugate between paclitaxel and the new brain delivery vector Angiopep-2. Br J Clin Pharmacol 97:155–185

51. Xin H, Jiang X, Gu J et al (2011) Angiopep-conjugated poly(ethylene glycol)-co-poly (e-caprolactone) nanoparticles as dual-targeting drug delivery system for brain glioma. Biomaterials 32:4293–4305

52. Feng X, Gao X, Kang T et al (2015) Mammary-derived growth inhibitor targeting peptide-modified PEG–PLA nanoparticles for enhanced targeted glioblastoma therapy. Bioconjug Chem 26:1850–1861

53. Guo J, Gao X, Su L et al (2011) Aptamer-functionalized PEG-PLGA nanoparticles for enhanced anti-glioma drug delivery. Biomaterials 32:8010–8020

54. Wang B, Lv L, Wang Z et al (2015) Improved anti-glioblastoma efficacy by IL-13Rα2 mediated copolymer nanoparticles loaded with paclitaxel. Sci Rep 5:16589

55. Xin H, Sha X, Jiang X et al (2012) Anti-glioblastoma efficacy and safety of paclitaxel-loading Angiopep-conjugated dual targeting PEG-PCL nanoparticles. Biomaterials 33:8167–8176

56. Xu Y, Shen M, Li Y et al (2016) The synergic antitumor effects of paclitaxel and temozolomide co-loaded in mPEG-PLGA nanoparticles on glioblastoma cells. Oncotarget 7:20890–20901

57. Zhang B, Sun X, Mei H et al (2013) LDLR-mediated peptide-22-conjugated nanoparticles for dual targeting therapy of brain glioma. Biomaterials 34:9171–9180

58. Zhang B, Wang H, Liao Z et al (2014) EGFP-EGF1-conjugated nanoparticles for targeting both neovascular and glioma cells in therapy of brain glioma. Biomaterials 35:4133–4145

59. Zhang P, Hu L, Yin Q et al (2012) Transferrin-modified c[RGDfK]-paclitaxel loaded hybrid micelle for sequential blood-brain barrier penetration and glioma targeting therapy. Mol Pharm 9:1590–1598

60. Vinchon-Petit S, Jarnet D, Paillard A et al (2010) In vivo evaluation of intracellular drug-nanocarriers infused into intracranial tumours by convection-enhanced delivery: distribution and radiosensitisation efficacy. J Neurooncol 97:195–201

61. Gu G, Xia H, Hu Q et al (2013) PEG-co-PCL nanoparticles modified with MMP-2/9 activatable low molecular weight protamine for enhanced targeted glioblastoma therapy. Biomaterials 34:196–208

62. Gao H, Zhang S, Cao S et al (2014) Angiopep-2 and Activatable cell-penetrating peptide dual functionalized nanoparticles for systemic glioma-targeting delivery. Mol Pharm 11:2755–2763

63. Shi K, Zhou J, Zhang Q et al (2015) Arginine-glycine-aspartic acid-modified lipid-polymer hybrid nanoparticles for docetaxel delivery in glioblastoma multiforme. J Biomed Nanotechnol 11:382–391

64. Steinger SCJ, Kreuter J, Khalansky AS et al (2004) Chemotherapy of glioblastoma in rats using doxorubicin-loaded nanoparticles. Int J Cancer 109:759–767

65. Madhankumar AB, Slagle-Webb B, Wang X et al (2009) Efficacy of interleukin-13 receptor–targeted liposomal doxorubicin in the intracranial brain tumor model. Mol cancer Ther 8:648–654

66. Petri B, Bootz A, Khalansky A et al (2007) Chemotherapy of brain tumour using doxorubicin bound to surfactant-coated poly(butyl cyanoacrylate) nanoparticles: revisiting the role of surfactants. J Control Release 117:51–58

67. Gelperina SE, Khalansky AS, Skidan IN et al (2002) Toxicological studies of doxorubicin bound to polysorbate-80-coated poly(butyl cyanoacrylate) nanoparticles in healthy rats and rats with intracranial glioblastoma. Toxicol Let 126:131–141

68. Fisusi FA, Siew A, Chooi KW et al (2016) Lomustine nanoparticles enable both bone marrow sparing and high brain drug levels–A strategy for brain cancer treatments. Pharm Res 33:1289–1303

69. Qian L, Zheng J, Wang K et al (2013) Cationic core-shell nanoparticles with carmustine contained within O^6-benzylguanine shell for glioma therapy. Biomaterials 34:8968–8978

70. Jain D, Bajaj A, Athawale R et al (2016) Surface-coated PLA nanoparticles loaded with temozolomide for improved brain deposition and potential treatment of gliomas: development, characterization and in vivo studies. Drug Del 23:999–1016

71. Chooi KW, Simão CM, Soundararajan R et al (2014) Physical characterisation and long-term stability studies on quaternary ammonium palmitoyl glycol chitosan (GCPQ)· a new drug delivery polymer. J Pharm Sci 103:2296–2306

72. Kim S, Rait A, Kim E et al (2015) Encapsulation of temozolomide in a tumor-targeting nanocomplex enhances anti-cancer efficacy and reduces toxicity in a mouse model of glioblastoma. Cancer Let 369:1–9

Chapter 2

Passive and Active Targeting of Brain Tumors

Amr Hefnawy and Ibrahim M. El-Sherbiny

Abstract

Delivery of the pharmaceutically active agents towards the central nervous system (CNS) is usually challenging due to the presence of the blood-brain barrier (BBB). Brain tumors, compared to other types of CNS diseases and other types of tumors, remain difficult to treat with very little improvement in the survival rate in the last decades (Huse JT, Holland EC, Nat Rev Cancer 10:319, 2010). In this case, it is not only required to deliver a chemotherapeutic through the BBB to the brain but also to deliver it selectively to the tumor site rather than the healthy brain tissues. Nonselective distribution of the chemotherapeutics usually causes severe side effects that limit the efficiency of the treatment. This chapter summarizes some selected recent methods to prepare carrier systems capable of both efficient delivery through the BBB and active or passive targeting of the tumors within the brain tissues. This includes the description of methods to prepare the carrier systems along with the needed characterization techniques as well as in vitro and in vivo testing of its efficiency. Additional aim of this chapter is to provide the basic knowledge and skills needed for preparation of such systems and discusses the potentials or limitations of the described systems.

Key words Blood-brain barrier, Brain tumors, Glioblastoma, Ligand-mediated targeting, Stimuli responsive targeting, Passive targeting

1 Introduction

1.1 Types of Brain Tumors and Their Molecular Hallmarks

Primary brain tumors are classified according to the type of cells affected into astrocytomas, oligodendrogliomas, and oligoastrocytomas [1]. Gliomas occur more commonly in adults while medulloblastoma is the type of brain tumors more commonly affecting children [2]. This type of cancer is still posing a significant challenge with no available satisfactory treatment. This is obvious from the low median survival rate reaching 15 months only [3]. Glioblastoma may be a primary tumor which is the case in most of the tumors. In other cases, glioblastoma may be secondary arising from lower grade astrocytic neoplasms.

Studying the molecular pathology of tumors did not only help in developing animal models of brain tumors, but it is also critical for developing actively targeted drug carrier systems. For instance,

Vivek Agrahari et al. (eds.), *Nanotherapy for Brain Tumor Drug Delivery*, Neuromethods, vol. 163,
https://doi.org/10.1007/978-1-0716-1052-7_2, © Springer Science+Business Media, LLC, part of Springer Nature 2021

primary glioblastomas are associated with upregulation and activation of epidermal growth factor receptor (EGFR) [4]. Other molecular characteristics of glioma include inactivating mutations of the retinoblastoma 1 (RB1 signaling pathway) and upregulation of cyclin-dependent kinase 4 and 6 (CDK4 and CDK6) which are negative regulators of RB1 [5]. Single-nucleotide polymorphism (SNP) mutations of certain genes as regulator of telomere elongation helicase 1 (RTEL1) and telomerase reverse transcriptase (TERT) can also be determinant in the development of gliomas. Less common oncogenic molecular pathways include upregulation of the platelet-derived growth factor receptor-α (PDGFRα) or the co-expression of the receptor and its ligand [6]. Similarly, hepatocyte growth factor (HGF) and its receptor RTK MeT can also be upregulated in glioma tumors [7].

1.2 Blood-Brain Barrier

The central nervous system is protected by the blood-brain barrier (BBB) which consists of a single layer of endothelial cells associated with pericytes and astrocytes separating the blood from the cerebral parenchyma. The barrier is characterized by the presence of tight junctions and the absence of fenestration [8]. Moreover, the BBB is characterized by its high metabolic capacity, high level of expression of efflux pumps as ATP binding cassette transporters, and the deficiency of pinocytic vesicles [9, 10]. Accordingly, the BBB represents a major challenge for treatment of brain tumors as it prevents the delivery of most of the intravenous drugs to the brain limiting the choices among available chemotherapeutic agents. It prevents the delivery of all large molecules and around 98% of small pharmaceutically active molecules (<5000 Da). Passage by diffusion to the brain is limited to lipophilic electrically neutral drugs and weak bases [11]. It should be noted that in the presence of brain tumors, the permeability towards the brain may be enhanced as a result of mediators secreted by the cancer cells as arachidonic acid, leukotrienes, thromboxane B2, and prostaglandin E [12]. Additionally, invasion and proliferation of cancer cells may lead to disruption of the barrier layer [13]. However, such enhancement of permeability is highly variable and dependent on the subtype of the tumors and their localization. This variability makes it difficult to determine the accurate dose for treatment.

1.3 Drug Delivery to the Brain

Several approaches have been developed to deliver chemotherapeutics through the BBB. For example, invasive methods were used for direct delivery of the chemotherapeutics to the cerebral parenchyma. The drug was loaded into polymeric nanoparticles to allow its sustained delivery following a single injection. Gliadel® is an example for a commercially available system using this approach [14–16]. Although Gliadel® showed significant increase in the survival rate for around 2 months, this invasive approach is associated with several side effects including infections and cerebral edema. The drawbacks of invasive modes of treatment motivated the search for noninvasive therapeutic approaches. One approach is to induce transient disruption of the BBB for few hours increasing the permeability of the chemotherapeutics towards the brain. This can be achieved by injection of hyperosmotic solution in the carotid artery or through administration of inflammatory cytokines as leukotrienes or vasoactive peptides [17, 18]. Other methods to enhance the permeability of drugs are to modify them in such a way to increase their lipophilicity. This method is very challenging since modifying the structure of the drug in most cases affects its activity and toxicity which in turn requires careful reassessment including clinical trials. Several trials in this approach were not successful for treatment of cancer as the resulting modified molecules did not show better activity compared to the unmodified molecules [19]. This failure to enhance the activity can be explained by the low solubility of the modified drugs in the interstitial fluids.

1.4 Intranasal Brain Delivery

Recently, there has been an increasing interest in the intranasal route of delivery which was reported to provide a chance for bypassing the BBB. [20] This route of administration provides the advantage of completely avoiding the permeability limiting effect of the BBB. Accordingly, the delivery through this route is not limited to certain class of drugs or specific degree of drug lipophilicity. Nevertheless, using this route may result in a significant damage of the nasal tissues particularly when the drug cargo is a potent general cytotoxic agent. Moreover, formulation of efficient and acceptable dosage forms for this route is challenging taking into consideration the limited space of the nasal cavity and the inconvenience of its blockage for repeated or prolonged periods. There are also several concerns regarding the distribution of the drugs after intranasal delivery as they would be expected to accumulate more in the hippocampus, the olfactory bulb, and the cerebrum while tumors in other brain regions as the cerebral cortex may remain unaffected [21]. However, drug delivery systems provide a promising tool in this aspect. Loading the active drug into a carrier system can allow efficient delivery to the brain with selectivity towards the tumor site compared to other healthy tissues. The advantage of such approach is that, in most cases, the activity and mechanism of action of the loaded drug remains unaffected.

Passive targeting may represent a promising solution to this limitation of brain delivery to direct the carrier systems more specifically towards the tumor site. This phenomenon depends mainly on the differences between the vascular structures of the normal tissues and the tumor tissues and its surrounding environment. Normal vascular epithelial cells are connected by tight junctions that prevent the leakage of molecules and particles larger than 11.5 nm [22]. On the other hand, tumor vasculature is characterized by defects and large fenestrations that allow the extravasation of larger particles that may sometimes reach around 1 μm [23]. This phenomenon is known as the enhanced permeability and retention effect which was first reported in 1986, and it is commonly employed as an efficient targeting method [24]. Accordingly, loading cytotoxic drugs in carrier systems in the size range of 100–500 nm will result in accumulating the drug selectively at the tumor site. For example, Brigger and his colleagues prepared poly (ethylene glycol)-coated hexadecylcyanoacrylate nanospheres. The coating helped the nanosphere escape the phagocytosis by the mononuclear macrophages allowing the nanosphere to remain in the circulation time sufficient to accumulate passively at the tumor site [25].

Active targeting is another alternative method that depends on the specific binding between the ligand and the target receptor which may be a mutant-form of receptor or a normal receptor that is overexpressed by the tumor cells. Ligand-receptor binding may result in cellular uptake of the carrier, or it may only result in accumulating the carrier at the tumor site so that the released drug can affect the tumor directly. Recently, there has been some successful approaches reported utilizing active targeting for delivery to brain cancer cells. For example, Dixit and his colleagues prepared gold nanoparticles decorated with transferrin peptide which can bind specifically to transferrin receptor. This receptor is highly expressed by glioblastoma multiform primary tumors [26]. The targeted nanoparticles were able to accumulate by two- to three-folds more compared to the non-targeted nanoparticles. The developed carrier system was used to deliver the prodrug photosynthesizer, phthalocyanine 4, achieving the same killing efficiency as the non-targeted drug using only 1/10 of the concentration. Moreover, this system can also be used for noninvasive imaging of brain tumor. [27] Another study reported the use of Angiopep-2, a peptide derived from aprotinin to target the delivery of doxorubicin to brain cells. The peptide binds to low-density lipoprotein receptor-related protein 1 which is abundant on the surface of brain capillaries. Binding of the peptide to the receptors induces transcytosis and parenchymal accumulation. The carrier system involved binding of the Angiopep-2 to the surface of poly (ethylene glycol)-co-poly(ε-caprolactone) polymersomes loaded with the drug. This carrier system significantly improved the survival time of mice with glioma model compared to non-targeted particles [28]. More advanced carrier system was introduced by McNeeley combining both passive and active targeting in a stepwise process. They prepared liposomes decorated with folic acid as targeting agents which were in turn coated with polyethylene glycol (PEG) 5000 linked through cysteine bonds. The PEG coating helps the nanoparticles escape immune clearance keeping them in the circulation for time sufficient to passively accumulate at the tumor site. The nanoparticles can then be uncoated using cysteine injection. The folic acid ligand can then actively enhance the intracellular uptake of the nanoparticles in cancer cells.

This chapter will introduce some of the most recent drug delivery systems developed for active and passive targeting of chemotherapeutics towards brain tumors.

2 Materials

Alginate (Alg), low molecular weight chitosan (Cs), doxorubicin (Dox), gelatin, glycine, mannitol, iron oxide magnetic nanoparticles (MNPs), gum Arabic, glutaraldehyde.

3 Methods

3.1 Intranasal Delivery of Doxorubicin (Dox) Using Nanoparticles Loaded into Intranasal Inserts [29]

This method describes the preparation of a platform for delivery of chemotherapeutics towards the brain through the intranasal route. In brief, Dox, as a model anticancer drug, was loaded into Alg/Cs nanoparticles based on counter ion complexation. The positively charged Dox was complexed with negatively charged alginate and the resulting complex was then coated with positively charged Cs. The formation of the nanoparticles provided a sustained release profile, diminished toxicity on healthy tissues including the nasal membrane, and also provided particle size suitable for passive targeting of tumors. The nanoparticles were then loaded into intranasal inserts making it suitable for administration through the intranasal route.

3.1.1 Preparation of Dox-Loaded Nanoparticles

1. Separately, prepare solutions of Dox.HCl (5 mg), Alg (5 mg), and Cs (10 mg) at the concentration of 1 mg/ml. Dox.HCl and Alg are dissolved in deionized water (dH$_2$O) while Cs is dissolved in 1% w/v aqueous acetic acid solution.

2. Neutralize the Dox.HCl solution with 3 drops of 10% w/v aqueous NaHCO$_3$ to obtain the free form of Dox (*see* **Notes 1–3**).

3. Dox solution is then added dropwise to the Alg solution with stirring to form Dox/Alg complex.

4. The formed complex is then sonicated for 1 min using a probe sonicator. Sonication should be done using 60 W power in cycles of 5 s on and 5 s off.

5. After sonication, the complex is coated by Cs through its dropwise addition to the Cs solution under homogenization (10,000 rpm). The final solution is left under homogenization for additional 5 min.

6. The formed nanoparticles were then separated by centrifugation at 14,000 rpm ($20380 \times g$) for 30 min, and the nanoparticles are finally freeze-dried for complete removal of water.

3.1.2 Loading of Dox Nanoparticles into Intranasal Inserts

1. The resulting Dox-loaded nanoparticles are dispersed in distilled H_2O with stirring to obtain a final concentration of 3.75% w/v equivalent to free Dox.

2. After dispersion of the nanoparticles, gelatin (1% w/v), glycine (1% w/v), and mannitol (5% w/v) are added to the solution.

3. The obtained solution is then poured into a suitable mold for formation of inserts of suitable size to be placed in the intranasal cavity.

4. The inserts are finally freeze-dried to obtain them in a form ready for in vivo assessment.

3.1.3 Characterization

1. Formation of the nanoparticles is confirmed using high-resolution transmission electron microscope (TEM) imaging. The size of the nanoparticles and the surface charge are measured using dynamic light scattering (DLS).

2. Drug entrapment efficiency is measured indirectly by UV-Vis spectrometer of the supernatant after centrifugation of the nanoparticles.

3. The morphology of the nasal inserts is evaluated using scanning electron microscopy (SEM).

4. Drug targeting to the brain is evaluated *in vivo* using rabbit animal model. Drug targeting is expressed in terms of exposure ratio, drug targeting percentage, and direct transport percentage.

3.1.4 Notes

1. Neutralization of Dox.HCl into the free form is a critical step affecting the entrapment efficiency of the drug which increased from 4% upon using the salt to around 80% upon using the free form. The difference in the entrapment efficiency can be explained based on the effect of pH on the electrostatic interactions. In brief, Dox.HCl has positively charged protonated amino groups masked by interaction with negatively charged chloride anions (Cl^-). Addition of $NaHCO_3$ makes the amino groups free for interaction with the carboxylic acid groups of Alg.

2. The solubility of Dox decreases by neutralization and it may result in precipitation of the drug with time. Accordingly, it is recommended that the neutralization step is done directly before complexation with Alg.

3. Dox color differs according to the ionization state which is affected by protonation and deprotonation of its amino groups and acidic phenolic groups (Fig. 1). In acidic pH (5.5), the

(a)

(b)

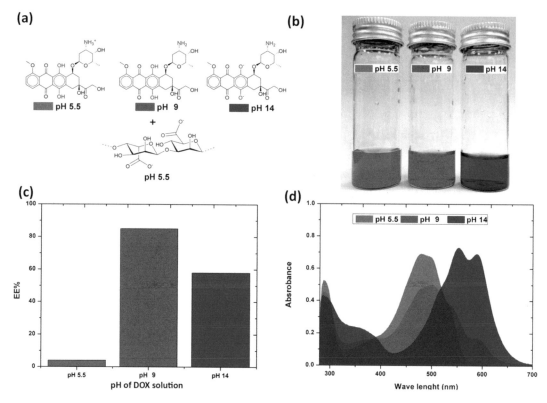

(c)

(d)

Fig. 1 Effect of pH on (**a**) Dox ionization, (**b**) color, (**c**) entrapment efficiency, and (**d**) UV absorbance [22]

drug has an orange color. In moderately alkaline pH (8–9), the color changes to dark red as the lone pair of electrons of the amino group becomes more available for participation in the aromatic resonance. At highly alkaline pH (14), the color turns into violet indicating the deprotonation of the phenolic groups and involvement of their electrons in the aromatic resonance.

3.2 Preparation of Poly(Ethyleneimine) (PEI)-Grafted Magnetic Nanoparticles for Magnetic Targeting of Brain Tumors

This study describes the preparation of a carrier system that is suitable for delivery of both chemotherapeutics and genes to the brain due to the presence of an external coating of PEI. The carrier system is composed of magnetic nanoparticles in the core coated with gum Arabic which is in turn conjugated to the PEI.

3.2.1 Synthesis of Magnetic Nanoparticles (MNP)

1. Prepare a solution of $FeCl_2/FeCl_3$ at the ratio of 1:2 in deionized water under nitrogen atmosphere to obtain a final concentration of 0.05 g/ml (*see* **Note 4**).

2. Prepare 2–3% w/v ammonia solution.

3. Add the ammonia solution dropwise with vigorous stirring to the $FeCl_2/FeCl_3$ with the rate adjusted to 1 drop every 30 s. Add volume sufficient to change the color of the solution from

yellow to black. Half of the required volume to achieve the color change is then added as excess to ensure complete conversion of the Fe salts to magnetic nanoparticles (*see* **Notes 5** and **6**).

4. Continue stirring the solution for additional 2 h.

5. Magnetic nanoparticles are separated using a strong magnet. The nanoparticles are then washed three times by dispersion in deionized water and then three times in acetone. The nanoparticles are finally dried in a vacuum oven.

3.2.2 Coating of MNP by Gum Arabic [30]

1. Prepare 14 ml suspension of 12 mg/ml MNPs in deionized H_2O.

2. (3-Aminopropyl)triethoxysilane (APTES) is then added to the MNPs to reach a final concentration of 10% v/v.

3. The mixture is then mixed by shaking followed by sonication at 10 W power for 10 min. The mixture is incubated at 70 °C for 1 h.

4. The amino-terminated MNPs are then washed six times using distilled water and the final volume is adjusted and kept constant using distilled water.

5. Aldehyde-terminated MNPs were obtained by addition of glutaraldehyde to the MNPs suspension to reach a final concentration of 5% v/v.

6. The solution is mixed again by shaking followed by sonication for 10 min and then incubation at room temperature for 1 h.

7. Aldehyde functionalized nanoparticles were then separated and washed six times with distilled water. The volume is finally adjusted to 14 ml using deionized water.

8. The suspension prepared in the previous step is divided into 1 ml aliquots. The MNPs are then left to settle down. Then, 1 ml of the gum Arabic solution at concentration of 70 mg/ml is added to the nanoparticles.

9. The samples are then shaken and sonicated at 10 W for 10 min and then incubated at room temperature for 1 h.

3.2.3 Preparation of Polyethylenimine (PEI)-Coated Nanoparticles [31]

1. A mixture of gum Arabic-coated MNPs (GA-MNPs), PEI, EDC, and sulfo-NHS with the ratio 1:1:2:2 was prepared. The molarity of GA-MNPs is calculated as equivalent to Fe.

2. The mixture is then incubated for 48 h.

3. The nanoparticles are then separated using a strong magnet and washed by dispersion in deionized water. Washing is repeated till complete removal of excess PEI in solution.

3.2.4 Characterization

1. Evaluation of the magnetic properties of the prepared MNPs and the final particles should be done to ensure that the carrier maintains its magnetic-responsive activity. This can be done using superconducting quantum interference device (SQUID) magnetometer.

2. Quantification of the amount of gum Arabic adsorbed onto the surface of the MNPs can be done using bicinchoninic acid test [32]. Samples of 50 µl were withdrawn from the supernatant and added to 96-well plate. Freshly prepared test reagent in 200 µl (mixture 50:1 of reagent A and B, respectively) was added to each sample. The well plate is then incubated for 20 min in dark environment before measuring the absorbance at 562 nm. The concentration can be obtained using a standard calibration curve for gum Arabic at concentrations 0–70 mg.

3. Quantification of terminal amino groups either in the amino-terminated MNPs or the PEI can be determined calorimetrically using the ninhydrin assay (also known as the Kaiser test). Briefly, 500 µl of 2% w/v ninhydrin solution dissolved in 0.1 M phosphate buffer at pH 9 is added to 200 µl of the sample. The mixture is then heated in a boiling water bath for 15 min. The MNPs are separated using a magnet or centrifuge before measuring absorbance at 570 nm. The concentration is obtained from a standard calibration curve measured for a serial dilution of ethanolamine or glycine.

3.2.5 Notes

1. The presence of inert atmosphere during the process is critical to avoid uncontrolled oxidation which would produce undesired other types of iron oxides. The inert atmosphere can be achieved by purging the deionized water with inert gas as argon or nitrogen before dissolving the iron chlorides and then replacing the air in the flask with inert gas using a gas inlet and air outlet through flask sealing.

2. The rate of ammonia addition affects the formation of the nanoparticles and their size. Accordingly, a fixed addition rate is needed to achieve low polydispersity of the nanoparticles which was best achieved using a syringe pump for addition at rate of 0.015 ml/min.

3. The color changes of the solution are indicative of the formation of the magnetic nanoparticles. The solution starts by a yellow color for the iron chloride salts which changes by addition of the basic solution to brown (hematite) before finally reaching the black color of the magnetite nanoparticles (Fe_3O_4).

4. Ninhydrin reagent is unstable and needs to be freshly prepared. It is also recommended to perform the assay with a suitable protection from light.

5. The route of administration is a critical factor towards successful targeting of the carrier system towards the brain tumors. It was found that injection of these particles through the carotid artery (intra-carotid route) achieved almost 30-fold higher accumulation at the tumor site after magnetic targeting compared to intravenous (IV) injection.

3.3 Actively Targeted Liposomes with Cysteine-Triggered Unmasking of the Targeting Ligands [33]

Ligand-based targeting has long been studied as a method for selective delivery of drugs towards target sites. Although the binding of these carrier systems to the target is crucial to targeting, their binding ability is also the reason for their rapid clearance and poor pharmacokinetic profile. These nanoparticles are more liable to opsonization by the immune system followed by engulfment and removal from the circulation by macrophages. Escaping the immune system was achieved by coating with polymers as polyethylene glycol 5000 (PEG_{5000}) which provides steric hindrance that prevents opsonization and cell engulfment making the nanoparticles stealth to the immune system. This steric effect, however, hinders the ability of the ligands on the surface of the nanoparticles to bind to their targets. Accordingly, studies aimed to achieve a balance between both effects.

This protocol describes the preparation of liposomes decorated with folate as targeting ligand that is in turn masked using cleavable PEG_{5000} chains which can be removed using cysteine injection. This approach provides a long circulation time of the PEG_{5000}-coated liposomes allowing passive targeting of tumors. After accumulation at the tumor site, the targeting ligand can be unmasked using cysteine dose (1 mmol/kg) allowing efficient and selective uptake into cancer cells. It should be noted that this concentration is generally considered safe.

3.3.1 Synthesis of Cysteine Cleavable Lipid PEG Conjugates

1. Dissolve 790 mg of 1,2-distearoyl-snglycero-3-phosphoethanolamine (DSPE) in 22 ml chloroform with the addition of 900 μl triethylamine at 55 °C (*see* **Note 9**).

2. Dissolve 263 mg of *N*-succinimidyl 3-[2-pyridyldithio]-propionamido (SPDP) in 3 ml of chloroform, which is then added to the previous solution of DSPE (Scheme 1). The reaction mixture is kept under stirring for 5 h at room temperature and the reaction progress is monitored using thin layer chromatography (TLC) (*see* **Note 10**).

3. Dissolve 1.75 g of PEG_{5000} in 9 ml of chloroform before adding this solution to the previous solution reaction. The mixture is kept under stirring overnight at room temperature. The reaction progress is monitored using TLC and by measuring UV absorbance at 343 nm (*see* **Note 11**).

4. After the reaction is completed, the organic solvent is removed using rotatory evaporator.

Scheme

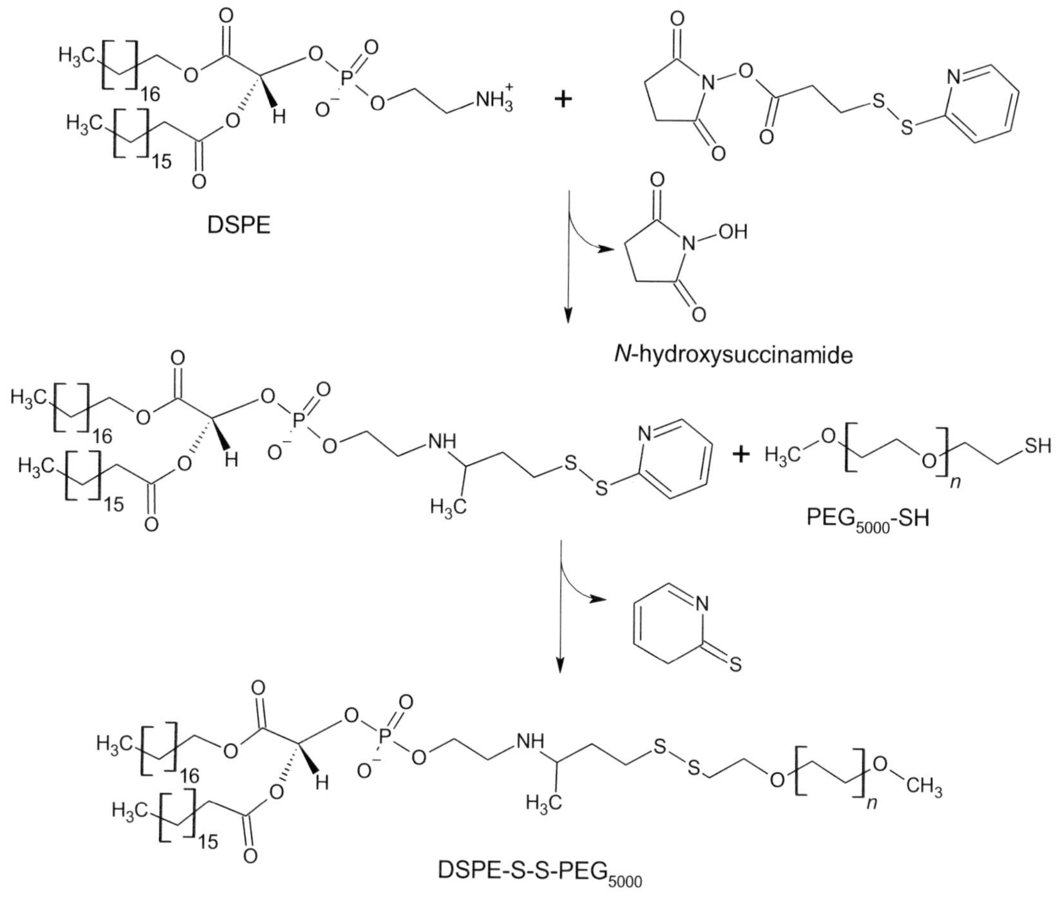

Scheme 1 Synthesis of DSPE-S-S-PEG5000

5. The remaining solid is then dissolved in acetonitrile which separates excess DSPE as a precipitate that should be removed by centrifugation.

6. The supernatant is then recovered and the organic solvent is evaporated using rotatory evaporator.

7. The solid product is then dissolved in the least amount of dichloromethane (DCM) to be applied to silica gel for column chromatography purification. The mobile phase used for separation consists of 200 ml fractions of gradually increasing concentration of methanol in DCM (4%, 6%, 9%, 12%, and 15%).

8. Solution coming out from the column is collected as 4 ml fractions. Fractions containing the desired product were detected using TLC and then evaporated to dryness using rotatory evaporator.

3.3.2 Synthesis of DSPE-PEG2000-Folate Conjugates [34]

1. Dissolve 76 mg DSPE-PEG$_{2000}$-amine and 16.7 folate (molar ratio 5:8) in 667 µl dimethylsufloxide (DMSO).

2. The reaction is then catalyzed by the addition of 333 µl pyridine and 21.7 mg dicyclohexylcarbodiimide (DCC).

3. The reaction is then kept under stirring for 4 h before removal of the pyridine using rotatory evaporator.

4. The product is suspended in 16.7 ml of distilled water and the suspension is dialyzed in 300,000 MWCO dialysis tube twice against 2 L of 50 mM NaCl solution followed by three times against 2 L of distilled water.

5. The final product is obtained in the dry form using lyophilization.

3.3.3 Preparation of Targeted and Non-targeted Liposomes [33, 34]

1. Liposomes were prepared using 65:35 mixtures of 1,2-dipalmitoyl-*sn*-glycerophosphocholine (DPPC) and cholesterol.

2. The lipid mixture (100 mg) is then dissolved in the least amount of ethanol at 60 °C followed by hydration using 400 mM ammonium sulfate buffer.

3. The liposome solution was extruded to 100 nm by extrusion five times through 0.2 µm filters and then ten times through 0.1 µm filters at 60 °C.

4. The liposomes are then dialyzed against 0.9% sodium chloride.

5. Folate conjugates were inserted in the lipid bilayer at concentration of 0.15% to form folate-targeted liposomes. The DSPE-PEG$_{2000}$-folate conjugate formed into micelles by dissolving into DMSO at 60 °C at the concentration of 28 mM. The conjugate solution is then diluted tenfolds by the addition of distilled water. The DMSO is removed by dialysis twice against 1 L of water.

6. The micelles are mixed at the concentration of 0.15% with the liposomes for 1 h at 60 °C to allow insertion of the folate conjugate into the micelles. Unincorporated folate conjugates were then removed using dialysis (MWCO 300,000).

7. Liposomes with controlled masking and demasking were prepared by repeating the step while replacing 8% of the lipid mixture used in **step 1** with cysteine cleavable DSPE–PEG$_{5000}$ conjugates.

8. Loading of Dox is achieved by dispersing liposomes in PBS buffer containing Dox at concentration of 15 mg/ml. The ratio between Dox and liposomes is 4:25. The mixtures is then heated at 60 °C for 1 h. The loading process is terminated by immediate cooling of the solution on ice followed by removing unloaded Dox using dialysis (*see* **Note 12**).

3.3.4 Characterization

1. The synthetic steps and the structure of the final product should be confirmed using various characterization methods such as matrix-assisted laser desorption/ionization time-of-flight (MALDI-TOF) mass spectroscopy, ^{1}H-NMR, and ^{13}C-NMR.

2. The thiolytic cleavability of the nanoparticles can be measured by incubating 1 mM of the nanoparticles with 10 mM of cysteine in PBS buffer at 37 °C for 30 min. The cleavage can then be detected by comparing polarity by TLC or HPLC. It can also be confirmed using MALDI-TOF mass spectrometry.

3.3.5 Notes

1. Triethylamine is used as a sterically hindered base making it poor nucleophile but keeping its ability to abstract protons from the reaction medium elevating its pH. Other examples for sterically hindered bases include *N,N*-diisopropylethylamine.

2. The product of the first reaction between SPDP and DSPE contains a pyridine ring which results in UV absorbance at 204 and 256 nm and accordingly can be visualized using a shortwave UV lamp. The final product, however, does not contain any conjugated system and will not be visible by color or UV absorbance. As a result, the TLC used for monitoring the reaction needs to be stained using appropriate coloring agents as cupric sulfate solution which is used for colorimetric detection of lipids through interaction with the phosphate bonds. (The staining solution consists of 3% w/v cupric sulfate and 15% phosphoric acid.)

3. The UV absorbance measured at 343 nm for this reaction is related to the formation of the pyridyl-2-thione side product of the reaction. Accordingly, it is expected that absorbance increases with the progress of the reaction till it reaches a plateau indicating that the reaction is completed.

4. Loading of Dox into the liposome is achieved based on concentration gradient (Fig. 2). The Dox is dissolved in a buffer in its original form as HCl salt. This salt is present in equilibrium with its free base form. While the free base form can pass through the lipid bilayer of the liposome, the salt form is too hydrophilic to pass through. When the Dox reaches the core of the liposome in the presence of ammonium sulfate buffer, it is converted to its sulfate salt which forms a gel that is entrapped within the liposome. This method can achieve up to 20-folds higher concentration of Dox inside the liposome compared to the surrounding environment [35].

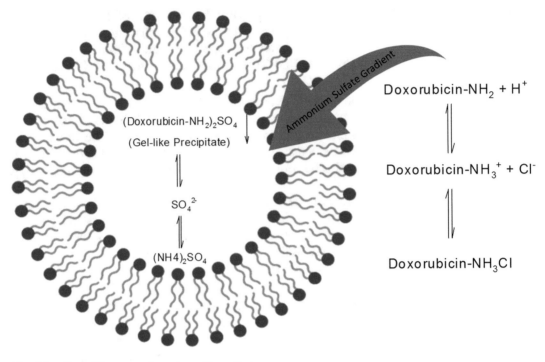

Fig. 2 Loading of liposomes based on pH gradient

4   Conclusion

This chapter described several methods to deliver chemotherapeutic agents to the brain with overcoming the limiting effect of the BBB. Using the intranasal route of delivery was beneficial in bypassing the BBB but it did not provide a selective delivery to cancer cells. On the other hand, using ligand-based carrier system can selectively target the cancer cells but it must first pass the BBB to reach the brain and interact with the targeted receptors. Accordingly, it is highly recommended to develop a hybrid carrier system that combines the advantages of the previously mentioned carrier systems. Future work will also involve the investigation of more specific targeting systems that are able to target subcellular structures as the mitochondria. It has been shown that mutations of the mitochondria are responsible for some CNS diseases such as Parkinson's disease. Such system would be expected to be a promising approach and an important step towards reaching a satisfactory treatment of brain tumors.

References

1. Behin A, Hoang-Xuan K, Carpentier AF, Delattre J-Y (2003) Primary brain tumours in adults. Lancet 361:323–331

2. Huse JT, Holland EC (2010) Targeting brain cancer: advances in the molecular pathology of malignant glioma and medulloblastoma. Nat Rev Cancer 10:319

3. Stupp R, Mason WP, Van Den Bent MJ et al (2005) Radiotherapy plus concomitant and adjuvant temozolomide for glioblastoma. N Engl J Med 352:987–996

4. Libermann TA, Nusbaum HR, Razon N et al (1985) Amplification, enhanced expression and possible rearrangement of EGF receptor gene in primary human brain tumours of glial origin. Nature 313:144

5. Henson JW, Schnitker BL, Correa KM et al (1994) The retinoblastoma gene is involved in malignant progression of astrocytomas. Ann Neurol 36:714–721

6. Clarke ID, Dirks PB (2003) A human brain tumor-derived PDGFR-α deletion mutant is transforming. Oncogene 22:722

7. Abounader R, Laterra J (2005) Scatter factor/hepatocyte growth factor in brain tumor growth and angiogenesis. Neuro-Oncology 7:436–451

8. Brightman MW, Reese TS (1969) Junctions between intimately apposed cell membranes in the vertebrate brain. J Cell Biol 40:648–677

9. Reese TS, Karnovsky MJ (1967) Fine structural localization of a blood-brain barrier to exogenous peroxidase. J Cell Biol 34:207–217

10. Golden PL, Pollack GM (2003) Blood–brain barrier efflux transport. J Pharm Sci 92:1739–1753

11. Abraham MH, Chadha HS, Mitchell RC, Bonding H (1994) Hydrogen bonding. 33. Factors that influence the distribution of solutes between blood and brain. J Pharm Sci 83:1257–1268

12. Wahl M, Schilling L, Unterberg A, Baethmann A (1993) Mediators of vascular and parenchymal mechanisms in secondary brain damage. In: Mechanisms of Secondary Brain Damage. Springer, New York, pp 64–72

13. Gururangan S, Friedman HS (2002) Innovations in design and delivery of chemotherapy for brain tumors. Neuroimaging Clin N Am 12:583–597

14. Menei P, Capelle L, Guyotat J et al (2005) Local and sustained delivery of 5-fluorouracil from biodegradable microspheres for the radiosensitization of malignant glioma: a randomized phase II trial. Neurosurgery 56:242–248

15. DiMeco F, Li KW, Tyler BM et al (2002) Local delivery of mitoxantrone for the treatment of malignant brain tumors in rats. J Neurosurg 97:1173–1178

16. Westphal M, Ram Z, Riddle V et al (2006) Gliadel® wafer in initial surgery for malignant glioma: long-term follow-up of a multicenter controlled trial. Acta Neurochir 148:269–275

17. Cosolo WC, Martinello P, Louis WJ, Christophidis N (1989) Blood-brain barrier disruption using mannitol: time course and electron microscopy studies. Am J Phys Regul Integr Comp Phys 256:R443–R447

18. Cloughesy TF, Black KL (1995) Pharmacological blood-brain barrier modification for selective drug delivery. J Neuro-Oncol 26:125–132

19. Greig NH, Genka S, Daly EM et al (1990) Physicochemical and pharmacokinetic parameters of seven lipophilic chlorambucil esters designed for brain penetration. Cancer Chemother Pharmacol 25:311–319

20. Hanson LR, Frey WH (2008) Intranasal delivery bypasses the blood-brain barrier to target therapeutic agents to the central nervous system and treat neurodegenerative disease. BMC Neurosci 9:S5

21. Zhao Y, Yue P, Tao T, CHEN Q (2007) Drug brain distribution following intranasal administration of Huperzine a in situ gel in rats 3. Acta Pharmacol Sin 28:273–278

22. Sukriti S, Tauseef M, Yazbeck P, Mehta D (2014) Mechanisms regulating endothelial permeability. Pulm Circ 4:535–551

23. Hashizume H, Baluk P, Morikawa S et al (2000) Openings between defective endothelial cells explain tumor vessel leakiness. Am J Pathol 156:1363–1380

24. Matsumura Y, Maeda H (1986) A new concept for macromolecular therapeutics in cancer chemotherapy: mechanism of tumoritropic accumulation of proteins and the antitumor agent smancs. Cancer Res 46:6387–6392. https://doi.org/10.1021/bc100070g

25. Brigger I, Morizet J, Aubert G et al (2002) Poly (ethylene glycol)-coated hexadecylcyanoacrylate nanospheres display a combined effect for brain tumor targeting. J Pharmacol Exp Ther 303:928–936

26. Recht L, Torres CO, Smith TW et al (1990) Transferrin receptor in normal and neoplastic brain tissue: implications for brain-tumor immunotherapy. J Neurosurg 72:941–945

27. Dixit S, Novak T, Miller K et al (2015) Transferrin receptor-targeted theranostic gold nanoparticles for photosensitizer delivery in brain tumors. Nanoscale 7:1782–1790

28. Lu F, Pang Z, Zhao J et al (2017) Angiopep-2-conjugated poly (ethylene glycol)-*co*-poly (-ε-caprolactone) polymersomes for dual-targeting drug delivery to glioma in rats. Int J Nanomedicine 12:2117

29. Hefnawy A, Khalil IA, El-Sherbiny IM (2017) Facile development of nanocomplex-in-nanoparticles for enhanced loading and selective delivery of doxorubicin to brain. Nanomedicine 12:2737–2761

30. Roque ACA, Bicho A, Batalha IL et al (2009) Biocompatible and bioactive gum Arabic coated iron oxide magnetic nanoparticles. J Biotechnol 144:313–320

31. Chertok B, David AE, Yang VC (2010) Polyethyleneimine-modified iron oxide nanoparticles for brain tumor drug delivery using magnetic targeting and intra-carotid administration. Biomaterials 31:6317–6324

32. Roque ACA, Wilson OC Jr (2008) Adsorption of gum Arabic on bioceramic nanoparticles. Mater Sci Eng C 28:443–447

33. McNeeley KM, Karathanasis E, Annapragada AV, Bellamkonda RV (2009) Masking and triggered unmasking of targeting ligands on nanocarriers to improve drug delivery to brain tumors. Biomaterials 30:3986–3995

34. Saul JM, Annapragada A, Natarajan JV, Bellamkonda RV (2003) Controlled targeting of liposomal doxorubicin via the folate receptor in vitro. J Control Release 92:49–67

35. Bolotin EM, Cohen R, Bar LK et al (1994) Ammonium sulfate gradients for efficient and stable remote loading of amphipathic weak bases into liposomes and ligandoliposomes. J Liposome Res 4:455–479

Chapter 3

Strategies to Modulate the Blood-Brain Barrier for Directed Brain Tumor Targeting

Paula Schiapparelli, Montserrat Lara-Velazquez, Rawan Al-kharboosh, Hao Su, Honggang Cui, and Alfredo Quinones-Hinojosa

Abstract

Glioblastoma (GBM) is the most common and aggressive primary brain tumors in adults. Despite being one of the most genetically and molecularly characterized tumors, patient prognosis remains dismal. Many factors have contributed to the challenges of developing effective clinical therapies, such as tumor heterogeneity, immune-suppressive microenvironment, and therapy resistance. From the perspective of therapeutics delivery, one of the main obstacles is to get newly discovered therapeutic agents across the blood-brain barrier (BBB) to penetrate the brain parenchyma. In this context, this book chapter provides an overview of three different strategies of delivering antitumor agents intended to circumvent the BBB and highlight the strengths and weakness of each approach.

Key words Blood-brain barrier, BBB permeability, Brain drug delivery, Glioma, Convection-enhanced delivery (CED), Focused ultrasound (FUS), Supramolecular hydrogels

1 Overview: Brain Tumor Therapy (Obstacles and Alternatives)

The unique anatomical, physiological, and pathological features of GBM greatly limit the effectiveness of conventional radio- and chemotherapy. Due to the high infiltrative nature of GBM, maximal surgical resection cannot be achieved and tumor cells left behind are believed to play a key role in tumor recurrence and patient relapse. The blood-brain barrier (BBB) presents the major obstacle for entrance of many therapeutically active compounds and nanoparticle-based drug delivery systems into brain. [1]The high level of P-glycoproteins in the BBB has also been reported to inhibit brain penetration and pharmacological activities of many drugs [2]. For these reasons, most clinical trials and practices have

Author ContributionsConceptualization: PS, MLV, RA, HS, HC, AQH. Formal Analysis: PS, MLV, RA, HS. Writing: PS, MLV, RA, HS. Visualization: PS, MLV, RA. Supervision: AQH, HC. Funding Acquisition: AQH, HC. Resources: AQH, HC.

Vivek Agrahari et al. (eds.), *Nanotherapy for Brain Tumor Drug Delivery*, Neuromethods, vol. 163, https://doi.org/10.1007/978-1-0716-1052-7_3, © Springer Science+Business Media, LLC, part of Springer Nature 2021

been focusing on the use of BBB penetrable chemotherapeutics which are mainly alkylating agents that kill cells by attaching an alkyl group to DNA [3]. After standard radiotherapy, 90% of the tumors recur at the original site, [4] and no significant benefit in survival is observed with increased dosages [5, 6]. Furthermore, systemic adjuvant chemotherapy such as with temozolomide (TMZ) only prolongs the survival of glioma patients by only a couple of months [7]. However, due to systemic exposure, TMZ dosage is limited by hematological toxicity, specifically thrombocytopenia and neutropenia. Local application of Gliadel® wafer, the carmustine-loaded polymer implants as an adjunct to surgery and radiation therapy, has been proven to extend the survival of patients with malignant gliomas, [8–10] strongly suggesting that the combined use of surgical resection and local drug delivery offers an effective treatment strategy to extend the survival of brain tumor patients [11]. Despite its clinical success, the encapsulated drug carmustine is completely released from the wafer in approximately 1 week [11]. There is an increased impetus for the advent of new strategies for treating malignant gliomas, allowing for ease of implantation at the time of surgery, sustained drug release, high tissue penetration, and effective mechanisms to overcome multi-drug resistance. In this chapter we discuss techniques for effective drug delivery to the brain: convection-enhanced delivery, MR-guided focused ultrasound, and self-assembling hydrogels.

2   The Blood-Brain Barrier

The ability to deliver diagnostic and therapeutic agents to the CNS has been a long-standing issue for the delivery of systemic therapy as the anatomical features of the BBB, which is comprised of unusually abundant and structurally unique tight junctions between the vascular endothelial and basement membrane [12]. The BBB offers several barriers to the entry of molecules into diseased and normal brain parenchyma. Physical barrier incorporates several cell types, which includes the first restriction, endothelial cells lining the vessels and stitched together by tight junctions; in the event that molecules are able to pass this restricted barrier, they must overcome other highly dense cellular barrier such as astrocytes and pericytes. The BBB also has biochemical barriers that would screen molecules that have managed to successfully cross the physical barrier; these biochemical barriers include transmembrane efflux transporters and pumps that would enable the active transporting of molecules and drugs out of the BBB limiting the accumulation of small molecules and drugs for therapy. Even if diagnostic and therapeutic compounds managed to cross this barrier, regulations via the physical and biochemical barriers maintain a tight hold preventing the passage of the vast majority of therapeutics. There has been increasing interest in specific BBB disruption permitting molecules and cells to pass through an otherwise uncompromised loci [13]. One hundred percent of large molecules and drugs over 500 Da will not pass through, and 98% of drugs and antibodies are incapable of crossing the barrier [14]; lipid-soluble small molecules are capable of diffusing through the brain if they can squeeze through the endothelial membrane; however, very few drugs fall into this category. There is a dire need for a system that penetrates the BBB transiently, focally, with minimal risk and increased clinical therapeutically utility.

3 Convection-Enhanced Delivery

Convection-enhanced delivery (CED) technology is an emerging strategy to maximize local drug-delivery into the central nervous system (CNS). The device is composed of a catheter inserted into a target area in the brain and a reservoir where drugs are pushed through hydrostatic pressure by a syringe pump to infuse substances into the interstitial space (Fig. 1). Due to the gradient created by the pump, CED allows targeted administration of a

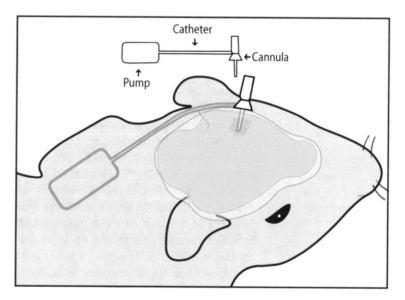

Fig. 1 Convention-enhanced delivery (CED): An intratumoral therapy option for brain cancer. The CED technology is a novel strategy that maximizes local drug administration into the central nervous system (CNS). The system is composed of a surgically implanted catheter inserted through a hole into the skull, where an agent of interest can be delivered through an active flow created by a pump which is connected to a catheter. The sustained and control administration rate mediated by the pump allows an extended and uniform distribution of the infuse agents in the tumor and peritumoral area

range of different molecular weight particles at specific locations in the CNS. CED has successfully been applied for the treatment of several pathologies such as brain tumors, neurodegenerative diseases (Parkinson and Alzheimer), epilepsy, and neuropathic disorders [15]. Recent clinical trials have shown the beneficial effect of multidrug administration through CED in glioma patients [16].

3.1 Background

CNS selective penetration of polarized components is a well-known function of the BBB. [17] Local therapy for cancer is a strategy to overcome this selectivity by placing the substance of interest (growth factor, antibody, protein conjugates, enzymes, or hydrogels) in close contact to the cancer cells [18]. CED is an approach for intratumoral therapy that allows the continuous administration of small or large compounds into the brain through a positive pressure gradient into the interstitial space. In normal conditions, as well as in abnormal circumstances (such as stroke or edema), the flow of the interstitial fluid in the brain regulates the distribution of compounds while limiting the diffusion of drugs into the tissue [19]. Thus, a homogenous distribution of reagents into the CNS after external administration is still a challenge [20]. The CED system is composed by a surgically implanted infusion catheter that allows drug delivery into targeted areas of the brain in a sustained and controlled rate [21]. This delivery technique is based on one or more catheters inserted through a hole into the skull to reach specific areas. Once in the desired location, the agent of interest is delivered by an active flow created by a pump, which is connected to a catheter. Depending on the catheter and the pump, infusion rates vary from 0.1 to 10 μL/min to enable spherical distribution form [19, 22].

Despite many advantages described for CED such as decrease toxicity, BBB penetration, and control rate of diffusion, this technique has several limitations (Table 1). Tumor mass and edema can contribute to microenvironmental changes that can create fluid-reflux though the catheter, eventually leading to a decreased dose release and target of unspecific areas. Furthermore, disturbed flow patterns secondary to air bubbles (that may contribute to flow-reflux) have been described by several authors [23, 24]. In brain cancer, specifically for gliomas, CED has been actively used to treat lesions located in inaccessible areas for the surgeons, allowing an active infusion of chemotherapeutics to treat these lesions. In preclinical studies, CED has been successfully used to treat GBM. Immunotoxins, drugs, and viruses infused directly into the tumor bed resulted in a decrease rate of tumor growth with no recurrence rate in the animals [25–27]. Additionally, radiographic analysis from these studies demonstrated an extended and uniform distribution of the infused agents all over the tumor and peritumoral area [28].

Table 1
Advantages and disadvantages of CED

Advantages	Disadvantages
Diffusion rate based in pressure gradients	Drug backflow secondary to an increase intratumoral pressure
Homogenous intratumoral distribution of the infused agent	Disturbance in flow patterns secondary to air bubbles
Infusion of larger volumes and higher rates of an agent	Brain tissue inflammation secondary to catheter placement
Decrease neurotoxicity	Volume distribution limited by special tissue characteristics
Drug administration is independent of an agent's molecular weight or diffusivity	
Potential monitoring of the distribution agent in real time	

Table 2
Clinical trials using CED for gliomas

Tumor	Drug	Infusion rate	Duration of treatment	References
GBM	TP-38	0.4 mL/h	50 h	[33]
GBM	IL-13 PE38QQR	0.75 mL/h	96 h	[34]
GBM Anaplastic astrocytoma Anaplastic oligoastrocytoma	Flucytosine	350 μL/h	5 days	[35] ClinicalTrials.gov Identifier: NCT01156584
GBM Anaplastic astrocytoma	Topotecan	200 μL/h	100 h	[36]
GBM Anaplastic astrocytoma	Trabedersen	4 μL/min	7 days	[37]

In humans, several clinical trials have shown beneficial impacts of CED to treat high-grade gliomas (Table 2) [16]. The clinical trial, conducted by Edward Oldfield's group, was the first study to show the efficacy of an agent infused by CED. TF-CRM107, a transferrin receptor, for the treatment of malignant tumors resulted in 50% tumor size reduction in half of the patients treated with CED (9 of 15 patients) [29]. Another study, a phase III clinical trial, PRECISE trial (NCT00076986), analyzed the outcomes after CED administration of IL13–PE38QQ (a recombinant bacteria exotoxin) in recurrent GBM patients, demonstrated a progression free survival of 17.7 weeks versus 11.4 weeks over carmustine wafers. Recent data from Cleveland Clinic using a novel design of a multiport catheter TM (four microcatheters joint by a central guide) showed a widespread distribution of the chemotherapeutic drug topotecan into the tumor mass and in the peritumoral area early post-administration of the drug (24 h) [30]. Currently, administration of agents such as topotecan, IL-13, paclitaxel, carboplatin, muscimol, TGF-β, and IL-4 are under investigation to be used as treatment for recurrent or progressive GBM. [22]. In this section we will provide the protocols for preclinical models using osmotic pumps to deliver drugs into the brain [31, 32].

3.2 Materials

1. Osmotic minipumps (ALZET Model 1007D for a 7-day delivery period, 0.5 μL/h).

2. Syringe and blunt 27-G needle for filling minipumps (ALZET).

3. Brain infusion assembly (ALZET Brain Infusion Kit 3).

4. Brain infusion cannula.

5. 6–8-week-old mice.

6. Drug of interest.

7. 0.9% (w/v) NaCl sterile solution.

8. Buprenorphine analgesic.

9. Surgical glue.

10. 100% isoflurane.

11. 50 mL conical polypropylene centrifuge tubes (e.g., BD Falcon).

12. Small pointed scissors.

13. Hemostatic forceps.

14. Bard Parker scalpels stainless steel surgical blade #10 (catalog number 371610).

15. Ideal micro-drill (catalog number1685).

16. Ideal micro-drill burr set (catalog number 60-1000) (working diameter 0.6 mm, 0.8 mm, 1.00 mm, 1.20 mm, and 1.60 mm).

17. Stereotaxic apparatus with mouse adaptor (Kopf Model 922 Small Animal Stereotaxic Instrument with Digital Display Console) catalog number 900864A).

18. Olympus microscope S2X7.

19. Thermal blanket (Sunbeam EA, catalog number SLA103).

20. Patterson scientific induction chamber (12 × 6 × 8) for isoflurane administration (catalog number 07-8917760).

3.3 Methods

3.3.1 Preparation of the Infusion Pumps

Before assembling the device, determine the desired catheter length (2.5–3 mm catheter length is recommended for 25–30 g mice) and cannula, following a re-sterilization process under UV light (*see* Subheading 3.3.4, **Note 1**). Under sterile conditions, assemble the brain infusion device (composed of a catheter tube, flow moderator, and cannula, Fig. 1). The cannula length can be modified shaving the tip of the cannula with a dremel. Alternately, length can be modified using the rings provided by the manufacturer.

3.3.2 Pump Loading

Dissolve the factor/drug under study in sterile saline (0.9% NaCl) solution or check for pump compatible vehicles such as DMSO, PG, PEG, etc. Under the hood, fill the osmotic pump using an insulin syringe and the blunt 27-G needle provided by the manufacturer (ALZET). Carefully, remove the needle tip and connect the infusion device to the pump. After the pumps are loaded, fill the brain infusion device with the desired drug or vehicle so that the final assembled system does not contain any air bubbles (*see* Subheading 3.3.4, **Note 2**). Drop the loaded pumps into a sterile Falcon tube with 0.9% NaCl sterile solution, allowing pumps to hydrate overnight at 37 °C into the hood. After 24 h, take the Falcon tube into the surgery room and keep the infusion pumps submerged in NaCl solution until use (*see* Subheading 3.3.4, **Note 2**).

3.3.3 Osmotic Pump Implantation

Place a mouse in an isoflurane chamber to induce anesthesia, and once the mouse is fully anesthetized, the animal is accommodated on a stereotactic platform, and open the skin with a vertical incision to expose the skull (*see* Subheading 3.3.4, **Note 3**).

Set up the desire coordinates for the pump placement using the Kopf stereotaxic instrument. Once at the desire location, make a hole with a drill through the animal's skull (X: 1.5 mm, Y:1.34 mm, and work well to target the caudate nuclei) (*see* Subheading 3.3.4, **Note 4**).

With a rounded-tip scissors, dissect a cavity in the subcutaneous space in the back of the animal to create a space that will hold the osmotic pump. Using the forceps, hold the pump connected to the brain infusion assembly, and place the pump in the animal's back pouch under the skin. Hold the infusion cannula with a cannula holder and slowly descend the cannula until it reaches the skull surface. Add glue to the base pedestal of the cannula and lower it to the cannula until it is placed in the skull. Close the mouse's back skin incision with stitches and reinforce with surgical glue (*see* Subheading 3.3.4, **Note 5**). Remove the animal from the frame and allow it to recover on a pre-warmed thermal blanket before returning it to a cage.

3.3.4 CED Notes

1. There are multiple options for different sizes and volumes of pumps that have different rates of drug release and duration. When choosing the pump, take into account the size of the

mouse, and use the smaller size to provide more comfort. Given that the pump is connected to a catheter in the brain, ensure that the animals can reach food and water comfortably.

2. Filling of the pump system must always be done in vertical position to avoid bubbles into the system; after filling the pump, leave a small amount of liquid outside the pump to prevent bubbles going in the infusion system before connecting the catheter. If there are any air bubbles trapped into the system, evacuate it completely and refill the system. The pumps have to be vertically placed in the conical tubes, with the catheter tube facing up.

3. Before opening the animal's skin, a back subcutaneous injection of buprenorphine (10 μL per 10 g animal weight) is recommended for pre-surgical analgesia.

4. Verify the coordinates for cannula placement using a mouse brain atlas (the mouse brain in stereotaxic coordinates, [38]); the coordinates may vary slightly depending on the animal strain. Confirmation of the cannula placement can be performed by hematoxylin/eosin staining to detect the grove surrounded by glial scar after the traumatic injury caused by the cannula. Another option is to inject directly to the cannula a dye such as Evans blue following the manufacturer's instructions (http://www.alzet.com/resources/technical.tips.html).

5. The cannula must be always held perpendicular to the skull. When stitching and gluing the skin, be careful not to bend or detach the cannula. Monitor the mice in the days following the surgery. In the event that the wound open, re-stich the skin.

4   MR-Guided Focused Ultrasound

Preclinical interventions to the CNS often require direct cranial administration of drugs to reach relevant therapeutic concentrations and reduce systemic toxicity. Despite promising strategies, the effectiveness of systemically administered therapeutics has been hindered by the presence of the blood-brain barrier (BBB). There is an increased impetus to deliver therapeutics across the blood-brain barrier in a focally enhanced way in order to reduce off-target toxicity. To address the fundamental limitation of brain drug delivery, MR-guided focused ultrasound (*MRgFUS*) is a noninvasive technique made to circumvent this very issue via short pulses of low power, low frequency directed focused ultrasound (*FUS*) guided by MRI. This technique is an innovative physical drug delivery paradigm that provides a novel approach for focal enhanced delivery by transiently increasing BBB permeability and allowing increased delivery of your therapeutic vehicle.

4.1   Background

The ability to deliver therapeutic agents to the CNS has been a long-standing issue for the delivery of systemic therapy [12]. Even if diagnostic and therapeutic compounds managed to cross this barrier, regulations via astrocytes and pericytes maintain a tight hold preventing the passage of the vast majority of therapeutics. There has been increasing interest in using low-energy frequency for BBB disruption to create a transient permeable pore allowing for bigger molecules and cells to pass through an otherwise uncompromised barrier [13]. The BBB works in several ways: 100% of large molecules and drugs over 500 Da and 98% of drugs and antibodies are incapable of crossing the barrier [14]; lipid-soluble small molecules are capable of diffusing through the brain if they can squeeze through the endothelial membrane; however, very few drugs fall into this category. There is a dire need for a system that penetrates the BBB transiently, focally, with minimal risk and increased clinical therapeutic utility.

A critical barrier to systemic delivery of therapeutics, whether it is cell or macromolecule, is the ability of cargo to reach areas of insult and injury in areas of the penumbra where the BBB is intact. MRgFUS-induced BBB opening is a noninvasive method of generating the critical transient regional blood-barrier disruption that is necessary to increase agent penetration into the brain. Localized BBB disruption is mediated by the systemic delivery of pre-formed microbubbles in combination with short bursts of low energy frequency; this low-energy FUS creates negligible thermal elevations with ultimate cytoarchitecture preservation [39]. The FUS applies concentrated energy in a specific spot in incremental diameter; the delivery of microbubbles allows their oscillation to compromise the tight junctions of the endothelial cells lining the vessels causing a transient opening for bigger molecules and smaller cells, to pass through [40]. To mediate BBB penetration, MRI is used to scan and guide your target of interest allowing for focally enhanced region to induce pore creation; online monitoring is done dynamically with contrast-enhanced scans [41].

Typically, FUS is applied with parallel administration of microbubbles that are 1–10 μm; there presence in circulation reduces the acoustic energy required to open the BBB by two orders of magnitude [42]. Microbubbles are important not only in BBB permeabilization, but it abolishes any risks encountered with potentially heating the skull and causing thermal ablation of tissue. Furthermore, low pulses and sequential bursts of FUS allows for the heat of FUS to disperse in the focal area further reducing risk. When FUS is applied with the parallel administration of microbubbles, the oscillation of the microbubbles produces the mechanical shear force needed for pore creation in the BBB.

Clinical applicability of MRIgFUS in the context of brain cancer has yet to be applied; however, its applicability in Alzheimer's is just making way as patients are now being enrolled in noninvasive

Table 3
Advantages and disadvantages of MRIgFUS

Advantages	Disadvantages
Immediate delivery to site after sonication	Transient pore closes within 6 h of opening
Low energy focused US—No thermal ablation	Systemic injection of drugs and cells required (safety concern)
Stable cavitation	Mechanical defects due to acoustic energy
Limited risk of skull heating	Must be optimized for disease/pathology/structure and location for each patient/acoustic energy

ultrasound procedures to permeabilize the BBB as of 2017 (NCT02986932). This prospective single-arm non-randomized phase I will evaluate BBB sonication parameters measured by gadolinium enhancement. Increasing sonication energy will be done in patients with mild Alzheimer's until BBB is observably opened by MRI to establish the minimum required sonication energy. This technique allows for the efficient localization of your treatment cargo to the precise brain structure and location of intent while sparing adjacent tissue.

Low energy bursts of focused ultrasound will deliver acoustic sonication parameters that would allow for a noninvasive non-thermal cavitation of brain at MRI-guided site of interest. MRIgFUS is a revolutionary, safe, and novel method for noninvasive systemic delivery of larger therapeutics into the brain, transiently and non-destructively bypassing the BBB. The global hurdle of systemic therapy due to the BBB makes access of therapeutics, let alone cellular therapy to the brain parenchyma, nearly impossible. This approach allows first time the use of FUS to permeabilize the BBB by creating a transient pore big enough for larger entities to access, bypassing some of the BBB hurdles otherwise seen in drug and cell delivery to the CNS in a very precise and focused manner with high precision and low off-target delivery (Table 3).

4.2   Materials

1. MRI-compatible three-axis FUS system.
2. Small animal MRI scanner >1.5T.
3. Ultrasound transducer.
   (a) Center frequency in the range of 25–1.5 MHz.
   (b) F-number or radius aperture/curvature of 1.
4. Ultrasound transducer with integrated hydrophone.
5. MRI-compatible PVDF receiver (for acoustic emission recordings).
6. Power amplifier.

7. Passive cavitation detector (PCD) and PCD cable.

8. Function generator.

9. Degassed, deionized water.

10. Gadolinium.

11. Microbubbles.

12. 18-gauge needle.

13. Ultrasound gel.

14. Saline.

15. Cells/drugs of interest to be localized in brain after systemic delivery.

16. Benchtop anesthesia chamber.

17. Isoflurane.

18. Electric razor.

19. Heparin saline.

20. Tail vein catheter/22-g needle.

21. Heat blanket.

4.3 Methods

4.3.1 Ultrasound and MRI Setup

To begin, set up an animal platform where degassed and deionized water is on the bed of the MRI and place the ultrasound transducer below the animal platform (where the water is set up). The ultrasound transducer must be connected to a power amplifier. Ensure that the focal length of the transducer is at the surface of the platform where the degassed water is. To ensure focal length optimization of transducer to water platform, sonicate the water to create a subtle fountain in the water. Alternatively, you could use a gel to measure the focal length of the transducer for precise sonication. The location of the focused ultrasound must be centered through a pinhole where, when placed over focus, it will allow the water to come through pinhole (*see* Subheading 5, **Note 1**). In order to record the location of the exact focus, set up an MRI 3-plane localizer sequence to record the location of the exact focus. To record acoustic emission through the hydrophone, mount the MRI-compatible hydrophone in the water bath directed at the focus of the transducer. Run a PCD cable to capture bursts of up to 10 ms at proton resonance frequency (PRF) up to 2 Hz.

4.3.2 Animal Prep

Anesthetize animals using isoflurane (*see* Subheading 5, **Note 2**). Shave the fur off the top of the animals' head to not interfere with imaging and focus. Flush saline/heparin mixture 33 U/mL to prevent clot formation in catheter. Place animal supine with the top of the head making contact with the water bath through the hole created in the platform (*see* Subheading 5, **Note 3**) Tape animal to the platform and place tape across chin. Place heat blanket on top of animal to keep it warm.

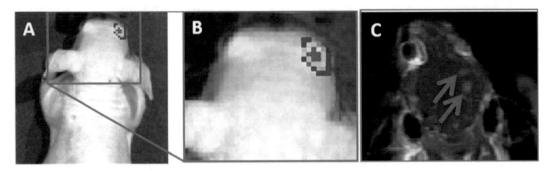

Fig. 2 MRI-guided focused ultrasound: (**a**) Systemic delivery of nanoparticles successfully localize brain region where BBB is disrupted by FUS. (**b**) Zoomed image. (**c**) Enhanced contrast MRI demonstrates BBB disruption

4.3.3 Target Selection

On the day of treatment, prior to sonication, baseline T1- and T2-weighted MRI should be acquired for targeting and as a baseline for contrast-enhanced imaging. Specify four targets (approximately 1.8 mm × 8 mm ellipsoids) in a 2 mm square, contained in a single hemisphere spanning and surrounding the area as identified on the planning images. Scheme should be as follows: (1) acquire baseline T1- and T2-weighted MR coordinates and (2) select target coordinates from T2-weighted scans (*see* Subheading 5, **Note 4**) (Fig. 2).

4.3.4 Microbubble Prep

Deliver microbubbles through IV at a dosage of 0.2 mL/kg of a 10:1 (*see* Subheading 5, **Note 5**) dilution (~108 bubbles/kg) followed by 0.1 mL saline flush through a tail vein catheter. During saline flush, begin sonication using 10-ms-long pulses at a pulse repetition frequency of 1 Hz for durations of 2 min. All of the targets should be sonicated with a single microbubble dose by interleaving the sonications. Target in situ pressures should be around 0.45–0.5 MPa at 1 MHz for an in situ correlated mechanical index of 0.45–0.5. Use a standard gadolinium-based MRI contrast agent at a dosage of 0.2 mL/kg delivered IV and confirm with T1-weighted MRI for immediate validation of BBB pore generation.

MRgFUS-mediated BBB permeabilization occurs when pulses of low intensity focused ultrasound (mechanical indices in the diagnostic range, though typically at a lower frequency) are applied to brain tissue that contains circulating microbubbles. Without the focused ultrasound, the microbubbles will be cleared from the blood with a half-life of about 1.3 min producing no effect; without the microbubbles, the ultrasound intensity is insufficient to affect the BBB; together, the interaction of the microbubbles and focused ultrasound is sufficient to locally and transiently permeabilize the BBB.

4.3.5   Ultrasound Delivery   Ultrasound parameters: low duty cycle bursts; however, bursts should be not continuous. Appropriate parameters of .5 MHz are around .23 MPa in situ pressure; use 10 ms bursts at 1 Hz PRF for 2 min (*see* Subheading 5, **Note 5**). Inject microbubble solution slowly into tail-vein catheter and flush 0.5 mL saline; sonication should begin simultaneously with start of microbubble injection.

4.3.6   Flow of MRI Visualized Cavitation   Image hyperintensity will indicate the locations where the small molecular gadolinium contrast agent have crossed the disrupted BBB mediated by the microbubbles. Concurrently with the administration of contrast, you should now deliver your cargo and experiment flow should run as follows (*see* Subheading 5, **Note 6**):

1. During sonication and while BBB is implicitly cavitation, inject gadolinium at 0.2 mL/kg.

2. Acquire T1-weighted images until contrast peak is diminished (*see* Subheading 5, **Note 6**).

3. Acquire T2-weighted image to check for any leaks or edema.

4. Inject cells/drugs.

5   MRgFUS Notes

1. Allowing the water to come through the pinhole will allow for a 2-plane system visualized by MRI and the transducer axial coordinates can be determined through the water surface coming through pinhole.

2. Since isoflurane can affect BBB disruption, the animals must be off of the gas or at low isoflurane 10 min prior to experiment.

3. Place ultrasound gel on top of the head while the animal is positioned supine to minimize air bubbles through experiment.

4. Avoid ventricles and midline of the brain.

5. Invert syringe until homogenous appearance of saline and microbubble (*see* **Note 7**).

6. At 1.5 MHz in situ pressure falls around 0.45–0.5 Mpa—at different frequencies can be estimated using 0.46 mechanical index.

7. Sites which have been opened by microbubble and sonication will show greater enhancement of the gadolinium.

6   Self-Assembling Hydrogels

The use of anticancer drugs to create a new class of supramolecular hydrogels has been developed over the past few years [43, 44]. This hydrogel-based drug delivery system is made by self-assembling drug amphiphiles (DAs) through conjugating a short peptide segment onto an anticancer drug via a biodegradable linker. Under physiologically relevant conditions, these DAs can be designed to spontaneously associate into discrete, stable supramolecular nanostructures that are potentially able to crosslink into 3D gel networks and possess the potential for self-delivery. Given that GBM tumors recur in close to the primary site, these injectable supramolecular hydrogels that cloud be in close proximity to the surrounding tissue are highly desired for local treatment.

6.1   Background

Many drug molecules are not suitable for clinical use due to their poor solubility, degradation, tissue penetration, and rapid clearance [45, 46]. An attractive approach is the use of a prodrug strategy to chemically modify a drug to improve its bioavailability and pharmacokinetic profile and achieve enhanced treatment efficacy. Conjugation of an auxiliary segment (small moiety) onto a drug of interest can improve its water solubility and bioavailability by preventing premature drug degradation and enhancing therapeutic efficacy [47]. Peptide-drug conjugates are formed through the covalent attachment of a specific peptide sequence and to a drug via a cleavable linker. The use of peptides provides great functionality; for example, it can allow for targeting specific receptors, allow higher tissue penetration, and can even be used for imaging [48–50]. Interestingly, it is sometimes unavoidable that the resulting prodrug can acquire an amphiphilic nature. This property can aid the prodrug to self-assemble into a variety of discrete stable supramolecular nanostructures [43, 51].

The unique features of these drug-made nanostructures are that they form their own drug delivery vehicles; they are capable of breaking down, either over time by hydrolysis or due to some specific stimulus, and release the active drug. The drug retains its inherent bioactivity but also acts as a building block in the construction of nano-sized assemblies [43, 44]. Furthermore, most of one-dimensional nanoassemblies are able to associate into discrete, stable supramolecular hydrogels under physiological environments that are potentially able to crosslink into 3D gel networks and possess the potential for self-delivery. Hydrogel-based therapeutics has been extensively explored in GBM (reviewed in [52]). In contrast with other kinds of hydrogels used in brain tumors (i.e., PLGA-based hydrogels like Oncogel™ [53]), the peptide-based supramolecular hydrogels are formed by physical entanglements of filamentous assemblies resulting from a series of non-covalent

Table 4
Advantages and disadvantages of supramolecular hydrogels

Advantages	Disadvantages
The great diversity of amino acid sequences provides the possibility to generate supramolecular interactions to form nanofibers and hydrogels of various properties	For local use, a procedure is necessary before implementation
The drug release rates can be precisely tuned through both the chemical linker design and the molecular design	Limited drug penetration in tumor tissue
The hydrogel form allows the delivered nanofiber to fill seamlessly the cavities left after tumor resection and to have a significantly prolonged retention time	The biocompatibility of peptide-based assemblies should be better elucidated
The choice of various functional peptides (e.g., tumor targeting and tissue penetrating) allows enhanced specificity of supramolecular hydrogels	The release rate over a long period of time is a critical factor to determine the clinical benefit of the hydrogels

interactions with the peptidic units. These entangled networks can be engineered to release therapeutics in a well-controlled manner (reviewed in [54]) Supramolecular hydrogels can be applied for the local treatment of brain tumors with several advantages to the techniques previously discussed (Table 4): (1) The hydrogel form will allow the delivered nanofiber to fill seamlessly the cavities left after tumor resection enabling a significantly prolonged retention time. (2) The nanofiber form allows for diffusion across larger areas relative to individual molecules while avoiding capillary loss. (3) The conjugated form offers an efficient strategy to overcome the MDR mechanisms that glioma cells possess or may develop over the course of the treatment. Conjugation of an anticancer drug with a short peptide is a well-proven strategy in the literature to address multidrug resistance mechanisms in cancer cells; thus it will be more effective than their respective free drug form in treating drug resistant glioma cells. (4) The drug release rates can be precisely tuned through the molecular design of the drug building blocks by tuning the linker chemistry and stiffness of the hydrogels. The release rate over a long period of time is a critical factor to determine the clinical benefit of the hydrogels. Next, we will describe the construction of prodrug-like therapeutic agents with self-assembling properties.

6.2 Materials

1. Fmoc amino acids, and coupling reagents (HBTU or HATU) from Advanced Automated Peptide Protein Technologies (AAPPTEC, Louisville, KY, USA).

2. Rink amide MBHA resins Novabiochem (San Diego, CA, USA).

3. Camptothecin AvaChem Scientific (San Antonio, TX, USA).

4. Varian ProStar Model 325 HPLC (Agilent Technologies, Santa Clara, CA).

5. Finnigan LDQ Deca ion-trap mass spectrometer (Thermo-Finnigan, Waltham, MA).

   In this example, we conjugated two camptothecin (CPT) molecules to a short hydrophilic segment of three different types to give three different DAs: dCPT-K2, DCPT-OEG$_5$-K2, and dCPT-Sup35-K2 (Figs. 3 and 4). Disulfanyl-ethyl carbonate linker (etcSS) was used as the biodegradable linker to bridge the peptide and the drug, which is expected to break down in the presence of a cancer-relevant intracellular reducing agent glutathione (GSH). Glioblastomas as well as several other cancer types present higher levels of GSH intracellularly [55]. Using linkers sensitive to GSH will allow activation of the drug at the specific tumor site.

   All peptides described are synthesized using standard Fmoc solid phase peptide synthesis protocols well established by our group and purified using reverse-phase preparative HPLC [51, 56–60]. The drug amphiphiles are prepared by reaction of the appropriate cysteine-containing peptide with an activated CPT disulfide derivative and again purified using reverse-phase HPLC [56, 59]. For all the studied DAs, the purity and identity of each DA is determined through analytic HPLC and mass spectrometric analysis (ESI or MALDI). Nanostructure morphology can be assessed using transmission electron microscopy (TEM). The internal structure/packing of the assemblies can be probed using circular dichroism spectroscopy.

6.2.1 Peptide Synthesis

All peptide conjugates were synthesized on the Rink Amide MBHA resins using standard 9-fluorenylmethoxycarbonyl (Fmoc) solid phase synthesis techniques at a 0.25 mmol synthesis scale. The Rink Amide MBHA resins were swelled with 10–15 mL of DCM in a 50 mL plastic tube and transferred to the synthesis flask of AAPPTec Focus peptide synthesizer after swelling. An automated peptide synthesizer (Focus XC, AAPPTec, Louisville, KY) was employed to build the (Ac-Cys)2KGN2Q2NYK2-NH2 (dCys-Sup35-K2) sequence before manual protocols were used to furnish the branching motif. Amino acids were properly weighed to be fourfold excess of the molar number of the resin for a single coupling step (*see* Subheading 7, **Note 1**). Prepare each Fmoc-amino acid/DMF solution at a concentration of 0.2 mmol/mL. Fill Fmoc deprotection solution (20% 4-methylpiperidine in DMF), HBTU coupling solution (0.4 M HBTU in DMF), DIEA solution (2 M DIEA in DMF/DCM = 4:1), DMF, DCM, and prepared amino acid solutions into the reagent reservoirs of the

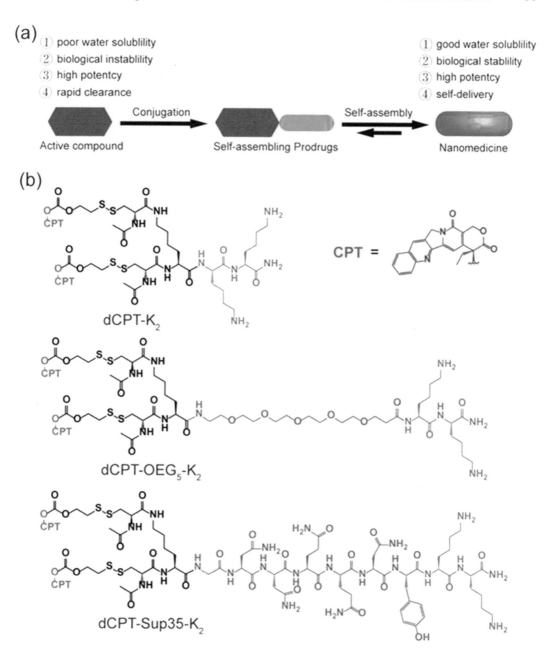

Fig. 3 (**a**) Schematic illustration of the concept for rationally designed self-assembling prodrugs (SAPD) with high potency. (**b**) Chemical structures of the studied three model drugs: dCPT-K2, dCPT-OEG5-K2, and dCPT-Sup35-K2. Two CPTs were conjugated to three different hydrophilic auxiliary segments through disulfanyl-ethyl carbonate linker (etcSS). (Reproduced from Theranostics 2016 with permission from Ivyspring International Publisher. Su H, Zhang P, Cheetham AG, Koo JM, Lin R, Masood A, Schiapparelli P, Quiñones-Hinojosa A, Cui H. Supramolecular Crafting of Self-Assembling Camptothecin Prodrugs with Enhanced Efficacy against Primary Cancer Cells. Theranostics 2016; 6(7):1065-1074. https:/doi.org/10.7150/thno.15420)

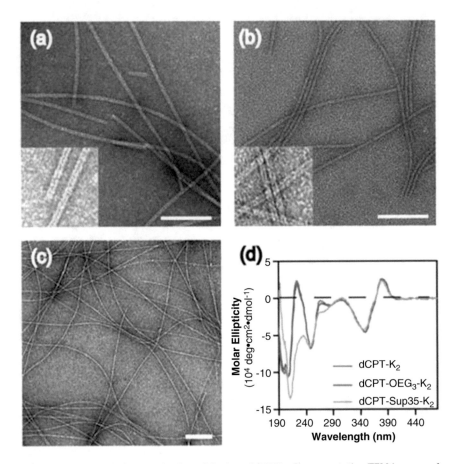

Fig. 4 Molecular assembly and characterization of designed SAPDs. Representative TEM images of nanotubes formed by dCPT-K2 with diameter of 8.2 ± 0.8 nm (**a**) and dCPT-OEG5-K2 with diameter of 7.5 ± 0.7 nm (**b**). (Insert) High-resolution TEM images display the tubular morphology. (**c**) TEM image of nanofibers formed by dCPT-Sup35-K2 with diameter of 6.9 ± 1.4 nm. (**d**) Normalized CD spectra of the three studied supramolecular prodrugs in water. All the samples were prepared at concentration of 100 μM in water and aged over 3 days before measurements. TEM samples were negatively stained with 2 wt% uranyl acetate and all scale bars = 100 nm). (Reproduced from Theranostics 2016 with permission from Ivyspring International Publisher: Su H, Zhang P, Cheetham AG, Koo JM, Lin R, Masood A, Schiapparelli P, Quiñones-Hinojosa A, Cui H. Supramolecular Crafting of Self-Assembling Camptothecin Prodrugs with Enhanced Efficacy against Primary Cancer Cells. Theranostics 2016; 6(7):1065-1074. https:/doi.org/10.7150/thno.15420)

synthesizer. Set up the program for the automated synthesis of the sequence Fmoc-GN2Q2NYK2-Rink. After automated synthesis, the resins (Fmoc-GN2Q2NYK2-Rink) were transferred into a peptide shaking vessel to start manual synthesis. After Fmoc deprotection with 20% 4-methylpiperidine/DMF, a ranched lysine Fmoc-Lys (Fmoc)-OH was conjugated to the end of the resin by adding a mixture of Fmoc-Lys (Fmoc)-OH, HBTU, and DIEA (4:4:6 molar equiv. to resin) in DMF for 2 h. Then, the last amino acid Fmoc-Cys(Trt)-OH was also added on by the similar method (*see* Subheading 7, **Note 2**). The final acetylation step was performed after deprotection of Fmoc group on the Cys with 20% acetic anhydride/DMF and 100 μL DIEA, shaken for 15 min, and the coupling was repeated twice (*see* Subheading 7, **Note 3**). The other two peptides (Ac-Cys)2KK2 (dCys-K2) and (Ac-Cys)2KOEG5 K2 (dCys-OEG5-K2) were both synthesized manually using the standard Fmoc chemistry protocols before the coupling of the camptothecin. Similar as the abovementioned manual protocols, Fmoc deprotections were performed using a 20% 4-methylpiperidine in DMF solution for 15 min and repeated once. The amino acid coupling cycle was performed after Fmoc deprotection, by adding a mixture of Fmoc-amino acids, HBTU, and DIEA (4:4:6 molar equiv. to resin) in DMF for 2 h. The addition of Fmoc-NH-OEG5-Propionic Acid was realized by the same protocols as amino acids coupling. The branching point was achieved by coupling Fmoc-Lys (Fmoc)-OH to yield two primary amine groups for the further coupling of Fmoc-Cys(Trt)-OH. Acetylation was performed on α-amino groups of N-terminus amino acids using a 20% acetic anhydride in DMF solution with 100 μL DIEA, shaken for 15 min, and the coupling was repeated twice. In all cases, reactions were monitored by the ninhydrin test (Kaiser test) (Anaspec Inc., Fremont, CA) for free amines. To perform a Kaiser test, take some [10–20] resins into a small test tube with addition of 2 drops of Kaiser test kit reagent A, 1 drop of reagent B, and 1 drop of reagent C, and heat the tube to 110 °C for 5 min. This Kaiser test is used to trace the presence or absence of free primary amino groups. After Fmoc deprotection, the resin beads and the solution should turn dark blue (positive result) since primary amine is present. Otherwise, repeat the deprotection process again until the Kaiser test indicates the presence of free amine. Meanwhile, after conjugation, the resins should remain their original color and the solution stay light yellow (negative result) because free amines are reacted in the coupling reaction. Otherwise, repeat the coupling one more time and check the Kaiser test again. Completed peptides were cleaved from the resins using a mixture of TFA/TIS/H_2O at a ratio of 95:2.5:2.5 for 3 h. Excess TFA was removed by rotary evaporation and the concentrated solution was precipitated in cold diethyl ether to get the crude product, which were collected, washed with more diethyl ether (to fully remove TFA), and dried under vacuum overnight.

6.2.2 Self-Assembling Prodrug (SAPD) Synthesis and Characterization

The synthesis of SAPDs was carried out by mixing CPT-etcSS-Pyr and the corresponding crude peptides in N_2-purged DMSO (2 mL) with a molar ratio of 2:1. After reacting for 5 days, the mixture was diluted to 10 mL with 0.1% TFA in acetonitrile/water and purified by preparative RP-HPLC using a Varian ProStar model 325 HPLC (Agilent Technologies, Santa Clara, CA, USA) equipped with a fraction collector. Separations were performed using a Varian PLRP-S column (100 Å, 10 μm, 150 × 25 mm) monitoring at 370 nm with the mobile phase starting from 25% MeCN (with 0.1% TFA) to 75% MeCN (with 0.1% TFA) at 30 min, and the flow rate is 20 mL/min. Collected fractions were analyzed by ESI-MS (LDQ Deca ion-trap mass spectrometer, Thermo Finnigan, USA), and the appropriate fractions were collected, concentrated, and lyophilized on a FreeZone −105 °C 4.5 L freeze dryer (Labconco, Kansas City, MO, USA). The powders obtained were then re-dissolved, calibrated, and aliquoted into cryo-vials before re-lyophilization.

The purity of the conjugates was analyzed by HPLC using the following conditions: Agilent Zorbax-C$_{18}$ column (5 μm, 4.6 × 150 mm); the flow rate was 1 mL/min, with the mobile phase starting from 15% MeCN (with 0.1% TFA) to 80% MeCN (with 0.1% TFA) at 25 min, hold for 5 min, and gradient back to the initial condition in 3 min; the monitored wavelength was 370 nm. Molecular masses were determined by ESI-MS.

6.2.3 Calibration of the Concentration

The concentrations of purified self-assembling prodrugs (SAPDs) were calibrated by analyzing the reduced product free CPT from the SAPDs. Briefly, a stock solution of the SAPD was prepared by dissolving in MeCN/H$_2$O (1:1). 5 μL of the stock solution was then diluted to 20 μL by addition of 15 μL MeCN/H$_2$O (1:1) and mixed with 20 μL of fresh prepared 1 M aqueous TCEP for 1 h with periodic vortexing. 25 μL solution was then injected into the HPLC (so as to completely fill the 20 μL loop) to measure the area of the peak due to free CPT at 370 nm. The CPT concentration was obtained by comparing the area under peak from treated solution with the standard calibration of CPT. Note that the standard calibration curve of CPT was predetermined by analytical HPLC and the equation of concentration C_{CPT} and A (absorption area under curve) was obtained. The SAPD concentration was calculated based on the applied dilutions and number of CPT molecules (*see* Subheading 7, **Note 4**). Finally, the stock solution was diluted to 100 μM according to the calibrated concentration and aliquoted into cryo-vials before re-lyophilization.

6.3 Techniques for Hydrogel Characterization

6.3.1 Transmission Electron Microscopy (TEM)

TEM can be used to characterize the nanostructure of self-assembled DAs.

DA solutions of assembled nanostructures (7–10 μL) are placed on grids and allowed to dry. Contrast-enhancing agents are often used to study the fine details of the nanostructure.

In our case 100 μM stock solutions of corresponding samples in water were prepared by direct dissolution of the respective lyophilized powders and allowed to age overnight. TEM samples were prepared by laying a thin layer of the appropriate solution onto carbon-coated copper grid (Electron Microscopy Services, Hatfield, PA, USA) and stained using 2 wt% aqueous uranyl acetate. Hydrogels were instantly diluted before sample preparation. The sample grid was then allowed to dry at room temperature prior to imaging. Bright-field TEM imaging was performed on FEI Tecnai 12 TWIN transmission electron microscope operated at an acceleration voltage of 100 kV. All TEM images were recorded by a SIS Megaview III wide-angle CCD camera.

6.3.2   Circular Dichroism (CD)

CD can be used to determine the secondary structure of peptides in the self-assembled state. Signatures for α-helical, β-sheet, and β-turn secondary structures as well as random-coil conformations can be determined. In addition, it allows for monitoring changes in the secondary structure of peptides in response to changes in solution conditions (e.g., pH, temperature, ionic strength, chaotropes, etc.).

All the CD spectra were recorded on a Jasco J-710 spectropolarimeter (JASCO, Easton, MD, USA) from 190 to 480 nm using a 1 mm (for 100 μM and 50 μM) or 10 mm (for 12.5 μM and 3.125 μM) path length quartz UV-Vis absorption cell (Thermo Fisher Scientific, Pittsburgh, PA, USA) (*see* Subheading 7, **Note 5**). Background spectra of the solvents (water here) were acquired and subtracted from the sample spectra. Collected data was normalized from ellipticity (mdeg) to molar ellipticity (deg cm^2 dmol^{-1}).

6.3.3   In Vitro Drug Release

To determine the release profile of the DAs, dCPT-K$_2$, dCPT-OEG$_5$-K$_2$, and dCPT-Sup35-K$_2$ with or without GSH was evaluated using RP-HPLC (Fig. 5). 50 μM stock solutions in deionized water were prepared and diluted to 25 μM with 20 mM PBS buffer with or without GSH (20 mM). The solutions were incubated at

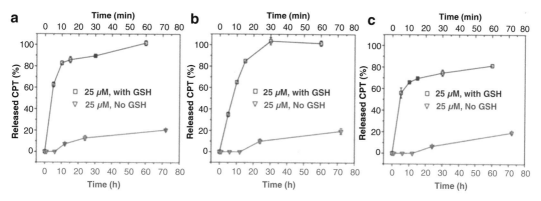

Fig. 5 Cumulative drug release from dCPT-K2 (**a**), dCPT-OEG5-K2 (**b**), and dCPT-Sup35-K2 (**c**) in PBS at 37 μL at 25 μM with (blue curve) or without (red curve) 10 mM GSH. Data were given as mean ± s.d. (*n* = 3). For samples with 10 mM GSH, they presented ultrafast release profiles (100% CPT released out for dCPT-K2 (**a**), dCPT-OEG5-K2 (**b**), and 81.4% CPT released out for dCPT-Sup35-K2 (**c**) within 1 h). For samples without GSH, around 20% CPT released out at 72 h. (Reproduced from Theranostics 2016 with permission from Ivyspring International Publisher: Su H, Zhang P, Cheetham AG, Koo JM, Lin R, Masood A, Schiapparelli P, Quiñones-Hinojosa A, Cui H. Supramolecular Crafting of Self-Assembling Camptothecin Prodrugs with Enhanced Efficacy against Primary Cancer Cells. Theranostics 2016; 6(7):1065–1074. https://doi.org/10.7150/thno.15420)

37 °C and samples were collected at 0 min, 5 min, 10 min, 15 min, 30 min, and 1 h for those treated with GSH, and 0 h, 1 h, 6 h, 12 h, 24 h, and 72 h for those without GSH. Note that samples with GSH will have very fast release because of the design linker in the DA (this mimics intracellular environment), while hydrolysis of ester is mainly responsible for the drug release without GSH (this mimics extracellular environment). The samples were acidified with addition of 0.2 μL of 2 M HCl, frozen with liquid nitrogen and stored at −30 °C until analysis. Note that acidification will quench the reaction of GSH and stop the release of CPT. The amounts of released CPT were monitored by RP-HPLC using the following conditions: Varian Pursuit XRs C_{18} (5 μm, 150 × 4.6 mm); 362 nm detection wavelength; 1 mL/min flow rate; the gradient began at 85% of mobile phase A (0.1% aqueous TFA) and 15% of mobile phase B (acetonitrile containing 0.1% TFA) to 45% mobile phase A and 55% mobile phase B at 10 min, then to 15% mobile phase A and 85% mobile phase B at 13 min and held for another 2 min. Selected time points were characterized and data were plotted as a percentage of the total CPT concentration.

6.3.4 Cell Viability in Response to Supramolecular Hydrogels

Cell viability was evaluated using the Sulforhodamine B (SRB) method [61]. Figure 6 tumor cells were seeded onto 96-well plates (5000 cells/well) and allowed to attach overnight. Stock solutions in 100 μM of three different SAPDs were prepared at the same time and aged overnight. The stock solutions were then diluted with fresh medium to achieve final CPT concentration of 0.1, 1, 10, 100,500, 1000, 5000, and 10,000 nM. After dilution, the SAPD-containing mediums were incubated with cells immediately. Medium containing the same concentration of free CPT ranging from 0.1 to 10,000 nM was also used to incubate the cells, with non-treated cells (medium only) as the control group. After 72 h incubation, the cell viability was evaluated using the SRB method according to the manufacturer's protocols (TOX-6, Sigma, St. Louis, MO).

6.4 Delivery Methods for Hydrogels in Vivo

GBM are highly infiltrative brain tumors, in 90% of the cases tumor recurrence after treatment appears within 2 cm of the resection cavity [4]. For this reason, applying hydrogels locally to the cavity increases the drug concentration at the potential recurrence site while minimizing systemic toxicities. In recent years, several laboratories (including ours) have developed surgical resection models in mice implanted with human-derived GBM cells [62–65]. The surgical cavity allows hydrogel deposition, typically between 2 and 5 μL of material. In addition, supramolecular hydrogels can be delivered by stereotactic injection to the brain parenchyma. For example, Lock LL et al. [48] developed a pemetrexed-peptide conjugate (Pem-FE, combining pemetrexed with a short peptide sequence containing two glutamic acids and two phenylalanines)

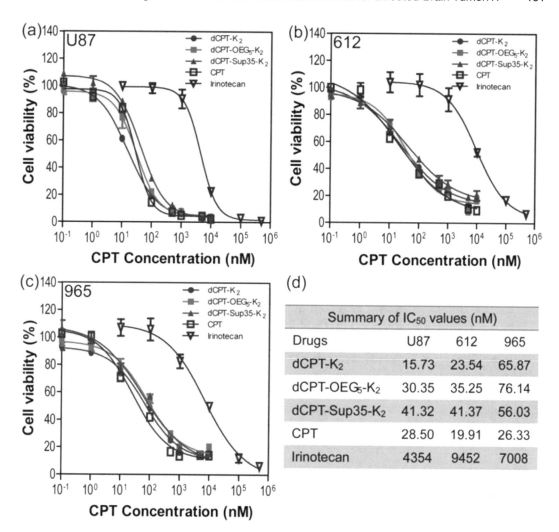

Fig. 6 In vitro cytotoxicity study of the SAPDs against human brain cancer cell lines U87 MG (**a**) and intraoperative derived human cancer cell lines (612 (**b**) and 965 (**c**)). All cancer cells were incubated with the appropriate SAPDs for 72 h, and cell viability was determined by SRB assay. Data were given as mean ± s. d. ($n = 3$) and IC50 values were calculated according to CPT concentration. All the SAPDs exhibited greater efficacy against three human brain cancer cell lines than irinotecan. (Reproduced from Theranostics 2016 with permission from Ivyspring International Publisher: Su H, Zhang P, Cheetham AG, Koo JM, Lin R, Masood A, Schiapparelli P, Quiñones-Hinojosa A, Cui H. Supramolecular Crafting of Self-Assembling Camptothecin Prodrugs with Enhanced Efficacy against Primary Cancer Cells. Theranostics 2016; 6(7):1065–1074. https://doi.org/10.7150/thno.15420)

and tested it using the GL261 orthotopic brain tumor model [66]. Pem molecular structure allowed the use of chemical exchange saturation transfer (CEST) using MRI imaging to determine the hydrogel brain distribution at different time points after implantation. Another strategy to visualize the hydrogels can be using immunofluorescence imaging, labeling the peptides with fluorophores such as Cy5 or Cy7.

Table 5
Other drug-peptide conjugates for targeting receptors in brain tumors

Receptor target site	Peptide-drug conjugate	References
Integrin αv	iRGD-polymer-doxorubicin	[69]
Low-density lipoprotein receptor	Angiopeptin2-paclitaxtel	[70]
Tenascin C and neuropilin-1	Paclitaxel-loaded Ft nanoparticles (Ft-NP-PTX)	[71]
Interleukin-4 receptor	AP1 peptide functionalized doxorubicin-loaded nanoparticles	[72]

Recently, other hydrogel systems have been used combined with surgical resection for GBM. Bastiancich et al. used a lauroyl-gemcitabine lipid nanocapsule-based hydrogel (GemC12-LNC) using two different approaches: intratumoral injection or deposition after resection [67]. They observed that the hydrogel was well tolerated mid and long term and increased survival.

6.5 Other Types of Peptide Conjugate Systems to Target Brain Tumors

Specific targeting of brain tumor cells can be achieved using targeting peptide-drug conjugates. The Arg-Gly-Asp (RGD) peptide is among the most common peptide used, and there are multiple variants being developed and studied (reviewed in [68]). Table 5 illustrates different examples of peptide-drug conjugates evaluated in GBM:

7 Self-Assembling Hydrogels Notes

1. When we use 0.25 mmol of resin, the amount amino acids should be 1 mmol for each step.

2. We need to use eightfold excess of Cys to the resin, since the branched lysine doubled the conjugation site (from 1 to 2).

3. Wash the resin three times each with DCM and DMF after each deprotection and coupling.

4. Each prodrug has two CPTs in the molecular design; thus the final concentration of prodrug is $1/2$ of measured CPT concentration.

5. All the solutions here are diluted from one stock solution and aged over night to equilibrium. The use of longer path length cell enables higher resolution of low concentration samples and minimizes the noise.

8 Future Directions

The delivery of therapeutics across the CNS has been universally hindered by the inability of the BBB to allow active compounds to enter sites of injury to the brain. The insurmountable obstacle posed by the BBB on many promising therapeutics have called the need for enhanced approaches aiding in BBB permeability that assist in increasing bioavailability and uniform distribution across brain tissue. While the aforementioned approaches call for an exogenous supplement to treatment that allow for BBB permeability, there is a need to design therapeutics to reach the brain penumbra, in their own right, without a supplemental application of a technology to allow drug entry like CED or FUS. Novel formulation of nanoparticles have emerged as promising carriers for CNS therapeutics that could allow for the systemic administration of drugs without the concern of decreased bio-index that would have otherwise been compromised with the limited accessibility imposed by the BBB [73–77]. Nanoengineering particles with small diameter capable of cargo delivery seems to be the next development in drug transport across the BBB. The field of nanoparticle discovery and design is offering new frontiers in drug transport potentially opening up new avenues for delivery-based therapeutics that would be less invasive and carry less risk to patients.

Traversing the BBB requires nanoparticle formulation design that would account for reaching distant sites upon systemic delivery, enabling transcytosis across the microvasculature barrier of the BBB, and finally, intracellular delivery of cargo to compromised tissue across various CNS pathologies. Ultimately, an effective nanoparticle would (1) enhance pre-transcytosis circulation (this could be manifested with the use of bioreducible formulations for prolonged circulation), (2) enhance BBB transcytosis as well as post-transcytosis for nanoparticle-brain tissue interaction for cargo delivery, and lastly, and (3) have effective design that would exploit innate transport processes to enter the brain, in their own right, without the need to permeabilize brain microvasculature [78, 79].

The new wave of BBB permeability approaches will allow for the discovery of engineered cargo-delivery vehicles to the CNS capable of navigating and crossing BBB compartments, which take advantage of the innate physiology of the BBB and will work diplomatically with the microenvironment rather than compromising it. The future of nanotechnology is a forthcoming field in CNS pathologies and their emergence will overshadow conventional approaches in the next decades to come.

Acknowledgments

We will like to thank Nick Ellens, PhD, for optimizing sonication parameters for MRIgFUS. AQH, PS, RA, and MLV are supported by NIH grants CA183827, CA195503, CA216855, and CA200399. MLV is also supported by CONACYT and PECEM from the National Autonomous University of Mexico.

References

1. Chakroun RW, Zhang P, Lin R, Schiapparelli P, Quinones-Hinojosa A, Cui H (2018) Nanotherapeutic systems for local treatment of brain tumors. Wiley Interdiscip Rev Nanomed Nanobiotechnol 10(1). https://doi.org/10.1002/wnan.1479

2. Schinkel AH, Wagenaar E, Mol CAAM, van Deemter L (1996) P-glycoprotein in the blood-brain barrier of mice influences the brain penetration and pharmacological activity of many drugs. J Clin Invest 97(11):2517–2524. https://doi.org/10.1172/Jci118699

3. Stewart LA (2002) Chemotherapy in adult high-grade glioma: a systematic review and meta-analysis of individual patient data from 12 randomised trials. Lancet 359 (9311):1011–1018

4. Hochberg FH, Pruitt A (1980) Assumptions in the radiotherapy of glioblastoma. Neurology 30(9):907–911

5. Selker RG, Shapiro WR, Burger P, Blackwood MS, Arena VC, Gilder JC, Malkin MG, Mealey JJ Jr, Neal JH, Olson J, Robertson JT, Barnett GH, Bloomfield S, Albright R, Hochberg FH, Hiesiger E, Green S (2002) Brain tumor cooperative G. the brain tumor cooperative group NIH trial 87-01: a randomized comparison of surgery, external radiotherapy, and carmustine versus surgery, interstitial radiotherapy boost, external radiation therapy, and carmustine. Neurosurgery 51(2):343–355; discussion 55–7

6. Tsao MN, Mehta MP, Whelan TJ, Morris DE, Hayman JA, Flickinger JC, Mills M, Rogers CL, Souhami L (2005) The American Society for Therapeutic Radiology and Oncology (ASTRO) evidence-based review of the role of radiosurgery for malignant glioma. Int J Radiat Oncol Biol Phys 63(1):47–55. https://doi.org/10.1016/j.ijrobp.2005.05.024

7. Stupp R, Mason WP, van den Bent MJ, Weller M, Fisher B, MJB T, Belanger K, Brandes AA, Marosi C, Bogdahn U, Curschmann J, Janzer RC, Ludwin SK, Gorlia T, Allgeier A, Lacombe D, Cairncross JG, Eisenhauer E, Mirimanoff RO, Van Den Weyngaert D, Kaendler S, Krauseneck P, Vinolas N, Villa S, Wurm RE, MHB M, Spagnolli F, Kantor G, Malhaire JP, Renard L, De Witte O, Scandolaro L, Vecht CJ, Maingon P, Lutterbach J, Kobierska A, Bolla M, Souchon R, Mitine C, Tzuk-Shina T, Kuten A, Haferkamp G, de Greve J, Priou F, Menten J, Rutten I, Clavere P, Malmstrom A, Jancar B, Newlands E, Pigott K, Twijnstra A, Chinot O, Reni M, Boiardi A, Fabbro M, Campone M, Bozzino J, Frenay M, Gijtenbeek J, Brandes AA, Delattre JY, Bogdahn U, De Paula U, van den Bent MJ, Hanzen C, Pavanato G, Schraub S, Pfeffer R, Soffietti R, Weller M, Kortmann RD, Taphoorn M, Torrecilla JL, Marosi C, Grisold W, Huget P, Forsyth P, Fulton D, Kirby S, Wong R, Fenton D, Fisher B, Cairncross G, Whitlock P, Belanger K, Burdette-Radoux S, Gertler S, Saunders S, Laing K, Siddiqui J, Martin LA, Gulavita S, Perry J, Mason W, Thiessen B, Pai H, Alam ZY, Eisenstat D, Mingrone W, Hofer S, Pesce G, Curschmann J, Dietrich PY, Stupp R, Mirimanoff RO, Thum P, Baumert B, Ryan G, European Organisation for Research and Treatment of Cancer Brain Tumor and Radiotherapy Groups; National Cancer Institute of Canada Clinical Trials Group (2005) Radiotherapy plus concomitant and adjuvant temozolomide for glioblastoma. New Engl J Med 352(10):987–996. https://doi.org/10.1056/Nejmoa043330

8. McGirt MJ, Than KD, Weingart JD, Chaichana KL, Attenello FJ, Olivi A, Laterra J, Kleinberg LR, Grossman SA, Brem H, Quinones-Hinojosa A (2009) Gliadel (BCNU) wafer plus concomitant temozolomide therapy after primary resection of glioblastoma multiforme. J Neurosurg 110(3):583–588. https://doi.org/10.3171/2008.5.17557

9. Attenello FJ, Mukherjee D, Datoo G, McGirt MJ, Bohan E, Weingart JD, Olivi A, Quinones-

Hinojosa A, Brem H (2008) Use of Gliadel (BCNU) wafer in the surgical treatment of malignant glioma: A 10-year institutional experience. Ann Surg Oncol 15(10):2887–2893. https://doi.org/10.1245/s10434-008-0048-2

10. Chaichana KL, Zaidi H, Pendleton C, McGirt MJ, Grossman R, Weingart JD, Olivi A, Quinones-Hinojosa A, Brem H (2011) The efficacy of carmustine wafers for older patients with glioblastoma multiforme: prolonging survival. Neurol Res 33(7):759–764. https://doi.org/10.1179/1743132811Y.0000000006

11. Fleming AB, Saltzman WM (2002) Pharmacokinetics of the carmustine implant. Clin Pharmacokinet 41(6):403–419. https://doi.org/10.2165/00003088-200241060-00002

12. Greene C, Campbell M (2016) Tight junction modulation of the blood brain barrier: CNS delivery of small molecules. Tissue Barriers 4 (1):e1138017. https://doi.org/10.1080/21688370.2015.1138017

13. Burgess A, Hynynen K (2014) Drug delivery across the blood-brain barrier using focused ultrasound. Expert Opin Drug Deliv 11 (5):711–721. https://doi.org/10.1517/17425247.2014.897693

14. Pardridge WM (2005) The blood-brain barrier: bottleneck in brain drug development. NeuroRx 2(1):3–14. https://doi.org/10.1602/neurorx.2.1.3

15. Saito R, Tominaga T (2012) Convection-enhanced delivery: from mechanisms to clinical drug delivery for diseases of the central nervous system. Neurol Med Chir (Tokyo) 52 (8):531–538

16. Hall WA, Rustamzadeh E, Asher AL (2003) Convection-enhanced delivery in clinical trials. Neurosurg Focus 14(2):e2

17. Sampson JH, Raghavan R, Brady M, Friedman AH, Bigner D (2011) Convection-enhanced delivery. J Neurosurg 115(3):463–464.; ; discussion 5-6. https://doi.org/10.3171/2010.11.JNS101801

18. Mehta AM, Sonabend AM, Bruce JN (2017) Convection-enhanced delivery. Neurotherapeutics 14(2):358–371. https://doi.org/10.1007/s13311-017-0520-4

19. Lei Y, Han H, Yuan F, Javeed A, Zhao Y (2017) The brain interstitial system: anatomy, modeling, in vivo measurement, and applications. Prog Neurobiol 157:230–246. https://doi.org/10.1016/j.pneurobio.2015.12.007

20. Lieberman DM, Laske DW, Morrison PF, Bankiewicz KS, Oldfield EH (1995) Convection-enhanced distribution of large molecules in gray matter during interstitial drug infusion. J Neurosurg 82 (6):1021–1029. https://doi.org/10.3171/jns.1995.82.6.1021

21. Corem-Salkmon E, Ram Z, Daniels D, Perlstein B, Last D, Salomon S, Tamar G, Shneor R, Guez D, Margel S, Mardor Y (2011) Convection-enhanced delivery of methotrexate-loaded maghemite nanoparticles. Int J Nanomedicine 6:1595–1602. https://doi.org/10.2147/IJN.S23025

22. Lonser RR, Sarntinoranont M, Morrison PF, Oldfield EH (2015) Convection-enhanced delivery to the central nervous system. J Neurosurg 122(3):697–706. https://doi.org/10.3171/2014.10.JNS14229

23. Casanova FCP, Sarntinoranont M (2014) Effect of needle insertion speed on tissue injury, stress, and backflow distribution for convection-enhanced delivery in the rat brain. PLoS One 9(4):e94919

24. Tykocki T, Miekisiak G (2016) Application of convection-enhanced drug delivery in the treatment of malignant gliomas. World Neurosurg 90:172–178. https://doi.org/10.1016/j.wneu.2016.02.040

25. Platt S, Nduom E, Kent M, Freeman C, Machaidze R, Kaluzova M, Wang L, Mao H, Hadjipanayis CG (2012) Canine model of convection-enhanced delivery of cetuximab-conjugated iron-oxide nanoparticles monitored with magnetic resonance imaging. Clin Neurosurg 59:107–113. https://doi.org/10.1227/NEU.0b013e31826989ef

26. Pollina J, Plunkett RJ, Ciesielski MJ, Lis A, Barone TA, Greenberg SJ, Fenstermaker RA (1998) Intratumoral infusion of topotecan prolongs survival in the nude rat intracranial U87 human glioma model. J Neuro-Oncol 39 (3):217–225. https://doi.org/10.1023/A:1005954121521

27. Saito R, Bringas JR, Panner A, Tamas M, Pieper RO, Berger MS, Bankiewicz KS (2004) Convection-enhanced delivery of tumor necrosis factor-related apoptosis-inducing ligand with systemic administration of temozolomide prolongs survival in an intracranial glioblastoma xenograft model. Cancer Res 64 (19):6858–6862. https://doi.org/10.1158/0008-5472.CAN-04-1683

28. Raghavan R, Brady ML, Rodriguez-Ponce MI, Hartlep A, Pedain C, Sampson JH (2006) Convection-enhanced delivery of therapeutics for brain disease, and its optimization. Neurosurg Focus 20(4):E12. https://doi.org/10.3171/foc.2006.20.4.7

29. Laske DW, Youle RJ, Oldfield EH (1997) Tumor regression with regional distribution of the targeted toxin TF-CRM107 in patients

with malignant brain tumors. Nat Med 3 (12):1362–1368

30. Vogelbaum MA, Brewer C, Barnett GH, Mohammadi AM, Peereboom DM, Ahluwalia MS, Gao S First-in-human evaluation of the Cleveland Multiport Catheter for convection-enhanced delivery of topotecan in recurrent high-grade glioma: results of pilot trial 1. J Neurosurg:1–10. https://doi.org/10.3171/2017.10.jns171845

31. Gonzalez-Perez O, Romero-Rodriguez R, Soriano-Navarro M, Garcia-Verdugo JM, Alvarez-Buylla A (2009) Epidermal growth factor induces the progeny of subventricular zone type B cells to migrate and differentiate into oligodendrocytes. Stem Cells 27 (8):2032–2043. https://doi.org/10.1002/stem.119

32. Moreno-Estelles M, Diaz-Moreno M, Gonzalez-Gomez P, Andreu Z, Mira H (2012;Chapter 2:Unit 2D 10) Single and dual birthdating procedures for assessing the response of adult neural stem cells to the infusion of a soluble factor using halogenated thymidine analogs. Curr Protoc Stem Cell Biol. https://doi.org/10.1002/9780470151808.sc02d10s21

33. Sampson JH, Akabani G, Archer GE, Berger MS, Coleman RE, Friedman AH, Friedman HS, Greer K, Herndon JE 2nd, Kunwar S, McLendon RE, Paolino A, Petry NA, Provenzale JM, Reardon DA, Wong TZ, Zalutsky MR, Pastan I, Bigner DD (2008) Intracerebral infusion of an EGFR-targeted toxin in recurrent malignant brain tumors. Neuro-Oncology 10(3):320–329. https://doi.org/10.1215/15228517-2008-012

34. Kunwar S, Chang S, Westphal M, Vogelbaum M, Sampson J, Barnett G, Shaffrey M, Ram Z, Piepmeier J, Prados M, Croteau D, Pedain C, Leland P, Husain SR, Joshi BH, Puri RK, Group PS (2010) Phase III randomized trial of CED of IL13-PE38QQR vs Gliadel wafers for recurrent glioblastoma. Neuro-Oncology 12(8):871–881. https://doi.org/10.1093/neuonc/nop054

35. Ostertag D, Amundson KK, Lopez Espinoza F, Martin B, Buckley T (2012) Galvao da Silva AP, Lin AH, Valenta DT, Perez OD, Ibanez CE, Chen CI, Pettersson PL, Burnett R, Daublebsky V, Hlavaty J, Gunzburg W, Kasahara N, Gruber HE, jolly DJ, Robbins JM. Brain tumor eradication and prolonged survival from intratumoral conversion of 5-fluorocytosine to 5-fluorouracil using a nonlytic retroviral replicating vector. Neuro-Oncology 14(2):145–159. https://doi.org/10.1093/neuonc/nor199

36. Bruce JN, Fine RL, Canoll P, Yun J, Kennedy BC, Rosenfeld SS, Sands SA, Surapaneni K, Lai R, Yanes CL, Bagiella E, DeLaPaz RL (2011) Regression of recurrent malignant gliomas with convection-enhanced delivery of topotecan. Neurosurgery 69(6):1272–1279.; ; discussion 9-80. https://doi.org/10.1227/NEU.0b013e3182233e24

37. Bogdahn U, Hau P, Stockhammer G, Venkataramana NK, Mahapatra AK, Suri A, Balasubramaniam A, Nair S, Oliushine V, Parfenov V, Poverennova I, Zaaroor M, Jachimczak P, Ludwig S, Schmaus S, Heinrichs H, Schlingensiepen KH (2011) Trabedersen glioma study G. targeted therapy for high-grade glioma with the TGF-beta2 inhibitor trabedersen: results of a randomized and controlled phase IIb study. Neuro-Oncology 13(1):132–142. https://doi.org/10.1093/neuonc/noq142

38. Paxinos G, Franklin KJ (2001) The Mouse Brain in stereotaxic coordinates, 2nd edn. Academic press

39. Choi JJ, Selert K, Vlachos F, Wong A, Konofagou EE (2011) Noninvasive and localized neuronal delivery using short ultrasonic pulses and microbubbles. Proc Natl Acad Sci U S A 108 (40):16539–16544. https://doi.org/10.1073/pnas.1105116108

40. Marquet F, Tung YS, Teichert T, Ferrera VP, Konofagou EE (2011) Noninvasive, transient and selective blood-brain barrier opening in non-human primates in vivo. PLoS One 6(7): e22598. https://doi.org/10.1371/journal.pone.0022598

41. Wang F, Cheng Y, Mei J, Song Y, Yang YQ, Liu Y, Wang Z (2009) Focused ultrasound microbubble destruction-mediated changes in blood-brain barrier permeability assessed by contrast-enhanced magnetic resonance imaging. J Ultrasound Med 28(11):1501–1509

42. Hynynen K, Jolesz FA (1998) Demonstration of potential noninvasive ultrasound brain therapy through an intact skull. Ultrasound Med Biol 24(2):275–283

43. Su H, Koo JM, Cui H (2015) One-component nanomedicine. J Control Release 219:383–395. https://doi.org/10.1016/j.jconrel.2015.09.056

44. Ma W, Cheetham AG, Cui H (2016) Building nanostructures with drugs. Nano Today 11 (1):13–30. https://doi.org/10.1016/j.nantod.2015.11.003

45. Rautio J, Kumpulainen H, Heimbach T, Oliyai R, Oh D, Jarvinen T, Savolainen J (2008) Prodrugs: design and clinical applications. Nat Rev Drug Discov 7(3):255–270. https://doi.org/10.1038/nrd2468

46. Venkatesh S, Lipper RA (2000) Role of the development scientist in compound lead selection and optimization. J Pharm Sci 89 (2):145–154. https://doi.org/10.1002/(SICI)1520-6017(200002)89:2<145::AID-JPS2>3.0.CO;2-6

47. Shen Y, Jin E, Zhang B, Murphy CJ, Sui M, Zhao J, Wang J, Tang J, Fan M, Van Kirk E, Murdoch WJ (2010) Prodrugs forming high drug loading multifunctional nanocapsules for intracellular cancer drug delivery. J Am Chem Soc 132(12):4259–4265. https://doi.org/10.1021/ja909475m

48. Lock LL, Li Y, Mao X, Chen H, Staedtke V, Bai R, Ma W, Lin R, Li Y, Liu G, Cui H (2017) One-component supramolecular filament hydrogels as Theranostic label-free magnetic resonance imaging agents. ACS Nano 11 (1):797–805. https://doi.org/10.1021/acsnano.6b07196

49. Ryppa C, Mann-Steinberg H, Fichtner I, Weber H, Satchi-Fainaro R, Biniossek ML, Kratz F (2008) In vitro and in vivo evaluation of doxorubicin conjugates with the divalent peptide E-[c(RGDfK)2] that targets integrin alphavbeta3. Bioconjug Chem 19 (7):1414–1422. https://doi.org/10.1021/bc800117r

50. Dal Pozzo A, Ni MH, Esposito E, Dallavalle S, Musso L, Bargiotti A, Pisano C, Vesci L, Bucci F, Castorina M, Fodera R, Giannini G, Aulicino C, Penco S (2010) Novel tumor-targeted RGD peptide-camptothecin conjugates: synthesis and biological evaluation. Bioorg Med Chem 18(1):64–72. https://doi.org/10.1016/j.bmc.2009.11.019

51. Cheetham AG, Zhang PC, Lin YA, Lock LL, Cui HG (2013) Supramolecular nanostructures formed by anticancer drug assembly. J Am Chem Soc 135(8):2907–2910. https://doi.org/10.1021/Ja3115983

52. Bastiancich C, Danhier P, Preat V, Danhier F (2016) Anticancer drug-loaded hydrogels as drug delivery systems for the local treatment of glioblastoma. J Control Release 243:29–42. https://doi.org/10.1016/j.jconrel.2016.09.034

53. Tyler B, Fowers KD, Li KW, Recinos VR, Caplan JM, Hdeib A, Grossman R, Basaldella L, Bekelis K, Pradilla G, Legnani F, Brem H (2010) A thermal gel depot for local delivery of paclitaxel to treat experimental brain tumors in rats. J Neurosurg 113(2):210–217. https://doi.org/10.3171/2009.11.JNS08162

54. Li Y, Wang F, Cui H (2016) Peptide-based supramolecular hydrogels for delivery of biologics. Bioeng Transl Med 1(3):306–322. https://doi.org/10.1002/btm2.10041

55. Schafer C, Fels C, Brucke M, Holzhausen HJ, Bahn H, Wellman M, Visvikis A, Fischer P, Rainov NG (2001) Gamma-glutamyl transferase expression in higher-grade astrocytic glioma. Acta Oncol 40(4):529–535

56. Lin R, Cheetham AG, Zhang PC, Lin YA, Cui HG (2013) Supramolecular filaments containing a fixed 41% paclitaxel loading. Chem Commun 49(43):4968–4970. https://doi.org/10.1039/C3cc41896k

57. Zhang PC, Cheetham AG, Lock LL, Cui HG (2013) Cellular uptake and cytotoxicity of drug-peptide conjugates regulated by conjugation site. Bioconjug Chem 24(4):604–613. https://doi.org/10.1021/bc300585h

58. Zhang PC, Lock LL, Cheetham AG, Cui HG (2014) Enhanced Cellular Entry and Efficacy of Tat Conjugates by Rational Design of the Auxiliary Segment. Mol Pharm 11 (3):964–973. https://doi.org/10.1021/mp400619v

59. Cheetham AG, Zhang P, Lin YA, Lin R, Cui H (2014) Synthesis and self-assembly of a mikto-arm star dual drug amphiphile containing both paclitaxel and camptothecin. J Mater Chem B 2 (42):7316–7326. https://doi.org/10.1039/c4tb01084a

60. Lin YA, Cheetham AG, Zhang P, Ou YC, Li Y, Liu G, Hermida-Merino D, Hamley IW, Cui H (2014) Multiwalled nanotubes formed by cationic mixtures of drug amphiphiles. ACS Nano 8(12):12690–12700. https://doi.org/10.1021/nn505688b

61. Skehan P, Storeng R, Scudiero D, Monks A, McMahon J, Vistica D, Warren JT, Bokesch H, Kenney S, Boyd MR (1990) New colorimetric cytotoxicity assay for anticancer-drug screening. J Natl Cancer Inst 82(13):1107–1112

62. Sheets KT, Bago JR, Paulk IL, Hingtgen SD (2018) Image-guided resection of glioblastoma and intracranial implantation of therapeutic stem cell-seeded scaffolds. J Vis Exp 137. https://doi.org/10.3791/57452

63. Hingtgen S, Figueiredo JL, Farrar C, Duebgen M, Martinez-Quintanilla J, Bhere D, Shah K (2013) Real-time multimodality imaging of glioblastoma tumor resection and recurrence. J Neuro-Oncol 111 (2):153–161. https://doi.org/10.1007/s11060-012-1008-z

64. Okolie O, Bago JR, Schmid RS, Irvin DM, Bash RE, Miller CR, Hingtgen SD (2016) Reactive astrocytes potentiate tumor aggressiveness in a murine glioma resection and recurrence model. Neuro-Oncology 18

(12):1622–1633. https://doi.org/10.1093/neuonc/now117

65. Kut C, Chaichana KL, Xi J, Raza SM, Ye X, McVeigh ER, Rodriguez FJ, Quinones-Hinojosa A, Li X (2015) Detection of human brain cancer infiltration ex vivo and in vivo using quantitative optical coherence tomography. Sci Transl Med 7(292):292ra100. https://doi.org/10.1126/scitranslmed.3010611

66. Kim JE, Patel MA, Mangraviti A, Kim ES, Theodros D, Velarde E, Liu A, Sankey EW, Tam A, Xu H, Mathios D, Jackson CM, Harris-Bookman S, Garzon-Muvdi T, Sheu M, Martin AM, Tyler BM, Tran PT, Ye X, Olivi A, Taube JM, Burger PC, Drake CG, Brem H, Pardoll DM, Lim M (2017) Combination therapy with anti-PD-1, anti-TIM-3, and focal radiation results in regression of murine gliomas. Clin Cancer Res 23 (1):124–136. https://doi.org/10.1158/1078-0432.CCR-15-1535

67. Bastiancich C, Bianco J, Vanvarenberg K, Ucakar B, Joudiou N, Gallez B, Bastiat G, Lagarce F, Preat V, Danhier F (2017) Injectable nanomedicine hydrogel for local chemotherapy of glioblastoma after surgical resection. J Control Release 264:45–54. https://doi.org/10.1016/j.jconrel.2017.08.019

68. Arosio D, Casagrande C (2016) Advancement in integrin facilitated drug delivery. Adv Drug Deliv Rev 97:111–143. https://doi.org/10.1016/j.addr.2015.12.001

69. Wang K, Zhang X, Liu Y, Liu C, Jiang B, Jiang Y (2014) Tumor penetrability and anti-angiogenesis using iRGD-mediated delivery of doxorubicin-polymer conjugates. Biomaterials 35(30):8735–8747. https://doi.org/10.1016/j.biomaterials.2014.06.042

70. Drappatz J, Brenner A, Wong ET, Eichler A, Schiff D, Groves MD, Mikkelsen T, Rosenfeld S, Sarantopoulos J, Meyers CA, Fielding RM, Elian K, Wang X, Lawrence B, Shing M, Kelsey S, Castaigne JP, Wen PY (2013) Phase I study of GRN1005 in recurrent malignant glioma. Clin Cancer Res 19 (6):1567–1576. https://doi.org/10.1158/1078-0432.CCR-12-2481

71. Kang T, Zhu Q, Jiang D, Feng X, Feng J, Jiang T, Yao J, Jing Y, Song Q, Jiang X, Gao X, Chen J (2016) Synergistic targeting tenascin C and neuropilin-1 for specific penetration of nanoparticles for anti-glioblastoma treatment. Biomaterials 101:60–75. https://doi.org/10.1016/j.biomaterials.2016.05.037

72. Sun Z, Yan X, Liu Y, Huang L, Kong C, Qu X, Wang M, Gao R, Qin H (2017) Application of dual targeting drug delivery system for the improvement of anti-glioma efficacy of doxorubicin. Oncotarget 8(35):58823–58834. https://doi.org/10.18632/oncotarget.19221

73. Jensen SA, Day ES, Ko CH, Hurley LA, Luciano JP, Kouri FM, Merkel TJ, Luthi AJ, Patel PC, Cutler JI, Daniel WL, Scott AW, Rotz MW, Meade TJ, Giljohann DA, Mirkin CA, Stegh AH (2013) Spherical nucleic acid nanoparticle conjugates as an RNAi-based therapy for glioblastoma. Sci Transl Med 5 (209):209ra152. https://doi.org/10.1126/scitranslmed.3006839

74. Zhang Y, Zhang YF, Bryant J, Charles A, Boado RJ, Pardridge WM (2004) Intravenous RNA interference gene therapy targeting the human epidermal growth factor receptor prolongs survival in intracranial brain cancer. Clin Cancer Res 10(11):3667–3677. https://doi.org/10.1158/1078-0432.CCR-03-0740

75. Karlsson J, Vaughan HJ, Green JJ (2018) Biodegradable polymeric nanoparticles for therapeutic cancer treatments. Annu Rev Chem Biomol Eng. https://doi.org/10.1146/annurev-chembioeng-060817-084055

76. Clark AJ, Davis ME (2015) Increased brain uptake of targeted nanoparticles by adding an acid-cleavable linkage between transferrin and the nanoparticle core. Proc Natl Acad Sci U S A 112(40):12486–12491. https://doi.org/10.1073/pnas.1517048112

77. Lin T, Zhao P, Jiang Y, Tang Y, Jin H, Pan Z, He H, Yang VC, Huang Y (2016) Blood-brain-barrier-penetrating albumin nanoparticles for biomimetic drug delivery via albumin-binding protein pathways for Antiglioma therapy. ACS Nano 10(11):9999–10012. https://doi.org/10.1021/acsnano.6b04268

78. Cai Q, Wang L, Deng G, Liu J, Chen Q, Chen Z (2016) Systemic delivery to central nervous system by engineered PLGA nanoparticles. Am J Transl Res 8(2):749–764

79. Saraiva C, Praca C, Ferreira R, Santos T, Ferreira L, Bernardino L (2016) Nanoparticle-mediated brain drug delivery: overcoming blood-brain barrier to treat neurodegenerative diseases. J Control Release 235:34–47. https://doi.org/10.1016/j.jconrel.2016.05.044

Chapter 4

Drug Delivery Approaches and Imaging Techniques for Brain Tumor

Mark Bell, Christine Pujol Rooks, and Vibhuti Agrahari

Abstract

Brain tumor is an abnormal growth of tissue in the CNS that can interrupt the brain function and proved challenging to treat, largely owing to the biological characteristics which often conspire to limit progress. These tumors are located in one of the body's most crucial organs and often beyond the reach due to BBB. The transport of substances across the BBB is strictly limited through both physical specialized connections (tight junctions) and metabolic barriers (enzymes and transport systems). Therefore, therapeutics have to pass through BBB before reaching the targeted sites in the brain tumor. This book chapter covered the types of brain tumors based on the diagnosis, strategies to improve the accumulation of anticancer drugs in the brain and brain tumor imaging. Numerous drug delivery approaches such as nanotechnology, focused ultrasound, hyperthermia, enhanced permeability and retention (EPR) effect, cell-penetrating peptides (CPP), ligand-mediated delivery, etc. have been discussed briefly to overcome the BBB and its advantages and limitations including other delivery system such as vaccines, stem cell therapy, etc. However, the main focus of this book chapter is on nanoparticle-based drug delivery system to overcome major obstacle in current brain cancer treatments. The different groups of nanoparticles that have been modified for brain tumor targeted drug delivery and brain targeted imaging have been discussed. Advances in these techniques suggest optimism for the future management of glioblastoma. Indeed, no single strategy is powerful enough to offer a substantial breakthrough for glioma treatment, so the future application of combined efforts and therapeutic agents might lead to a successful resolution.

Key words CNS, BBB, Glioblastoma, Nanoparticles, Targeted delivery

1 Introduction

Brain tumors are among the most feared of all forms of cancer. More than two-thirds of adults diagnosed with glioblastoma—the most aggressive type of brain cancer—will die within 2 years of diagnosis [1]. According to the National Brain Tumor Society (http://braintumor.org/brain-tumor-information/brain-tumor-facts/), in 2019, approximately 700,000 Americans are living with a brain tumor. It is estimated that approximately 86,970 will be diagnosed, and 16,830 people will die from malignant brain tumors in 2019 (http://braintumor.org/brain-tumor-information/brain-

Vivek Agrahari et al. (eds.), *Nanotherapy for Brain Tumor Drug Delivery*, Neuromethods, vol. 163,
https://doi.org/10.1007/978-1-0716-1052-7_4, © Springer Science+Business Media, LLC, part of Springer Nature 2021

tumor-facts/). Brain tumors have proved challenging to treat, largely owing to the biological characteristics of these cancers, which often conspire to limit progress. These tumors are located in one of the body's most crucial organs and often beyond the reach. These tumors are positioned behind the blood–brain barrier (BBB)—a system of tight junctions and transport proteins that protect delicate neural tissues from exposure to factors in the general circulation, thus also impeding exposure to systemic chemotherapy [1]. The current treatment consists of surgical resection, followed by immunotherapy, radiotherapy, and chemotherapy [2]. However, the benefits of current treatment are quite insignificant to patients. Based on the aforementioned facts, approaches in central nervous system (CNS) drug delivery to the brain tumor using nanocarrier systems and imaging techniques are briefly discussed in this chapter.

1.1 Types of Brain Tumor

Brain tumor is an abnormal growth of tissue in the CNS that can interrupt the brain function. The typical human brain has more than 100 billion cells consisting of various neuronal type of cells as well as support cells known as glial cells [3]. With this large abundance of cell differentiation, the number of specific brain tumor is immense, often including several subtypes for each cell. There are over 120 types of brain cancer which are classified based on the original cells [3, 4]. Few examples of diagnosed brain cancers based on the cell type are listed in Table 1 [4]. Brain tumors, those originating from glial cells, are called gliomas. The three subtypes of cells from which gliomas arise are astrocytes, oligodendrocytes, and ependymal cells generating astrocytoma, oligodendroglioma, and ependymoma, respectively [5, 6].

The World Health Organization (WHO) classifies gliomas within four grades (grade I to grade IV) [4]. Low-grade tumors (grade I/II) are generally treated with careful monitoring and/or surgery. Higher-grade tumors (grade III/IV) (malignant gliomas) are difficult to treat and additional approaches such as radiotherapy, chemotherapy, or targeted therapy are required. Malignant gliomas such as glioblastoma multiforme (GBM) are often difficult to treat because of their invasiveness and resistance to surgical procedures as well as chemo−/radiotherapy, poor prognosis, dismal survival rates, increased rate of angiogenesis, resistance to apoptosis, and limited delivery of chemotherapeutics across the CNS [7]. Under the standard treatment regimen (resection followed by radiotherapy and/or chemotherapy), the average survival rate for malignant brain tumor patients is ~34.4%, whereas, for GBM, the 5-year relative survival rate is ~5.1% (http://braintumor.org/brain-tumor-information/brain-tumor-facts/).

Table 1
Types of brain tumor

Type	Description	Current treatment	References
Astrocytic tumors	Astrocytes are a type of glial cell that plays a support role within the brain neurology	Current treatment includes surgical excision. Based on cancer progression, radiation and chemotherapy are viable options	[8]
Oligodendroglial tumors	This cell type is very similar to astrocytes with fewer starlike projections. These cells work in the production of myelin sheaths within nerve cells. Most often occurs within the frontal or temporal lobes	Based on the grade of tumor, there are several treatment options. For lower-grade tumors surgical removal may be adequate. As the grade increases radiation and chemotherapy is often added after the surgical procedure	[9]
Glioblastoma	Most common primary brain tumor begins in glial tissue and mainly includes astrocytoma, oligodendroglia, and ependymal, based on the glial cells associated	Chemotherapy, radiation therapy, surgical removal, targeted drug delivery	[10]
Ependymal tumors	Ependymal cell line cerebrospinal fluid areas within the central nervous system. Cancer can occur in these cells and occur throughout different layers of the ependymal	Typically surgery performed followed by radiation therapy	[11]
Pineal region tumors	Pineal tumors can produce excess of melatonin	Surgery is the standard treatment for these types of cancer. Sometimes radiation therapy can be used as a solo treatment	[12]
Meningioma	The meninges are the tissue that surrounds the central nervous tissue. Depending on the location of the cancer, this can cause pressure on this portion of the central nervous system	Surgery is the standard treatment but hard to detect. Chemotherapy for these cancers is being researched	[13]
Metastatic tumors	Stems from a previous existing cancer. The most common cancers that cause metastatic brain tumors are lung, breast, colon, kidney, and skin cancers	Surgery, radiosurgery, and chemotherapy to targeted areas	[14]

1.2 The Blood–Brain Barrier

The brain is divided into the right and left hemispheres and is made of four distinct parts that include the cerebrum, cerebellum, brainstem, and diencephalon. Each part of the brain contains myelinated and unmyelinated neurons which compose white matter and gray matter, respectively. These neurons line the millions of pathways that enable the brain to integrate incoming information and regulate autonomic and motor functions. Among the neurons are various glial cells to support and insulate neurons. Deep within the brain are inner ventricles filled with cerebrospinal fluid for nourishment, hormone transport, tissue cushioning, and waste removal. Surrounding the brain are three layers of meninges that serve to contain cerebrospinal fluid and to protect the brain and its numerous blood vessels. Despite the rich blood supply, entry to the brain is limited to small nutrients, ions, and lipophilic substances [15]. The majority of small molecule drugs and macromolecule agents, such as proteins, peptides, and antibodies, do not readily permeate into the brain parenchyma, which is one of the most significant challenges of an effective CNS drug delivery [16]. Transport of substances across the BBB is strictly limited through both physical specialized connections (tight junctions) and metabolic barriers (enzymes and transport systems). Tight junctions between epithelial cells compose the blood–brain barrier and prevent entry of harmful toxins into brain tissue. This can be explained by the fact that before reaching the targeted sites in the CNS tumor, therapeutics have to pass through BBB [17].

The function of the BBB is to separate the brain extracellular fluid from the circulating blood, transport beneficial endogenous molecules and essential nutrients into the brain, and filter harmful compounds from the brain back to the bloodstream. The components of the BBB are monolayer of capillary endothelial cells, basement membrane, vascular endothelium, pericytes, astrocytes, and the intracellular space between the membranes [18]. Unfortunately, over 95% of substances never reach the brain in therapeutically relevant concentrations. Therefore, the blood–brain barrier is one of the key reasons for its selective permeability and being challenging in the treatment of brain tumor. So how do molecules get through the BBB? The hallmark of the BBB is its utilization of transporters to shuttle key molecules to the brain [19]. There are several differentiated transporter types that are expressed in the BBB. However, the most important class of transporters concerning drug movement is the ATP-binding cassette (ABC) and solute carrier transporter (SLC). Utilizing these transporters or disrupting the endothelial tight junctions is one of the main goals in delivering chemotherapeutics in the treatment of brain tumors [20].

2   Strategies for Enhanced Brain Drug Delivery

The BBB is universally considered as both the target and the barrier for systemic drug delivery to the brain [19]. The three strategies that have been widely investigated to improve the accumulation of anticancer drugs in the brain includes bypassing the BBB, disrupting the BBB, and harnessing the endogenous transport system of the BBB [21]. Numerous approaches such as nanotechnology, focused ultrasound, hyperthermia, enhanced permeability and retention (EPR) effect, cell-penetrating peptides (CPP), ligand-mediated delivery, etc. have been developed to overcome the BBB as described below. The advantages and disadvantages of these methods are discussed in Table 2. However, the main focus of this book chapter is on nanoparticle-based drug delivery system.

Table 2
Strategies to enhance the permeability of the BBB for the treatment of glioma

Strategies	Advantages	Potential drawbacks
Nanotechnology	• Sustained/controlled release • Tumor specific • Surface hydrophilic polymers can be added to delay clearance • Variety of nanocarrier types to transport different drugs	• Rapid clearance • Particles aggregation • Potential toxicity • Potential immunogenicity (nanocarrier specific)
FUS	• Easy to execute • Compatible with drugs/nanoparticles	• Nonspecific targeting • Possible tissue damage • Strict control on FUS exposure required
Hyperthermia	• Easy to execute • Compatible with drugs/nanoparticles	• Particle size/concentration limitation • Possible tissue damage • Possible increase in intracranial pressure
EPR effects	• Tumor specific • Safe medication-induced methods • Effective for highly perfused tumors	• Limited efficacy under normal condition • Limited toxicity
CPPs	• Great BBB penetrating ability • No specific receptors needed to cross BBB	• Potential toxicity. • Nonspecific targeting initiate immune response
Ligand-mediated delivery	• Tumor specific • Low cytotoxicity • Highly effective BBB crossing	• Rapid degradation • Rely on the binding efficacy • Compete with endogenous ligands for binding • Possible receptor downregulation • Immunogenic response

2.1 Nanotechnology   Nanotechnology is a broad term that includes nanocarriers such as liposomes, polymersomes, nanomicelles, nanogels, dendrimers, exosomes, carbon nanotubes, and a variety of nanoparticles (NPs) such as polymeric NPs and magnetic NPs. These nanocarriers are able to transport small molecules and large macromolecules across the BBB, and some nanocarriers can control the release of drug once across. Targeting ligands can be attached to nanocarriers to deliver molecules to specific parts of the brain, which offers a unique advantage compared to other strategies to enhance BBB permeability [20, 22].

2.2 Focused Ultrasound   Focused ultrasound (FUS) is the usage of microtubules and heat to form a local cavitation in the BBB, essentially creating a temporary opening for molecules to cross. However, this method carries the risk of permanent brain tissue damage, especially with higher mechanical indexes of applied FUS. FUS exposure should be tightly controlled and used with caution if at all [23, 24].

2.3 Hyperthermia   Hyperthermia is another strategy in which elevated temperatures (41–43 °C) are used to temporarily and locally disrupt the BBB to allow molecules to cross. Modalities include electrohyperthermia, laser, and magnetic hyperthermia. FUS has also been used as a heat source to induce localized hyperthermia, though magnetic energy, microwaves, laser, and radiofrequency are also being studied. Due to BBB disruption and protein and DNA denaturation, hyperthermia carries the risk of brain tissue damage [25].

3   EPR Effects

The enhanced permeability and retention (EPR) effect is a process in which macromolecules permeate tissue through extravasation in the presence of permeability factors such as nitrous oxide (NO), bradykinin, and prostaglandins. Several medications have been identified that are effective in augmenting the EPR effect for drug delivery to brain tumors. Nitroglycerin and isosorbide dinitrate increase permeability via NO release. Angiotensin-converting enzyme (ACE) inhibitors combined with angiotensin II (AT-II) have been used in animal models to enhance drug delivery by inducing hypertension [26, 27].

4   Cell-Penetrating Peptides

Cell-penetrating peptides (CPPs) were originally isolated from a transcription activator in human immunodeficiency virus (HIV). CPPs have cationic residues that allow their uptake into cells so that CPPs can deliver biological molecules across the BBB. The exact mechanism of uptake of CPPs is not fully understood, but more CPP designs are being explored to increase the amount of drug delivered and prevent drug degradation before reaching the target site [25, 28].

5   Ligand-Mediated Delivery

Several receptors are abundant in the BBB including LDL receptors (LDLRs), Tf receptor 1 (TfR1), insulin receptors, glutathione transporters, and lactoferrin (Lf) receptors, among others. Ligand-mediated delivery involves attaching ligands to nanocarriers so that they can bind to specific receptors on the BBB such as LDLR and thus cross the BBB [29]. Various ligands have been studied and applied clinically including monoclonal antibodies [30], aptamers [31], and Si-RNA [32]. Ligands can be designed to have higher affinity for the receptors to further improve drug delivery [25].

5.1   Strategy I: Bypassing the BBB

The BBB can be bypassed by local delivery. For instance, drug-encapsulated wafers can be implanted in the tumor cavity during surgical resection of brain tumors. A Gliadel® (MGI Pharma, Inc.) wafer, a polifeprosan 20 polymer matrix containing carmustine (~1.45 cm in diameter and 1 mm thick), is the first implant approved by the FDA for the treatment of gliomas. Preclinical and clinical studies showed that Gliadel® wafer modestly prolonged the median survival in newly diagnosed and recurrent high-grade gliomas with minimal systemic side effects [21]. However, it was discovered that most recurrent tumors appeared adjacent to the implanted wafers, indicating limited drug penetration. Hence, convection-enhanced delivery (CED), defined as a continuous injection of therapeutic fluids under positive pressure, has been adopted to improve the penetration of drugs or drug encapsulated nanoparticles [33].

The nose-to-brain transport offers a noninvasive drug delivery to bypass the BBB. Drugs and nanoparticles can be directly transported to the brain paracellularly or transcellularly along the olfactory and trigeminal nerve without significant peripheral exposure. Thus, intranasal delivery has the potential to bypass peripheral clearance, decrease systemic toxicity, and lower the dose needed. For instance, intranasal delivery of 3H-5-fluorouracil (5FU) resulted in a significantly higher drug concentration in the CSF compared to intravenous injection. Intranasal administration of a 3′-fluorescein isothiocyanate (FITC)-labeled oligonucleotide N3′ → P5′ thio-phosphoramidate telomerase inhibitor (FITC-GRN163) successfully delivered the drug to the brain, inhibited tumor growth, and prolonged the median survival rates from 35 to 75 days compared to the control group. In conclusion, administrating nanoparticles through the nose-to-brain pathways provides a practical, noninvasive approach for delivering therapeutic agents to the brain and can potentially improve efficacy in the treatment of brain tumors due to their capabilities to reduce local toxicity caused by chemotherapy. Nevertheless, intranasal delivery is restricted by the capability to deliver concentrated drugs in a limited volume [21].

5.2 Strategy II: Disrupting the BBB

Another strategy for brain delivery involves disrupting the BBB using biochemical reagents such as cell-penetrating peptides, surfactants, and hyperosmotic agents or alternatively by physical methods such as magnetic nanoparticle-induced hyperthermia and focused ultrasound sonication (FUS). However, the disruption is tumor region specific and may lead to deleterious consequences. Alternatively, FUS in conjunction with microbubble contrast agents has been shown to disrupt the BBB locally. Under the guidance of magnetic resonance imaging, the acoustic waves can also be focused locally to desired regions in the brain and deep into soft tissues. The cavitation activity triggered by FUS leads to mechanical stress on the vasculature. Following cavitation, cytoplasmic channels were found to form along with a greater number of endocytosis/transcytosis vesicles and the opening of the BBB tight junctions, suggesting increased pinocytosis/endocytosis and improved BBB penetration [34]. Alternatively, FUS can also be used to trigger local drug release if the drugs are encapsulated or conjugated to the microbubbles [35]. However, potential side effects of BBB disruption such as hemorrhage, brain damage, and inflammation should be closely monitored, and a trade-off should be carefully considered in achieving maximum brain accumulation while minimizing possible side effects.

5.3 Strategy III: Penetrating the BBB

Endogenous transporters and receptors can be taken advantage of to improve neural penetration of drugs in a noninvasive manner with high efficiency. Nanoparticles decorated with transporter/receptor-specific ligands or antibodies have a higher chance to pass the BBB and target the tumor site more specifically than non-targeting nanoparticles. For example, Leamon et al. revealed that folate receptor (FR) functionalized liposomes delivered twofold more oligonucleotides to the FR-positive tumor cells than non-targeted formulations both in vitro and in vivo [36].

6   Nanoparticle-Based Drug Delivery and Imaging Techniques to Identify Brain Tumors

In recent years, there has been a great explosion in nanotechnology-based research for the diagnosis and treatment of brain tumors. By virtue of its small size and great target specificity,

Table 3
Nanotechnology to overcome major obstacle in current brain cancer treatments

Treatment strategy	Major obstacles	Solutions with nanotechnology
Surgery	• Difficult to identify tumor boundaries for maximal resection	• Intraoperative imaging guidance with nanoparticle-based imaging probes to differentiate diffused tumor margins from normal brain tissue with high specificity/sensitivity/resolutio
Radiation therapy	• Radioresistance	• Deliver nanoparticle-based high-Z element radiosensitizer • Deliver O_2-generating/containing nanoparticles to alleviate tumor hypoxia • Deliver gasotransmitters, such as NO and H_2S, with nanoplatforms
Chemotherapy	• BBB • Low tumor accumulation of drug • Tumor heterogeneity affecting sensitivity • Drug resistance	• Modify nanodrug systems with targeting ligands for BBB crossing and active tumor binding • Encapsulate therapeutic drugs into nanocarrier for prolonged circulation half-lives and on-demand drug release • Engineer the nanoparticles with cell-penetrating peptides for deep tumor penetration • Develop nanoparticle-based "all-in-one" drug cocktail for combinational therapy

Reproduced with permission [47]

nanotechnology offers tumor-specific drug delivery strategies to cross barriers and overcome the limitations of conventional drug delivery systems as described in Table 3. The so-called BBB-crossing nanotechnology is expected to make a revolutionary impact on conventional brain cancer management. The emerging role of functional neuroimaging techniques [37–39], such as magnetic resonance imaging (MRI), computed tomography (CT), positron emission tomography (PET), fluorescence (FL) imaging, photoacoustic (PA) imaging, Raman imaging, and dual modal imaging, in establishing diagnosis and for predicting prognosis in patients with brain tumor is remarkable. In particular, the design and synthesis of multifunctional BBB crossing nanoparticles (NPs) with excellent magnetic, optical, thermal, and TME-responsive performance promise great opportunities for improving the outcomes of current brain cancer treatment protocols as described in Tables 4 and 5.

Each imaging technique is uniquely designed to bypass the BBB defenses through a specific mechanism. Classically, the use of MRI has been the mainstay of brain tumor diagnostics and quantification of disease progression [40]. Iron-oxide nanoparticle (IONP) has been traditionally used as a contrast agent in MRI; however, it has largely been replaced by gadolinium [40]. The

Table 4
Examples of the different groups of nanoparticles that have been modified for targeted imaging and brain tumor targeted drug delivery

Class of nanoparticles	Targeting strategy	Model	Advantage	Application
Gadolinium nanoparticles	1. Interleukin (IL)-13 2. Anti-GD2 antibody 3. Angiopep-2	GBM	1. Local drug concentration increased. 2. Good persistence and distribution effect. 3. Reduced concentration required for imaging. 4. Large growth in the r_2/r_1 ratio.	MRI
Polymeric nanoparticles (1) dendrimer (2) PEG (3) PLGA	1. CTX. 2. Anti-PDGFRß. 3. CTB.	U87 mouse model/glioma bearing nude mice	1. Good persisting and distribution effect 2. Local drug concentration increased	Chemical Exchange Saturation Transfer (CEST)
Superparamagnetic iron oxide nanoparticles	1. Focused ultrasound. 2. Magnetic targeting. 3. cRGD peptide, 4. EGFRvIII antibody. 5. CTX. 6. Lactoferrin. 7. Tat.	Brain parenchymal/U87 xenograft model	1. Excellent T_1-weighted effects 2. Can be guided by focused ultrasound and magnetic targeting	MRI
Quantum dots (QDs)	1. RGD peptide 2. Aptamer 32	U251 glioma cells/glioma	1. Image more sensitively 2. Good light stability	FL
Au nanoparticles	1. TAT 2. Ala-Ala-Asn-Cys-Lys 3. RGD	1. Glioma cells 2. U87MG xenograft model	1. Enhance tumor killing effect 2. Enhanced retention 3. Can be tracked using a variety of imaging techniques	CT/MR/SERRS PDT
Liposome nanoparticles	1. VEGF-bevacizumab 2. CREKA peptide	GBM	1. Long imaging time 2. Improve imaging effect 3. High drug loading 4. Stability in body fluids	FL/Photothermal Therapy (PTT_/FL
Holotransferrin nanoparticles	Holotransferrin	Glioma models	1. Good biocompatibility 2. Has its own targeting effect	FL/PA/PTT

(continued)

Table 4
(continued)

Class of nanoparticles	Targeting strategy	Model	Advantage	Application
Mesoporous silica nanoparticles	Asn-Gly-Arg (NGR)	Glioma model	1. High drug loading 2. Stability in body fluids	Therapy
Carbon nanotube	Angiopep-2	Glioma model	1. Simple preparation method, cost effective, and environmentally friendly 2. Has both photothermal therapy and imaging function	PTT

Modified and Reproduced with permission [47]

Table 5
Representative nanoplatforms for brain tumor imaging

Types	Characteristics	Representative NPs
MR imaging	• High spatial resolution • Unlimited penetration • Limited sensitivity • Long acquisition time	SPION, Gd-DTPA dendrimer NP, $Gd@C_{82}O_{14}(OH)_{14}(NH_2)_6$
PET	• High sensitivity • Unlimited penetration • Lack of anatomical information • Hazardous ionizing radiation	^{64}Cu-labelled micelles and liposomes
FL imaging	• High sensitivity • Noninvasiveness • Limited penetration • Low spatial resolution	Semiconducting QDs, carbon dots, carbon nanotubules, rare-earth doped NPs, organic dye-based NPs
PA imaging	• High spatial resolution • High sensitivity • Noninvasiveness • Relatively deep penetration	Gold nanostructures, semiconducting polymer NPs, conjugated polymer NPs, carbon nanomaterials, transition-metal dichalcogenides/MXenen-based nanomaterials
Raman imaging	• High specificity • Ultrahigh sensitivity • Limited penetration • Limited spatial resolution	Gold nanostructures
Multimodal imaging	• Take the strengths of each modality • Compensate for respective inherent drawbacks	Gd-DOTA/Alexa-680 co-loaded NP, AuaMIL-88(A), Au-AZ/Au-AK pair, TB1-RGD AIE dots, SERRS-MOST-nanostar, Gd-DOTA/*trans*-1,2-bis(4-pyridyl)-ethylene co-loaded silica-coated Au NP

Reproduced with permission [29]

introduction of nanoplatforms to these two contrast agents has significantly increased their viability in clinical settings.

Polyethyleneimin-modified IONP (SPION), among others, is an example of an IONP platform that is utilized for brain tumor imaging [41]. Gadolinium's usage has increased dramatically due to the implementation of nanoparticle chelation techniques. CT/PET imaging does not currently have many nanoparticle systems that are being explored; however, there has been some promising results using Cu-labeled liposomes in PET scans [42]. By utilizing these nanoparticle platforms, the contrast agent is collected in higher volumes within the tumor region, increasing the length of visibility while imaging [27].

One of the most highly researcher areas of nanoparticle imaging techniques has been fluorescence imaging. While fluorescence imaging is highly sensitive and is a promising modality of imaging, the risk of adverse effects when used in vivo due to tissue absorption has been an issue. To overcome in vivo complications, the use of nanoparticle systems has been created. Quantum dots, rare-earth doped nanoparticles, carbon nanotubes, and organic-dye nanoparticles have all been developed to overcome the negative effects of raw fluorescence imaging [42, 43]. PA imaging utilizes a nonionizing technique that is similar to FL imaging in terms of its contrast ability. While there are currently no nanoparticle systems used in brain tumor imaging within humans, several experiments have shown positive results utilizing a variety of nanoparticle-based contrast agents in animal models [44].

Raman imaging commonly utilizes gold nanoparticles to assist in the development of this technique within clinical settings [45]. Surface-enhanced resonance raman spectroscopy (SERRS) has been used in glioma cell lines in order to properly image diffuse margins of tumors. The development of dual-modality nanoparticle systems is a relevant research topic. The development of a nanoparticle that utilizes multiple imaging modalities could potentially provide diagnosticians with a clearer diagnostic image to improve early detection and treatment of brain tumors.

One example of a dual-modality technology that is being researched is the use of gold nanoprobes as a single imaging agent for combination MR and Raman imaging [46].

7 Other Drug Delivery System

7.1 Stem Cell Therapy

Stem cells (SCs) have been investigated for treating brain tumors due to their ability to carry drugs across the blood–brain barrier and evade immuno-surveillance [48]. The main target of stem cell therapy is glioblastoma (GBM), an aggressive and persistent type of brain tumor with unregulated growth of cancer stem cells (CSCs) and no current cure [49]. Mesenchymal stem cells

(MSCs) and neural stem cells (NSCs) seems to be the safest and efficacious SCs as they can migrate directly to brain tumors and pose no ethical concerns [50]. These stem cells have been modified to carry and express genes for proteins that reduce tumor burden either by stopping proliferation, signaling apoptosis, or preventing angiogenesis. Tumor necrosis factor-related apoptosis inducing ligand (TRAIL) is the most widely used signal in stem cell therapy studies and NSCs implanted with TRAIL show promise for brain tumor treatment [48]. Other studies have explored stem cells implanted with interferon (IFN-b), bone morphogenetic proteins (BMP), parts of metalloproteinase-2 (PEX), antiangiogenic thrombospondin-1 (aaTSP-1), interleukin (IL)-12, IL-18, nanoparticles, and specific miRNA sequences [51].

Though stem cells effectively migrate to the brain, the route of administration of stem cells into the body is a critical aspect of therapy. Current routes of administration include intranasal delivery and surgical implantation [51]. Intranasal delivery is preferred for multifocal tumors as it is noninvasive, it can be used for repeated doses, and it can avoid vascular complications of implantation [48]. SCs delivered through the nasal passages migrate via olfactory nerves directly to the brain tumor in response to chemokines [52]. Thus, intranasal delivery of SCs limits systemic effects and complications. There has been much success of stem cell therapy of brain tumors in preclinical studies using murine models. One study transplanted cytosine deaminase (CD)-expressing NSCs and 5-fluorouracil prodrug into mice and demonstrated safety and efficacy in support of progressing into clinical trials for treating glioma [53]. Another study explored carboxylesterase (CE)-expressing NSCs with irinotecan prodrug and resulted in less medulloblastoma tumor growth compared to irinotecan alone [50].

In contrast, clinical trials are yielding less consistent efficacy. While research on stem cells for brain tumors in mice models is already directed toward combination therapies to prolong survival [51], studies on humans are still determining safety of stem cell therapy. In a phase I clinical trial, Portnow et al. (2017) found that CD-expressing NSCs did exhibit tropism to gliomas and were not tumorigenic [54]. Such studies are crucial since stem cell therapy can be useful for treating brain tumors, but not if the SCs form new tumors instead. More studies will be required to establish satisfactory levels of safety and efficacy for human subjects before progressing to subsequent phases of clinical trials.

7.2 Vaccines for Glioblastoma

Numerous trials of vaccines employing various strategies against glioblastoma are being conducted from phase I to phase III. Though some have shown promising results, none has come close to curing it. For instance, one of the ongoing phase II clinical trials, NCT02808364, is evaluating the safety and efficacy of

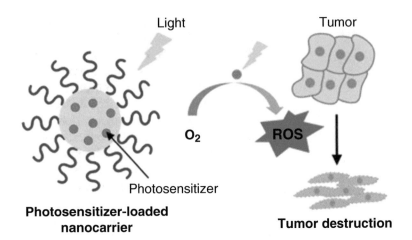

Fig. 1 Photodynamic therapy utilized for tumor destruction. (Reproduced with permission from [57])

"Personalized Cellular Vaccine for Recurrent Glioblastoma" (Per-CellVac2) in recurrent glioblastoma. This trial specifically uses immunization with autologous tumor cells, antigen pulsed DCsor allogeneic peripheral blood mononuclear cells. Results of this trial will be useful for determining the future course of action for cell-based immunotherapy in glioblastoma [55].

A randomized phase III trial NCT00045968 is evaluating long-term effects of addition of an autologous DCVax®-L vaccine to standard therapy for newly diagnosed glioblastoma, i.e., after surgery and chemoradiotherapy, patients receive temozolomide plus DCVax-L or temozolomide alone. Because of the crossover trial design, nearly 90% of the patients have received DCVax-L so far, and the vaccine has improved the survival rates of some patients, and those with median overall survival of 40.5 months are being analyzed further. The trial is ongoing to enable continued study of glioblastoma patients who are living beyond the expected length of survival [55].

7.3 Photodynamic Therapy

Photodynamic therapy (PDT) is based on release of singlet oxygen with toxic effects on the tumor that occurs when a photosensitizer at the tumor site is exposed to laser light of a certain wavelength as represented in Fig. 1 [56, 57]. PDT for cancer using talaporfin sodium with laser is approved in Japan. Progression-free survival of 1 year and an overall survival of 2 years and 8 months were shown in an open clinical trial of intraoperative PDT in glioblastoma [58]. A pilot clinical trial, the INDYGO (INtraoperative photoDYnamic Therapy for Glioblastomas), is planned as an addition to the standard of care of glioblastoma as a requirement prior to a randomized clinical trial [59]. In this clinical trial, PDT treatment following fluorescence-guided surgical resection will involve delivery of

5-aminolevulinic acid (ALA) processed by the cells to generate a photosensitizer protoporphyrin IX (PIX).

8 Conclusion and Future Perspective

Brain tumors such as glioblastoma multiforme are the most common and aggressive primary brain tumor. Even with aggressive treatment including surgical resection, radiation, and chemotherapy, patient outcomes remain poor. The complexity of the BBB complicates drug reaching to the tumor location, but also provides many unique opportunities for drug delivery to the brain. During the past few years, several approaches have been developed in attempts to overcome the BBB and improve the efficiency of pharmacological treatments. Among these approaches, the use of nanotechnology-based systems has revolutionized the field of drug delivery, offering the possibility to deliver therapeutic agents to local areas in the brain. Nonetheless, advanced methods for drug delivery of chemotherapy such as stem cell therapy, vaccines, photodynamic therapy, and nanoparticle-based therapy with diagnostic and therapeutic value for brain tumors were discussed in this chapter. The recent progress in other techniques such as focused ultrasound, hyperthermia, enhanced permeability factor, cell-penetrating peptides, and ligand-mediated delivery is powerful platform technologies for the creation of new methods to treat GBM which are briefly summarized in this chapter. The strategies described in this chapter share the ability to enhance the permeability and to deliver therapeutics across the BBB to reach the tumor. These strategies that have been widely investigated to improve the accumulation of anticancer drugs in the brain include bypassing the BBB, disrupting the BBB, and harnessing the endogenous transport system of the BBB.

Current therapy does not provide a long-term solution for brain tumor patients and failed in improving the quality of the patient's life. The nanotherapeutic approaches provide an attractive option for the treatment of brain tumors and has the potential to address the limitations of current therapeutic approaches. Suggestions for improvements in glioblastoma therapy include controlled/targeted delivery of anticancer therapy to glioblastoma through the BBB using polymer, direct introduction of genetically modified bacteria that selectively destroy cancer cells but spare the normal brain into the remaining tumor after resection, personalized/precision medicine approaches to therapy in clinical trials, and translation into practice of neurosurgery and neuro-oncology. Advances in these techniques suggest optimism for the future management of glioblastoma. Indeed, no single strategy is powerful enough to offer a substantial breakthrough for glioma treatment, so the future application of combined efforts and therapeutic agents might lead to a successful resolution. Overall, to achieve the goal of efficient drug delivery to brain tumor, a better understanding of the physiochemical properties of therapeutic molecules, pharmacokinetics of the delivery systems, molecular mechanisms involved in BBB regulation, and drug transportation is required.

References

1. Aldape K, Brindle KM, Chesler L, Chopra R, Gajjar A, Gilbert MR, Gottardo N, Gutmann DH, Hargrave D, Holland EC, Jones DTW, Joyce JA, Kearns P, Kieran MW, Mellinghoff IK, Merchant M, Pfister SM, Pollard SM, Ramaswamy V, Rich JN, Robinson GW, Rowitch DH, Sampson JH, Taylor MD, Workman P, Gilbertson RJ (2019) Challenges to curing primary brain tumours. Nat Rev Clin Oncol 16(8):509–520

2. Agrahari V (2017) The exciting potential of nanotherapy in brain-tumor targeted drug delivery approaches. Neural Regen Res 12 (2):197–200

3. von Bartheld CS, Bahney J, Herculano-Houzel S (2016) The search for true numbers of neurons and glial cells in the human brain: a review of 150 years of cell counting. J Comp Neurol 524(18):3865–3895

4. Louis DN, Perry A, Reifenberger G, von Deimling A, Figarella-Branger D, Cavenee WK, Ohgaki H, Wiestler OD, Kleihues P, Ellison DW (2016) The 2016 World Health Organization classification of tumors of the central nervous system: a summary. Acta Neuropathol 131(6):803–820

5. Gutkin A, Cohen ZR, Peer D (2016) Harnessing nanomedicine for therapeutic intervention in glioblastoma. Expert Opin Drug Deliv 13 (11):1573–1582

6. Woodworth GF, Dunn GP, Nance EA, Hanes J, Brem H (2014) Emerging insights into barriers to effective brain tumor therapeutics. Front Oncol 4:126–126

7. Pourgholi F, Hajivalili M, Farhad J-N, Kafil HS, Yousefi M (2016) Nanoparticles: Novel vehicles in treatment of Glioblastoma. Biomed Pharmacother 77:98–107

8. John Lin C-C, Yu K, Hatcher A, Huang T-W, Lee HK, Carlson J, Weston MC, Chen F, Zhang Y, Zhu W, Mohila CA, Ahmed N, Patel AJ, Arenkiel BR, Noebels JL, Creighton CJ, Deneen B (2017) Identification of diverse astrocyte populations and their malignant analogs. Nat Neurosci 20(3):396–405

9. Jaeckle KA (2014) Oligodendroglial Tumors. Semin Oncol 41(4):468–477

10. Wirsching H-G, Galanis E, Weller M (2016) Chapter 23 - Glioblastoma. In: Berger MS, Weller M (eds) Handb Clin Neurol. Elsevier, Amsterdam, pp 381–397

11. Rudà R, Reifenberger G, Frappaz D, Pfister SM, Laprie A, Santarius T, Roth P, Tonn JC, Soffietti R, Weller M, Moyal EC-J (2017) EANO guidelines for the diagnosis and treatment of ependymal tumors. Neuro-Oncology 20(4):445–456

12. Dahiya S, Perry A (2010) Pineal tumors. Adv Anat Pathol 17(6):419–427

13. Marosi C, Hassler M, Roessler K, Reni M, Sant M, Mazza E, Vecht C (2008) Meningioma. Crit Rev Oncol Hematol 67(2):153–171

14. Fox BD, Cheung VJ, Patel AJ, Suki D, Rao G (2011) Epidemiology of metastatic brain tumors. Neurosurg Clin N Am 22(1):1–6

15. Gabathuler R (2010) Approaches to transport therapeutic drugs across the blood–brain barrier to treat brain diseases. Neurobiol Dis 37 (1):48–57

16. Lu C-T, Zhao Y-Z, Wong HL, Cai J, Peng L, Tian X-Q (2014) Current approaches to enhance CNS delivery of drugs across the brain barriers. Int J Nanomedicine 9:2241–2257

17. Hladky SB, Barrand MA (2018) Elimination of substances from the brain parenchyma: efflux via perivascular pathways and via the blood-brain barrier. Fluids Barriers CNS 15(1):30–30

18. Teleanu DM, Chircov C, Grumezescu AM, Volceanov A, Teleanu RI (2018) Blood-brain delivery methods using nanotechnology. Pharmaceutics 10(4):269

19. van Tellingen O, Yetkin-Arik B, de Gooijer MC, Wesseling P, Wurdinger T, de Vries HE (2015) Overcoming the blood–brain tumor barrier for effective glioblastoma treatment. Drug Resist Updat 19:1–12

20. Meng J, Agrahari V, Youm I (2017) Advances in targeted drug delivery approaches for the central nervous system tumors: the inspiration of Nanobiotechnology. J Neuroimmune Pharmacol 12(1):84–98

21. Sun C, Ding Y, Zhou L, Shi D, Sun L, Webster TJ, Shen Y (2017) Noninvasive nanoparticle

strategies for brain tumor targeting. Nanomedicine 13(8):2605–2621

22. Jain KK (2012) Nanobiotechnology-based strategies for crossing the blood–brain barrier. Nanomedicine 7(8):1225–1233

23. Chu P-C, Chai W-Y, Tsai C-H, Kang S-T, Yeh C-K, Liu H-L (2016) Focused ultrasound-induced blood-brain barrier opening: association with mechanical index and cavitation index analyzed by dynamic contrast-enhanced magnetic-resonance imaging. Sci Rep 6:33264–33264

24. Jena L, McErlean E, McCarthy H (2019) Delivery across the blood-brain barrier: nanomedicine for glioblastoma multiforme. Drug Deliv Transl Res 10(2):304–318

25. Zhang F, Xu C-L, Liu C-M (2015) Drug delivery strategies to enhance the permeability of the blood-brain barrier for treatment of glioma. Drug Des Devel Ther 9:2089–2100

26. Fang J, Nakamura H, Maeda H (2011) The EPR effect: unique features of tumor blood vessels for drug delivery, factors involved, and limitations and augmentation of the effect. Adv Drug Deliv Rev 63(3):136–151

27. Maeda H, Nakamura H, Fang J (2013) The EPR effect for macromolecular drug delivery to solid tumors: improvement of tumor uptake, lowering of systemic toxicity, and distinct tumor imaging in vivo. Adv Drug Deliv Rev 65(1):71–79

28. Tian Y, Mi G, Chen Q, Chaurasiya B, Li Y, Shi D, Zhang Y, Webster TJ, Sun C, Shen Y (2018) Acid-induced activated cell-penetrating peptide-modified cholesterol-conjugated Polyoxyethylene sorbitol Oleate mixed micelles for pH-triggered drug release and efficient brain tumor targeting based on a charge reversal mechanism. ACS Appl Mater Interfaces 10 (50):43411–43428

29. Tang W, Fan W, Lau J, Deng L, Shen Z, Chen X (2019) Emerging blood–brain-barrier-crossing nanotechnology for brain cancer theranostics. Chem Soc Rev 48(11):2967–3014

30. Thom G, Hatcher J, Hearn A, Paterson J, Rodrigo N, Beljean A, Gurrell I, Webster C (2018) Isolation of blood-brain barrier-crossing antibodies from a phage display library by competitive elution and their ability to penetrate the central nervous system. MAbs 10 (2):304–314

31. Hays EM, Duan W, Shigdar S (2017) Aptamers and glioblastoma: their potential use for imaging and therapeutic applications. Int J Mol Sci 18(12):2576

32. Zheng M, Tao W, Zou Y, Farokhzad OC, Shi B (2018) Nanotechnology-based strategies for siRNA brain delivery for disease therapy. Trends Biotechnol 36(5):562–575

33. Jahangiri A, Chin AT, Flanigan PM, Chen R, Bankiewicz K, Aghi MK (2017) Convection-enhanced delivery in glioblastoma: a review of preclinical and clinical studies. J Neurosurg 126(1):191–200

34. Poon C, McMahon D, Hynynen K (2017) Noninvasive and targeted delivery of therapeutics to the brain using focused ultrasound. Neuropharmacology 120:20–37

35. Ye D, Zhang X, Yue Y, Raliya R, Biswas P, Taylor S, Tai Y-C, Rubin JB, Liu Y, Chen H (2018) Focused ultrasound combined with microbubble-mediated intranasal delivery of gold nanoclusters to the brain. J Control Release 286:145–153

36. Leamon CP, Reddy JA, Dorton R, Bloomfield A, Emsweller K, Parker N, Westrick E (2008) Impact of high and low folate diets on tissue folate receptor levels and antitumor responses toward folate-drug conjugates. J Pharmacol Exp Ther 327(3):918–925

37. Pope WB (2018) Brain metastases: neuroimaging. Handb Clin Neurol 149:89–112

38. Fink JR, Muzi M, Peck M, Krohn KA (2015) Multimodality Brain Tumor Imaging: MR Imaging, PET, and PET/MR Imaging. J Nucl Med 56(10):1554–1561

39. Pope WB, Djoukhadar I, Jackson A (2016) Chapter 3 - Neuroimaging. In: Berger MS, Weller M (eds) Handb Clin Neurol. Elsevier, Amsterdam, pp 27–50

40. Wang L, Tang W, Zhen Z, Chen H, Xie J, Zhao Q (2014) Improving detection specificity of iron oxide nanoparticles (IONPs) using the SWIFT sequence with long T(2) suppression. Magn Reson Imaging 32(6):671–678

41. Chertok B, David AE, Yang VC (2010) Polyethyleneimine-modified iron oxide nanoparticles for brain tumor drug delivery using magnetic targeting and intra-carotid administration. Biomaterials 31(24):6317–6324

42. Seo JW, Ang J, Mahakian LM, Tam S, Fite B, Ingham ES, Beyer J, Forsayeth J, Bankiewicz KS, Xu T, Ferrara KW (2015) Self-assembled 20-nm (64)Cu-micelles enhance accumulation in rat glioblastoma. J Control Release 220 (Pt A):51–60

43. Hong G, Zou Y, Antaris AL, Diao S, Wu D, Cheng K, Zhang X, Chen C, Liu B, He Y, Wu JZ, Yuan J, Zhang B, Tao Z, Fukunaga C, Dai H (2014) Ultrafast fluorescence imaging in vivo with conjugated polymer fluorophores in the second near-infrared window. Nat Commun 5(1):4206

44. Fan Q, Cheng K, Yang Z, Zhang R, Yang M, Hu X, Ma X, Bu L, Lu X, Xiong X, Huang W, Zhao H, Cheng Z (2015) Perylene-diimide-based nanoparticles as highly efficient photoacoustic agents for deep brain tumor imaging in living mice. Adv Mater 27(5):843–847

45. Saha K, Agasti SS, Kim C, Li X, Rotello VM (2012) Gold nanoparticles in chemical and biological sensing. Chem Rev 112 (5):2739–2779

46. Huang R, Harmsen S, Samii JM, Karabeber H, Pitter KL, Holland EC, Kircher MF (2016) High precision imaging of microscopic spread of glioblastoma with a targeted ultrasensitive SERRS molecular imaging probe. Theranostics 6(8):1075–1084

47. Wu X, Yang H, Yang W, Chen X, Gao J, Gong X, Wang H, Duan Y, Wei D, Chang J (2019) Nanoparticle-based diagnostic and therapeutic systems for brain tumors. J Mater Chem B 7(31):4734–4750

48. Zhang C-L, Huang T, Wu B-L, He W-X, Liu D (2017) Stem cells in cancer therapy: opportunities and challenges. Oncotarget 8 (43):75756–75766

49. Lathia JD, Mack SC, Mulkearns-Hubert EE, Valentim CLL, Rich JN (2015) Cancer stem cells in glioblastoma. Genes Dev 29 (12):1203–1217

50. Mariotti V, Greco SJ, Mohan RD, Nahas GR, Rameshwar P (2014) Stem cell in alternative treatments for brain tumors: potential for gene delivery. Mol Cell Ther 2:24–24

51. Shah K (2016) Stem cell-based therapies for tumors in the brain: are we there yet? Neuro-Oncology 18(8):1066–1078

52. Li G, Bonamici N, Dey M, Lesniak MS, Balyasnikova IV (2018) Intranasal delivery of stem cell-based therapies for the treatment of brain malignancies. Expert Opin Drug Deliv 15 (2):163–172

53. Aboody KS, Najbauer J, Metz MZ, D'Apuzzo M, Gutova M, Annala AJ, Synold TW, Couture LA, Blanchard S, Moats RA, Garcia E, Aramburo S, Valenzuela VV, Frank RT, Barish ME, Brown CE, Kim SU, Badie B, Portnow J (2013) Neural stem cell-mediated enzyme/prodrug therapy for glioma: preclinical studies. Sci Transl Med 5(184):184ra59

54. Portnow J, Synold TW, Badie B, Tirughana R, Lacey SF, D'Apuzzo M, Metz MZ, Najbauer J, Bedell V, Vo T, Gutova M, Frankel P, Chen M, Aboody KS (2017) Neural stem cell–based anticancer gene therapy: a first-in-human study in recurrent high-grade glioma patients. Clin Cancer Res 23(12):2951–2960

55. Jain KK (2018) A critical overview of targeted therapies for glioblastoma. Front Oncol 8:419–419

56. Dupont C, Vermandel M, Reyns N, Mordon S (2018) La thérapie photodynamique. Med Sci (Paris) 34(11):901–903

57. Hong EJ, Choi DG, Shim MS (2016) Targeted and effective photodynamic therapy for cancer using functionalized nanomaterials. Acta Pharm Sin B 6(4):297–307

58. Akimoto J (2016) Photodynamic therapy for malignant brain tumors. Neurol Med Chir (Tokyo) 56(4):151–157

59. Dupont C, Vermandel M, Leroy H-A, Quidet M, Lecomte F, Delhem N, Mordon S, Reyns N (2018) INtraoperative photoDYnamic therapy for GliOblastomas (INDYGO): study protocol for a phase I clinical trial. Neurosurgery 84(6):E414–E419

Chapter 5

Surface-Modified Nanodrug Carriers for Brain Cancer Treatment

Aniket S. Wadajkar, Nina P. Connolly, Christine P. Carney, Pranjali P. Kanvinde, Jeffrey A. Winkles, Graeme F. Woodworth, and Anthony J. Kim

Abstract

Brain cancers, especially glioblastomas (GBM), are among the most deadly human tumors due to high proliferation rates, invasion into functioning brain parenchyma, genomic instability, cellular and molecular heterogeneity, and more. The brain exhibits a uniquely complex physiology and utilizes specialized mechanisms, including blood-brain barrier formation, to prevent the influx of many molecules from the external environment, including most systemically administered chemotherapeutics. Nanoparticles are particularly promising for drug delivery to the brain due to their small size, ability to encapsulate drugs, controlled drug release profiles, and their potential to be actively targeted to cells and structures within the brain. In this chapter, we outline why targeted nanotechnology is poised to overcome the brain's specialized barriers to drug delivery and introduce some of the cell surface molecules, including the fibroblast growth factor-inducible 14 (Fn14) receptor, employed for targeting nanodrug carriers to GBM. Importantly, we review an advanced nanotherapeutic formulation developed by our team designed for optimal (1) brain tumor accumulation and penetration by reducing nonspecific interactions with blood plasma proteins and extracellular matrix proteins (off-target effects) and (2) brain tumor targeting by increasing specific interactions with targeting molecules that are overexpressed on cancer cells (on-target effects). These decreased adhesivity, receptor-targeted (DART) nanoparticles were developed for GBM treatment by balancing and maintaining low nonspecific adhesivity and high Fn14 receptor binding.

Key words Glioblastoma, Nonspecific binding, Receptor targeting, Fibroblast growth factor inducible-14, Nanotherapeutics

1 Introduction to Brain Cancer and Current Therapies

Primary intrinsic brain cancers represent a biologically diverse and clinically distinct group of intracranial tumors, many of which are highly aggressive diseases with dismal prognosis. Approximately 80% of these tumors are gliomas, which are derived specifically from glial cells including astrocytes, oligodendrocytes, ependymal cells, and microglia [1–4]. The most aggressive human brain

Vivek Agrahari et al. (eds.), *Nanotherapy for Brain Tumor Drug Delivery*, Neuromethods, vol. 163, https://doi.org/10.1007/978-1-0716-1052-7_5, © Springer Science+Business Media, LLC, part of Springer Nature 2021

tumor, glioblastoma (GBM), accounts for ~15% of all primary intrinsic brain tumors and 60–75% of all astrocytomas [5]. Patients diagnosed with GBM face a median survival of <6 months without intervention and only ~15 months with most aggressive treatment regimens [6].

GBM tumors exhibit genomic instability, cellular and molecular heterogeneity, extensive vascularization, and high cellular proliferation and invasion [1–4]. These characteristics, and brain-specific drug delivery challenges [7], limit treatment efficacy for GBM patients. Systemic corticosteroids, introduced in the 1960s, were among the first treatment options for GBM. These steroids alleviated the peritumoral edema and "mass effect" associated with a growing tumor by reducing compression of the surrounding normal tissue, which is the most common cause of neurological symptoms in patients with GBM [8]. Whole brain radiotherapy (WBRT) was then used as an adjuvant shortly after. Though limited by its potential for causing central nervous system (CNS) toxicity, WBRT was found to double the median survival time for GBM patients from 6 to 12 months [9]. Systemic chemotherapy was then introduced for GBM treatment in the 1990s. DNA alkylating agents, such as carmustine, further improved the median survival time by 2 months. Temozolomide (TMZ), another DNA alkylating agent, was introduced in the early 2000s, with the added benefit of oral administration [10]. The use of carmustine-loaded interstitial wafers (CIW) for local drug delivery was approved by the FDA for treatment of recurrent high grade gliomas in 1997 and primary high grade glioma in 2004 [11–15]. CIWs, made with non-inflammatory and biodegradable polymers, are implanted along the border of the tumor resected cavity in the brain. CIWs release the drug directly to the residual tumor cells, increasing the median survival by 2 months from 11.6 to 13.9 months [14].

In 2005, a treatment regimen of concurrent radiochemotherapy followed by adjuvant chemotherapy ("Stupp protocol") became the standard of care for GBM [6]. This treatment regimen involves TMZ administration and intensity modulated radiation therapy in fractionated doses (2 Gy daily) over the course of 6 weeks for a total dose of 60 Gy. In addition, after the radiation treatment, patients receive TMZ daily. This regimen resulted in a significant improvement in the 2-year survival rate. Increases in median survival, however, were modest: 14.6 months in the Stupp protocol compared to 12.1 months in patients receiving radiotherapy alone.

2 Barriers to Brain Cancer Treatment

Several factors conspire to make GBM prognosis notoriously poor, including the invasive nature of glioma cells. These invading cells are difficult to remove via surgical resection and cannot be safely targeted with radiation without damaging the surrounding healthy brain tissue. Tumors recur upon proliferation of these cells, which generally remain within centimeters of the original tumor [16, 17]. The ability to eradicate these cells, however, is complicated by the extremely complex physiology and biology of the brain. The three main barriers to therapeutic intervention in GBM patients include the blood-brain barrier (BBB), the glial lymphatic system (GLS), and the narrow extracellular spaces (ECS) within the brain parenchyma [7, 18].

The BBB is primarily composed of cerebral endothelial cells, pericytes, astrocytic end-feet processes, and a thick basement membrane, creating a specialized barrier that restricts more than 98% of systemically administered drugs or small molecules from entering the brain [19–22]. Thus, it becomes very difficult to achieve therapeutic concentrations of drugs within brain tumors [21]. The rapid proliferation and growth of tumors in the brain results in a compromised BBB at the tumor core. These fenestrations in the blood vessels allow some local accumulation of small particles. Unfortunately, the BBB remains relatively intact in the unresectable areas where tumor cells have invaded the healthy brain tissue, making it extremely difficult to deliver drugs targeting these invasive cells [22, 23]. The second barrier, the GLS, serves as a lymphatic drainage pathway for the clearance of drugs and other small molecules from the CNS [24]. In the GLS, astrocytic processes containing aquaporins make contact with blood vessels in the brain and pump cerebrospinal fluid (CSF) throughout the brain parenchyma. Despite serving as a clearance pathway to remove any waste products, the GLS also clears therapeutic agents and nanoparticles from the cerebral circulation [24, 25]. The third barrier, the ECS, comprises approximately 10–20% of the brain's volume and is filled with anisotropic, electrostatically charged extracellular matrix (ECM). By occupying spaces between neuronal and glial cells, the ECM presents a "brain penetration barrier" for systemically and locally delivered drugs [26, 27]. Any agent delivered to the brain must navigate through these narrow spaces while simultaneously interacting with the electrostatically charged components of the ECS. The result is a further reduction in drug delivery to the cancer cells.

Several strategies are under investigation to evade or bypass these barriers to CNS drug delivery, including the use of magnetic resonance-guided focused ultrasound (MRgFUS) to reversibly open the BBB, leveraging of alternative administrative routes such as local delivery or intranasal delivery, and development of nanoparticle drug delivery systems [7, 28–31]. In particular, surface modification of small molecules or drug-loaded nanoparticles have been studied by several groups in order to enhance brain penetration as well as specifically target the cancer cells. For instance, modifying the nanoparticle surface with hydrophilic polymers not only increases their blood circulation time but also enhances their diffusion through the ECS, resulting in improved dispersion in the neuronal tissue [32, 33]. Development of such nonadhesive nanoparticles in conjunction with conjugation of cancer cell-specific targeting molecules has great potential for enhancing drug delivery to GBM cells.

3   Drug Nanoparticle Formulations May Overcome Brain Tumor Treatment Barriers

Previous studies exploring drug delivery to brain tumors have revealed undesirable biological behavior or potential toxicities of the drugs caused by clearance and/or degradation mechanisms, distributing the drug to non-CNS or off-target sites [7, 34]. Encapsulating therapeutics in a nanoparticle system and then targeting them to specific disease components may decrease some of these limitations. In addition, nanoparticles have become a major focus in the field of drug delivery to the brain due to their small size, enhanced drug solubility, ability for multifunctional applications, and a controlled and sustained drug release profile [35].

Conventionally, systemic delivery of nanotherapeutics was widely investigated as a means of delivering drugs to the brain, leveraging the compromised BBB at the tumor core. The extent to which nanoparticles extravasate and remain within the tumor tissue is dictated by their physicochemical properties including, size, shape, charge, and hydrophobicity. For example, Householder et al. [36] observed that systemically administered ~200 nm camptothecin-loaded poly(lactic-*co*-glycolic acid) (PLGA) nanoparticles accumulated in GL261 tumors in C57BL/6 albino mice at levels ten times that of healthy brains. These nanoparticles effectively slowed GL261 tumor growth and imparted a significant survival benefit compared to mice receiving saline or free camptothecin.

Surface characteristics of nanoparticles play a vital role in the nonspecific binding to blood plasma proteins [32]. Ambruosi and colleagues [37] investigated the biodistribution of ~270 nm doxorubicin (Dox)-loaded poly(butyl-2-cyano [3-(14)C]acrylate) (PBCA) nanoparticles and Dox-loaded polysorbate 80-coated PBCA nanoparticles in rats harboring 101/8 GBM tumors following intravenous injection. Nanoparticles were found to accumulate at the tumor site and in the contralateral hemisphere in GBM-bearing rats. Moreover, surface coating of nanoparticles with poly(ethylene glycol) (PEG) increased blood circulation time, giving more time for nanoparticles to accumulate at the tumor site [38]. Despite the presence of compromised BBB at the tumor core, the mechanism of action by which passively targeted nanoparticles cross the BBB is not fully understood. In addition, the permeability of blood vessels varies throughout the tumor, preventing a uniform distribution of nanotherapeutics through blood vessel fenestrations into the tumor mass [33]. Nanoparticles, although PEGylated, are taken up by the normal healthy cells, causing off-target toxicities [30]. Therefore, nanoparticles designed to specifically target cancer cells without harming the surrounding healthy cells are of great interest, especially for drug delivery to invasive GBM tumor cells. Active targeting of nanoparticles involves conjugation of targeting molecules such as monoclonal antibodies, peptides, aptamers, growth factors, or hormones that recognize cell surface proteins preferentially expressed on tumor cells [39, 40]. Upon antibody binding to its target, nanoparticles are internalized via receptor mediated endocytosis for intracellular drug release.

A variety of materials including metals, polymers, and lipids have been used to formulate nanoparticles for targeted drug delivery to brain cancer cells. Metallic nanoparticles, made of inorganic materials such as gold, silver, or iron oxide, are more permeable to the BBB due to their small size compared to polymeric or lipid particles. Several in vivo studies have demonstrated that metallic nanoparticles are capable of leaving the systemic circulation and entering the brain through several mechanisms including passive diffusion, carrier-mediated transport (CMT), or trans-synaptic transport [41]. For example, Cheng et al. [42] demonstrated that ~5 nm gold nanoparticles modified with the trans-activator of transcription peptide can cross the BBB and deliver drugs and/or contrast agents in a murine intracranial glioma xenograft model. In contrast to metallic nanoparticles, polymeric nanoparticles are synthesized by organic chemistry from less rigid and less dense nanomaterials. Their "soft" physical nature enables encapsulation of a wide range of therapeutic molecules. Perhaps the greatest advantage polymeric nanoparticles offer is the ability to easily modify their physicochemical properties such as size, shape, charge, and surface chemistry. Previous investigations using nanoparticles composed of albumin or poly(butylcyanocrylate) have suggested that polymeric nanoparticles utilize a transcytosis mechanism to cross the BBB, such as CMT or receptor-mediated transport (RMT) [43–45]. These findings further supported the development of polymeric nanoparticles as carriers for therapeutic agents that are otherwise unable to cross the BBB. For example, Ren et al. [46] demonstrated the CMT of particles across the BBB using poly (lactic acid)-b-PEG nanoparticles coated with polysorbate 80. Loading these nanoparticles with amphotericin B, which cannot utilize CMT, resulted in increased delivery of the drug to the mouse brain. Finally, in the absence of CMT or RMT, molecules in the systemic circulation are capable of crossing the BBB via lipid-mediated free diffusion. Such transport, however, is limited to small molecules (<400 Da) with high lipid solubility, excluding most brain-targeting small molecule drug candidates.

Similar to the polymers, liposomes have been studied extensively for drug delivery to brain tumor. Liposomes are self-assembled vesicles composed of amphiphilic phospholipids that closely resemble the lipid bilayer of the cell membrane, enabling delivery of hydrophilic, hydrophobic, and amphoteric molecules across the BBB [47]. Liposomal doxorubicin (Dox) has been approved for use in patients with recurrent and/or refractory high grade glioma [48, 49], as well as in patients with brain metastases originating from non-CNS tumors [50]. Further, Qin et al. [51] recently demonstrated enhanced targeting and delivery of edaravone to the rat brain by conjugating cyclic RGD (cRGD) peptide to the liposome surface. Ying and colleagues [52] followed this by developing daunorubicin-loaded liposomes conjugated to both *p*-aminophenyl-a-d-mannopyranoside (MAN), a mannose analog, and transferrin in order to enhance their targeting of glioma cells across the BBB. This dual-targeting strategy improved the transport of daunorubicin across the BBB by up to 24.9% in in vitro and in vivo models.

4   Molecular Targets for Brain Cancer Nanotherapeutics

In general, nanoparticles typically utilize transcellular pathways to cross the BBB. By targeting various BBB-associated receptors expressed by brain endothelial cells, such as the transferrin receptor or lipoprotein receptor-related protein 1, nanoparticles can be delivered across the BBB by hijacking endogenous CMT and RMT mechanisms [31, 34]. Nanoparticles can also be engineered to recognize proteins preferentially expressed on target cancer cells, as opposed to healthy cells, diminishing drug side effects. Numerous cell surface proteins have been identified that are overexpressed in GBM cells, and several of these are discussed below (also *see* Table 1).

4.1 Epidermal Growth Factor Receptor (EGFR)

EGFR amplification is one of the most common genetic alterations found in patients with malignant gliomas, occurring in approximately 50% of cases [53]. Iron oxide nanoparticles (IONPs) conjugated to cetuximab, a monoclonal antibody to EGFR, showed a 60-fold increase in EGFR binding to EGFR-expressing glioma stem cells when compared to drug alone, increasing EGFR inhibition and eventually apoptosis of the EGFR-positive glioma cells [54, 55]. Also, gold nanoparticles coated with both PEG and anti-EGFR antibodies exhibited enhanced cell uptake in U251 and BT2012035 glioma cells and could be successfully delivered to a 9L rat glioma after BBB disruption using MRgFUS [28].

Table 1
Cell surface proteins overexpressed on GBM cells and the examples of target-specific nanoparticles

Target	Targeting moiety	Nanocarrier	Drug	References
Epidermal growth factor receptor (EGFR)	EGFR	Silica shell gold nanoparticle	Near-infrared surface enhanced Raman scattering (SERS)	Diaz et al. [28]
EGFR variant III (EGFRvIII)	EGFRvIII mAb	Iron oxide nanoparticle	N/A	Hadjipanayis et al. [58]
	Cetuximab(EGFR and EGFRvIII mAb)	Iron oxide nanoparticle	N/A	Bouras et al. [88]
$\alpha_v\beta_3$ integrin	Cyclic arginine-glycine-aspartic acid (RGD) peptide	Poly(trimethylene carbonate)-based micellar nanoparticulate system	Paclitaxel	Jiang et al. [89], Jiang et al. [62, 90]
Transferrin receptor	Transferrin	PEG-poly (ε-caprolactone) (PCL) polymersome	Doxorubicin	Pang et al. [65, 91]
	Transferrin	PEG-G4 poly (amidoamine) (PAMAM) dendrimer	Doxorubicin	Li et al. [92]
	Transferrin	Self-assembling PEGylated nanoparticle	Zoledronic acid	Porru et al. [93]
	Anti-transferrin receptor single chain antibody fragment (TfRscFV)	scL (immunoliposome nanocomplex)	wtp53 plasmid	Kim et al. [94]
	Transferrin	Polysorbate 80 coated PLGA nanoparticle	Methotrexate	Jain et al. [95]
	TfRscFV	scL (immunoliposome nanocomplex)	Temozolomide	Kim et al. [96]
	Transferrin	PEGylated liposomes	JQ1 (small molecule bromodomain inhibitor)	Lam et al. [66]
Lipoprotein receptor-related protein 1 (LRP1)	Lactoferrin	PEG-PCL	Doxorubicin	Pang et al. [68]
	Angiopep2 (ANG2)	PEGylated gold nanoparticle (PEG-AuNP)	Doxorubicin	Ruan et al. [69]
	ANG2	PEG-PCL	Paclitaxel	Zin et al. [97]
GLUT	2-deoxy-D-glucose	PEGylated poly (trimethylene carbonate) nanoparticles	Paclitaxel	Jiang et al. [73]

(continued)

Table 1
(continued)

Target	Targeting moiety	Nanocarrier	Drug	References
GFAP	GFAP mAb	PEGylated liposome	N/A	Chekhonin et al. [77]
Cx43	Recombinant E2 extracellular loop of Cx43 mAb	PEGylated liposome	N/A	Chekhonin et al. [77]
Fn14	ITEM-4 mAb	PEGylated polystyrene-based nanoparticle	N/A	Schneider et al. [101]
	ITEM-4 mAb	PEGylated PLGA nanoparticles	N/A	Wadajkar et al. [91]

4.2 Epidermal Growth Factor Receptor Variant III (EGFRvIII)

The EGFRvIII protein contains an in-frame deletion within the EGFR extracellular domain, creating a new glycine residue where the endogenous codon was split. This mutation produces a constitutively active tyrosine kinase and promotes GBM cell proliferation, migration, and resistance to chemotherapy and radiation treatments [56]. While EGFR gene amplification is found in ~50% of GBM patients, about half of these patients also express EGFRvIII [57]. The invasive properties endowed by EGFRvIII expression, combined with its prevalence in GBM tumors but not healthy brain tissue, make the EGFR variant a valuable target for GBM patient therapy. A study investigating the surface conjugation of 10 nm IONPs to an EGFRvIII-specific antibody showed a twofold increase in glioma cell targeting when applied to EGFRvIII-expressing U87MG cells compared to non-expressing U87 cells [58].

4.3 αvβ3 Integrin

Integrins are a family of transmembrane receptors expressed by all cells that are critical for cell-ECM adhesion [59]. Integrins commonly bind to the RGD (Arg-Gly-Asp) tripeptide sequence expressed in various ECM proteins, including fibronectin vitronectin, collagen, and fibrinogen [60, 61]. Jiang et al. [62] observed that paclitaxel-loaded RGD-functionalized micelles (cRGDyK-NP/PTX) showed enhanced cell penetration, tumor accumulation, and microtubule stabilization in U87MG gliomas in vivo, resulting in a 1.2- and 1.5-fold increase in mean survival time of mice compared to non-targeted-NP-PTX-treated and free taxol-treated mice.

4.4 Transferrin Receptor (TfR)

The TfR plays a major role in iron transport and metabolism and is expressed on the luminal membrane surface of brain endothelial cells and cancer cells [63]. TfR expression has also been correlated to tumor grade and/or prognosis in several cancers, including glioma [64]. In a study of orthotopic U87 tumor-bearing mice, treatment with Tf-targeted Dox-loaded polymersomes resulted in a 2.5- and five-fold decrease in tumor volume compared to non-targeted or free Dox treatment [65]. More recently, TFr-functionalized nanoparticles have been developed to deliver dual combination therapies. Using PEGylated TfR-targeted nanoparticles loaded with TMZ and the bromodomain inhibitor JQ1, Lam et al. demonstrated 1.5- to two-fold decrease in tumor size and survival in mice harboring U87MG or GL261 GBM tumors compared to free drug alone [66]. Several additional formulations targeting TfR are under active investigation for GBM (*see* Table 1).

4.5 Lipoprotein Receptor-Related Protein (LRP)

LRP is a member of the low-density lipoprotein receptor family and a similar target to TfR due to its overexpression on both brain endothelial cells and glioma cells. In contrast to TfR, LRP mediates the transport of several ligands, including lactoferrin (Lf), melano-transferrin, receptor-associated protein, and Angiopep-2 peptide [67]. A study investigating Lf-conjugated biodegradable polymersomes (Lf-PO-Dox/tetrandrine (Tet)) not only demonstrated enhanced delivery of Dox and Tet across the BBB and showed greater uptake and cytotoxicity against C6 glioma cells but also showed increased accumulation within the C6 glioma intracranial tumor model, resulting in a three-fold decrease in tumor volumes compared to that of controls [68]. Angiopep-2 is another specific ligand of the LRP receptor, which has been used as a targeting moiety for PEGylated doxycycline-loaded gold nanoparticles. In one such study, treatment of mice harboring C6 gliomas with these nanoparticles resulted in a 2.89-fold longer survival compared to mice treated with saline [69].

4.6 D-Glucose Transporter (GLUT)

Despite constituting only about 2% of total body weight, the normal adult brain consumes roughly 20% of the body's total glucose each day for its energy demands [70]. The endothelial cells of the BBB facilitate the transport of such large quantities of glucose via expression of sodium-independent facilitative GLUT transporters and sodium-dependent transporters [71]. Of these, GLUT-1 is especially enriched in the BBB and is thought to be the predominant mediator of facilitative glucose transport into the brain [72]. Because GLUT transporters are overexpressed in both brain microvessels and tumor cells, they represent promising targets for nanotherapeutics. For example, Jiang et al. [73] showed PTX-loaded nanoparticles conjugated with 2-deoxy-D-glucose accumulated specifically within the glioma tumor, evidenced by a 900-fold increased fluorescent signal in the tumor compared to that of non-targeted particles.

4.7 Glial Fibrillary Acidic Protein (GFAP) and Connexin 43 (Cx43)

The edges of high-grade gliomas are comprised of invasive glioma cells, whose movement into the surrounding non-tumor tissue creates a band of reactive astrocytes between the tumor and normal tissue [2, 7]. These astrocytes express two important structural proteins: the astroglial cell marker GFAP and the gap junction protein Cx43 [74, 75]. The gap junctions formed between glioma cells and GFAP-positive astrocytes assist the migration of Cx43-positive gliomas cells to the invasive edge of the tumor [74, 76]. One group tested Cx43- and GFAP-conjugated PEGy-lated liposomes in a C6 glioma model and observed the accumulation of fluorescent nanoparticles in the periphery of the tumor where expression of Cx43 and GFAP was highly concentrated, suggesting that GFAP and Cx43 are efficient targets for peritumoral area, primarily for the delivery of therapeutic agents to the glioma cells invading into healthy, non-tumor tissue [77].

4.8 Fibroblast Growth Factor-Inducible 14 (Fn14)

Fn14 is a member of the tumor necrosis factor (TNF) receptor superfamily (Fig. 1a) and is activated by the TNF-like weak inducer of apoptosis (TWEAK) ligand [78–81]. TWEAK-Fn14 signaling plays an important role in wound repair and inflammation; however, the TWEAK-Fn14 system has also been implicated in several pathological settings. Fn14, in particular, is thought to play a supportive role in the development of several cancers, including GBM [82]. TWEAK and Fn14 expression is low in normal, healthy tissue, but increases in many solid primary and metastatic tumors [78, 79]. In addition, Fn14 expression is preferentially upregulated in highly malignant and advanced brain tumors, most notably in the

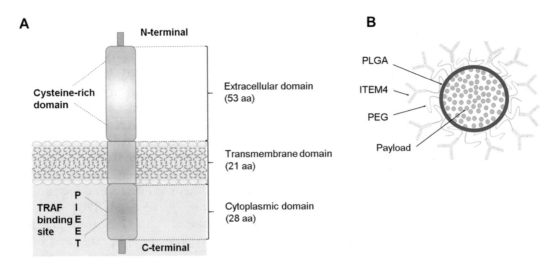

Fig. 1 Schematic of Fn14 receptor and DART nanoparticle formulation. (**a**) Human Fn14 contains only 102 amino acids and consists of three domains: an extracellular ligand binding domain, a transmembrane domain, and an intracellular domain that serves as a binding site for downstream effectors. (**b**) Schematic of DART nanoparticles conjugated with ITEM4 mAb specific for Fn14

GBM mesenchymal tumor subtype, which is a highly aggressive and invasive subtype [82–87]. Importantly, Fn14 is upregulated along the invasive edge of the tumors and in migrating glioma cell populations themselves [82, 84, 85]. These findings suggest that Fn14 is a valuable target for nanoparticle drug delivery to invasive GBM tumor cells, which is discussed below.

5 DART Nanoparticles: Tissue-Penetrating, Fn14-Targeted Nanotherapeutics

Regardless of a local or systemic delivery route, nanodrug particles for GBM treatment must diffuse through brain parenchyma and deliver drug specifically to the cancer cells (Fig. 2). The barrier properties of the ECS, including a dense layer of ECM proteins, is a major challenge for nanoparticle diffusion [7, 26, 27]. For example, Thorne and Nicholson [98] analyzed the diffusion of quantum dots and dextrans through the ECS and observed that navigation was limited to nanoparticles between 30 and 70 nm in

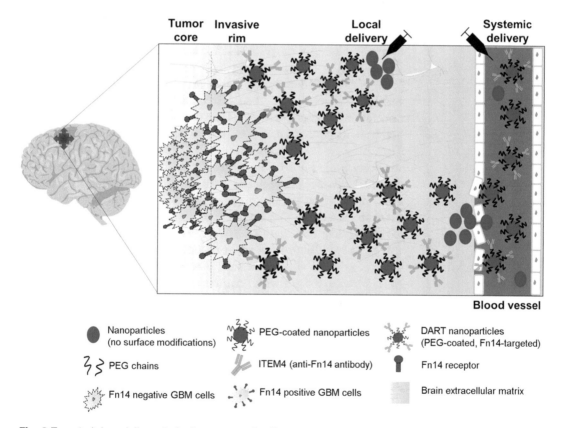

Fig. 2 Targeted drug delivery to brain cancer cells. Nanoparticle drug delivery systems must diffuse through brain parenchyma and deliver drug specifically to the cancer cells, regardless of a local or systemic delivery route. DART nanoparticles—designed to resist blood clearance, penetrate brain tissue, and target Fn14-positive GBM cells—have the potential to improve drug pharmacokinetic profile and therapeutic efficacy

diameter. In order to fully capitalize on the potential benefits of surface-modified, molecularly targeted nanoparticles, low levels of nonspecific adhesivity and off-target binding must be maintained while setting an effective level of target-specific binding. Achieving this balance between decreased nonspecific adhesivity to normal healthy tissue components and receptor-specific targeting (referred to as DART characteristics) is often quite challenging, as many targeting moieties (antibodies and related fragments, peptides, carbohydrates, and others) exhibit nonspecific binding to cellular, extravascular, or intravascular components. Coating nanoparticles with hydrophilic, low molecular weight PEG not only increases their blood circulation time [38] but also minimizes nonspecific binding to ECM components, thus enhancing brain penetration [99]. Using state-of-the-art techniques such as surface plasmon resonance (SPR), multiple particle tracking (MPT), and intravital microscopy, our group has demonstrated that polystyrene nanoparticles with dense PEG coatings have a larger size threshold for brain penetration [100]. In addition, we showed that the PEG density positively correlates with nanoparticle diffusion and the PEG densities below a specific value leads to ECM binding. Further, we modified the surface of these nanoparticles by conjugating a Fn14 monoclonal antibody (ITEM4) for active targeting to tumor cells, along with a PEG coating for minimal nonspecific binding [101]. The Fn14-targeted nanoparticles were selectively bound and internalized by Fn14-positive U87 glioma cells compared to non-targeted nanoparticles. The nanoparticles did not bind to brain ECM proteins and were able to diffuse freely through brain tissue despite increased nanoparticle size as a result of PEG and antibody conjugation. The targeted nanoparticles co-localized with the U87 GBM cells in the brain after intracranial injection, whereas the non-targeted penetrating nanoparticles did not. Results from these studies imply that Fn14-targeted nanoparticles may be useful for delivering therapeutics directly to the invading glioma cells while sparing the normal brain tissue. In more recent studies, our group designed and optimized biodegradable PLGA-PEG-based DART nanoparticles with balanced decreased nonspecific adhesivity and Fn14 receptor targeting (Fig. 1b) [91]. We utilized SPR to demonstrate minimal binding to brain ECM components and strong binding to Fn14. In addition, we employed MPT and fluorescent imaging to confirm enhanced diffusion, uptake, and retention in brain tumor cells. These results and nanoparticle design considerations offer promising new methods to optimize therapeutic nanocarriers for improving drug delivery and treatment for invasive brain tumors.

6 Future Perspectives

GBM is a highly aggressive and lethal form of brain cancer, largely due to the invasive nature of the cells, which prevents complete surgical removal of tumors and sufficient dosing of adjuvant therapies. Barriers to GBM treatment limit efficacy of treatments and highlight the need for new therapeutic strategies that can penetrate brain tissue and provide sustained drug concentrations by targeting invading GBM cells. Nanodrug formulations offer several advantages to that of other therapeutic platforms including increased drug solubility, drug stability, longer half-life in circulation, and the ability to specifically target tumor cells. The use of targeted nanotherapeutics, designed to kill GBM cells left behind after surgical resection, but not healthy brain cells, has emerged as promising strategy with the potential to improve pharmacokinetic profiles and therapeutic efficacy. Despite the promise of nanoparticles for the treatment of GBM, our knowledge of the molecular and microenvironmental cues driving GBM progression is lacking. Elucidating the complex biology of the GBM tumor microenvironment is a crucial first step toward designing optimal GBM-specific nanotherapeutics. Additionally, understanding mechanisms affecting the movement and diffusion kinetics of nanotherapeutics, in both normal and tumor tissues, is vital for designing and optimizing properties that enable particles to travel long distances. These important pieces of information will help us select the best target, as well as optimize physicochemical properties, so that nanotherapeutics can reach those targets for effective drug delivery.

Biodegradable polymeric nanoparticle drug formulations offer the potential to augment and control the biological behavior of multiple therapeutic agent types including chemotherapies, oligonucleotides, proteins, and others [102, 103]. Such formulations offer the ability to minimize off-target toxicities, clearance/degradation in circulation, and poor accumulation within the pathological site [34]. One strategy that shows promise in preclinical studies is to augment brain penetration of nanoparticles, which improved efficacy and safety of drugs such as PTX, cisplatin, and others [104–106]. Another strategy is to target nanotherapeutics directly to the diseased sites, thereby increasing the payload at the target to enhance treatment efficacy. Targeted delivery strategies have shown early evidence of improved delivery and therapeutic efficacy when nanotherapeutics were targeted to the cell surface markers overexpressed in GBM cells. Despite the ability to optimize nanoparticle penetration and targeting, nanotherapeutics may still be limited by rapid clearance or partitioning and other phenomena that can lead to off-target effects. In order to fully harness the potential benefits of nanotherapeutics, effectively balancing non-specific adhesivity with specific targeting (DART concept) to the disease is a crucial consideration.

By far, PEG is the best polymer for surface modification of nanoparticles for providing hydrophilicity, neutral charge, and non-adhesivity to the nanoparticle surface. In addition, Fn14 antibody conjugation to the PEG layer is a promising strategy for restricting drug delivery to the GBM cells. Our initial investigation revealed that PEG-coated, Fn14-targeted DART nanotherapeutics showed enhanced brain tissue dispersion, increased tumor cell uptake, and improved tumor retention [91]. The less expected finding related to the retention of nanoparticles at the target site suggests that a given therapeutic dose may have an even greater therapeutic effect than we had estimated previously. In addition, our group is currently investigating use of smaller Fn14-binding proteins (e.g., antigen-binding fragment (Fab) and single-chain variable fragment (scFv)) as Fn14 targeting moieties, which are expected to improve diffusion into tumors if engineered to maintain the half-life and binding affinity of their full-length counterpart. We anticipate that future applications of DART characteristics to other therapeutic formulations will lead to significant improvements in therapeutic ratio and the application of therapeutic agents that were not considered safe or feasible for brain cancer, and other diseases with challenging delivery considerations.

Acknowledgments

This work was supported, in part, by the National Institutes of Health (R37 CA218617 (A.J.K.), K08 NS09043 (G.F.W.), RO1 NS107813 (G.F.W.), an Institutional Research Grant IRG-97-153-10 (A.S.W.) from the American Cancer Society. A.S.W. and N.P.C. were supported in part by NIH Training Grant T32 CA154274.

References

1. Wen PY, Kesari S (2008) Malignant gliomas in adults. N Engl J Med 359(5):492–507

2. Gladson CL, Prayson RA, Liu WM (2010) The pathobiology of glioma tumors. Annu Rev Pathol 5:33–50

3. Cloughesy TF, Cavenee WK, Mischel PS (2014) Glioblastoma: from molecular pathology to targeted treatment. Annu Rev Pathol 9:1–25

4. Aldape K et al (2015) Glioblastoma: pathology, molecular mechanisms and markers. Acta Neuropathol 129(6):829–848

5. Louis DN et al (2016) The 2016 World Health Organization classification of tumors of the central nervous system: a summary. Acta Neuropathol 131(6):803–820

6. Stupp R et al (2005) Radiotherapy plus concomitant and adjuvant temozolomide for glioblastoma. N Engl J Med 352 (10):987–996

7. Woodworth GF et al (2014) Emerging insights into barriers to effective brain tumor therapeutics. Front Oncol 4:126

8. Galicich JH, French LA, Melby JC (1961) Use of dexamethasone in treatment of cerebral edema associated with brain tumors. J Lancet 81:46–53

9. Shibamoto Y et al (1990) Supratentorial malignant glioma: an analysis of radiation therapy in 178 cases. Radiother Oncol 18 (1):9–17

10. Stewart LA (2002) Chemotherapy in adult high-grade glioma: a systematic review and meta-analysis of individual patient data from 12 randomised trials. Lancet 359 (9311):1011–1018

11. Brem H et al (1995) Placebo-controlled trial of safety and efficacy of intraoperative controlled delivery by biodegradable polymers of chemotherapy for recurrent gliomas. The Polymer-brain Tumor Treatment Group. Lancet 345(8956):1008–1012

12. Brem H et al (1995) The safety of interstitial chemotherapy with BCNU-loaded polymer followed by radiation therapy in the treatment of newly diagnosed malignant gliomas: phase I trial. J Neuro-Oncol 26(2):111–123

13. Valtonen S et al (1997) Interstitial chemotherapy with carmustine-loaded polymers for high-grade gliomas: a randomized double-blind study. Neurosurgery 41(1):44–48; discussion 48-9

14. Westphal M et al (2003) A phase 3 trial of local chemotherapy with biodegradable carmustine (BCNU) wafers (Gliadel wafers) in patients with primary malignant glioma. Neuro-Oncology 5(2):79–88

15. Westphal M et al (2006) Gliadel wafer in initial surgery for malignant glioma: long-term follow-up of a multicenter controlled trial. Acta Neurochir 148(3):269–275. discussion 275

16. Giese A et al (2003) Cost of migration: invasion of malignant gliomas and implications for treatment. J Clin Oncol 21(8):1624–1636

17. Vehlow A, Cordes N (2013) Invasion as target for therapy of glioblastoma multiforme. Biochim Biophys Acta 1836(2):236–244

18. van Tellingen O et al (2015) Overcoming the blood-brain tumor barrier for effective glioblastoma treatment. Drug Resist Updat 19:1–12

19. Armulik A et al (2010) Pericytes regulate the blood-brain barrier. Nature 468 (7323):557–561

20. Pardridge WM (2005) The blood-brain barrier: bottleneck in brain drug development. NeuroRx 2(1):3–14

21. Obermeier B, Daneman R, Ransohoff RM (2013) Development, maintenance and disruption of the blood-brain barrier. Nat Med 19(12):1584–1596

22. Groothuis DR (2000) The blood-brain and blood-tumor barriers: a review of strategies for increasing drug delivery. Neuro-Oncology 2(1):45–59

23. Tate MC, Aghi MK (2009) Biology of angiogenesis and invasion in glioma. Neurotherapeutics 6(3):447–457

24. Plog BA, Nedergaard M (2018) The Glymphatic system in central nervous system health and disease: past, present, and future. Annu Rev Pathol 13:379–394

25. Louveau A et al (2015) Structural and functional features of central nervous system lymphatic vessels. Nature 523(7560):337–341

26. Bellail AC et al (2004) Microregional extracellular matrix heterogeneity in brain modulates glioma cell invasion. Int J Biochem Cell Biol 36(6):1046–1069

27. Nicholson C, Tao L (1993) Hindered diffusion of high molecular weight compounds in brain extracellular microenvironment measured with integrative optical imaging. Biophys J 65(6):2277–2290

28. Diaz RJ et al (2014) Focused ultrasound delivery of Raman nanoparticles across the blood-brain barrier: potential for targeting experimental brain tumors. Nanomedicine 10(5):1075–1087

29. Alonso MJ (2004) Nanomedicines for overcoming biological barriers. Biomed Pharmacother 58(3):168–172

30. Jain RK, Stylianopoulos T (2010) Delivering nanomedicine to solid tumors. Nat Rev Clin Oncol 7(11):653–664

31. Hersh DS et al (2016) Evolving drug delivery strategies to overcome the blood brain barrier. Curr Pharm Des 22(9):1177–1193

32. Calvo P et al (2001) Long-circulating PEGylated polycyanoacrylate nanoparticles as new drug carrier for brain delivery. Pharm Res 18 (8):1157–1166

33. Blanco E, Shen H, Ferrari M (2015) Principles of nanoparticle design for overcoming biological barriers to drug delivery. Nat Biotechnol 33(9):941–951

34. Wadajkar AS et al (2017) Tumor-targeted nanotherapeutics: overcoming treatment barriers for glioblastoma. Wiley Interdiscip Rev Nanomed Nanobiotechnol 9(4)

35. Pourgholi F et al (2016) Nanoparticles: novel vehicles in treatment of glioblastoma. Biomed Pharmacother 77:98–107

36. Householder KT et al (2015) Intravenous delivery of camptothecin-loaded PLGA nanoparticles for the treatment of intracranial glioma. Int J Pharm 479(2):374–380

37. Ambruosi A et al (2006) Biodistribution of polysorbate 80-coated doxorubicin-loaded

[14C]-poly(butyl cyanoacrylate) nanoparticles after intravenous administration to glioblastoma-bearing rats. J Drug Target 14(2):97–105

38. Nance E et al (2014) Non-invasive delivery of stealth, brain-penetrating nanoparticles across the blood-brain barrier using MRI-guided focused ultrasound. J Control Release 189:123–132

39. Byrne JD, Betancourt T, Brannon-Peppas L (2008) Active targeting schemes for nanoparticle systems in cancer therapeutics. Adv Drug Deliv Rev 60(15):1615–1626

40. Peer D et al (2007) Nanocarriers as an emerging platform for cancer therapy. Nat Nanotechnol 2(12):751–760

41. Yang Z et al (2010) A review of nanoparticle functionality and toxicity on the central nervous system. J R Soc Interface 7(Suppl 4): S411–S422

42. Cheng Y et al (2014) Blood-brain barrier permeable gold nanoparticles: An efficient delivery platform for enhanced malignant glioma therapy and imaging. Small 10(24):5137–5150

43. Barbu E et al (2009) The potential for nanoparticle-based drug delivery to the brain: overcoming the blood-brain barrier. Expert Opin Drug Deliv 6(6):553–565

44. Kreuter J et al (2002) Apolipoprotein-mediated transport of nanoparticle-bound drugs across the blood-brain barrier. J Drug Target 10(4):317–325

45. Zensi A et al (2009) Albumin nanoparticles targeted with Apo E enter the CNS by transcytosis and are delivered to neurones. J Control Release 137(1):78–86

46. Ren T et al (2009) Preparation and therapeutic efficacy of polysorbate-80-coated amphotericin B/PLA-b-PEG nanoparticles. J Biomater Sci Polym Ed 20(10):1369–1380

47. Samad A, Sultana Y, Aqil M (2007) Liposomal drug delivery systems: an update review. Curr Drug Deliv 4(4):297–305

48. Chastagner P et al (2015) Phase I study of non-pegylated liposomal doxorubicin in children with recurrent/refractory high-grade glioma. Cancer Chemother Pharmacol 76(2):425–432

49. Fabel K et al (2001) Long-term stabilization in patients with malignant glioma after treatment with liposomal doxorubicin. Cancer 92(7):1936–1942

50. Linot B et al (2014) Use of liposomal doxorubicin-cyclophosphamide combination in breast cancer patients with brain metastases: a monocentric retrospective study. J Neuro-Oncol 117(2):253–259

51. Qin J et al (2014) cRGD mediated liposomes enhanced antidepressant-like effects of edaravone in rats. Eur J Pharm Sci 58:63–71

52. Ying X et al (2010) Dual-targeting daunorubicin liposomes improve the therapeutic efficacy of brain glioma in animals. J Control Release 141(2):183–192

53. Padfield E, Ellis HP, Kurian KM (2015) Current therapeutic advances targeting EGFR and EGFRvIII in glioblastoma. Front Oncol 5:5

54. Eller JL et al (2002) Activity of anti-epidermal growth factor receptor monoclonal antibody C225 against glioblastoma multiforme. Neurosurgery 51(4):1005–1013; discussion 1013-4

55. Kaluzova M et al (2015) Targeted therapy of glioblastoma stem-like cells and tumor non-stem cells using cetuximab-conjugated iron-oxide nanoparticles. Oncotarget 6(11):8788–8806

56. An Z et al (2018) Epidermal growth factor receptor and EGFRvIII in glioblastoma: signaling pathways and targeted therapies. Oncogene 37(12):1561–1575

57. Frederick L et al (2000) Diversity and frequency of epidermal growth factor receptor mutations in human glioblastomas. Cancer Res 60(5):1383–1387

58. Hadjipanayis CG et al (2010) EGFRvIII antibody-conjugated iron oxide nanoparticles for magnetic resonance imaging-guided convection-enhanced delivery and targeted therapy of glioblastoma. Cancer Res 70(15):6303–6312

59. Kim C, Ye F, Ginsberg MH (2011) Regulation of integrin activation. Annu Rev Cell Dev Biol 27:321–345

60. Mattern RH et al (2005) Glioma cell integrin expression and their interactions with integrin antagonists: Research Article. Cancer Ther 3a:325–340

61. Reardon DA et al (2008) Cilengitide: an integrin-targeting arginine-glycine-aspartic acid peptide with promising activity for glioblastoma multiforme. Expert Opin Investig Drugs 17(8):1225–1235

62. Jiang X et al (2013) Integrin-facilitated transcytosis for enhanced penetration of advanced gliomas by poly(trimethylene carbonate)-based nanoparticles encapsulating paclitaxel. Biomaterials 34(12):2969–2979

63. Dufes C, Al Robaian M, Somani S (2013) Transferrin and the transferrin receptor for

the targeted delivery of therapeutic agents to the brain and cancer cells. Ther Deliv 4 (5):629–640

64. Rosager AM et al (2017) Transferrin receptor-1 and ferritin heavy and light chains in astrocytic brain tumors: expression and prognostic value. PLoS One 12(8):e0182954

65. Pang Z et al (2011) Enhanced intracellular delivery and chemotherapy for glioma rats by transferrin-conjugated biodegradable polymersomes loaded with doxorubicin. Bioconjug Chem 22(6):1171–1180

66. Lam FC et al (2018) Enhanced efficacy of combined temozolomide and bromodomain inhibitor therapy for gliomas using targeted nanoparticles. Nat Commun 9(1):1991

67. Lillis AP, Mikhailenko I, Strickland DK (2005) Beyond endocytosis: LRP function in cell migration, proliferation and vascular permeability. J Thromb Haemost 3 (8):1884–1893

68. Pang Z et al (2010) Lactoferrin-conjugated biodegradable polymersome holding doxorubicin and tetrandrine for chemotherapy of glioma rats. Mol Pharm 7(6):1995–2005

69. Ruan S et al (2015) Tumor microenvironment sensitive doxorubicin delivery and release to glioma using angiopep-2 decorated gold nanoparticles. Biomaterials 37:425–435

70. Benarroch EE (2014) Brain glucose transporters: implications for neurologic disease. Neurology 82(15):1374–1379

71. Patching SG (2017) Glucose transporters at the blood-brain barrier: function, regulation and gateways for drug delivery. Mol Neurobiol 54(2):1046–1077

72. Devraj K et al (2011) GLUT-1 glucose transporters in the blood-brain barrier: differential phosphorylation. J Neurosci Res 89 (12):1913–1925

73. Jiang X et al (2014) Nanoparticles of 2-deoxy-D-glucose functionalized poly(ethylene glycol)-co-poly(trimethylene carbonate) for dual-targeted drug delivery in glioma treatment. Biomaterials 35(1):518–529

74. Oliveira R et al (2005) Contribution of gap junctional communication between tumor cells and astroglia to the invasion of the brain parenchyma by human glioblastomas. BMC Cell Biol 6(1):7

75. Schulz R et al (2015) Connexin 43 is an emerging therapeutic target in ischemia/reperfusion injury, cardioprotection and neuroprotection. Pharmacol Ther 153:90–106

76. Bates DC et al (2007) Connexin43 enhances glioma invasion by a mechanism involving the carboxy terminus. Glia 55(15):1554–1564

77. Chekhonin VP et al (2012) Targeted delivery of liposomal nanocontainers to the peritumoral zone of glioma by means of monoclonal antibodies against GFAP and the extracellular loop of Cx43. Nanomedicine 8(1):63–70

78. Cheng E et al (2013) TWEAK/Fn14 Axis-targeted therapeutics: moving basic science discoveries to the clinic. Front Immunol 4:473

79. Winkles JA (2008) The TWEAK-Fn14 cytokine-receptor axis: discovery, biology and therapeutic targeting. Nat Rev Drug Discov 7(5):411–425

80. Burkly LC (2014) TWEAK/Fn14 axis: the current paradigm of tissue injury-inducible function in the midst of complexities. Semin Immunol 26(3):229–236

81. Burkly LC, Michaelson JS, Zheng TS (2011) TWEAK/Fn14 pathway: an immunological switch for shaping tissue responses. Immunol Rev 244(1):99–114

82. Perez JG et al (2016) The TWEAK receptor Fn14 is a potential cell surface portal for targeted delivery of glioblastoma therapeutics. Oncogene 35(17):2145–2155

83. Tran NL et al (2003) The human Fn14 receptor gene is up-regulated in migrating glioma cells in vitro and overexpressed in advanced glial tumors. Am J Pathol 162(4):1313–1321

84. Tran NL et al (2006) Increased fibroblast growth factor-inducible 14 expression levels promote glioma cell invasion via Rac1 and nuclear factor-kappaB and correlate with poor patient outcome. Cancer Res 66 (19):9535–9542

85. Fortin SP et al (2012) Cdc42 and the guanine nucleotide exchange factors Ect2 and trio mediate Fn14-induced migration and invasion of glioblastoma cells. Mol Cancer Res 10(7):958–968

86. Hersh DS et al (2018) Differential expression of the TWEAK receptor Fn14 in IDH1 wild-type and mutant gliomas. J Neuro-Oncol 138 (2):241–250

87. Hersh DS et al (2018) The TNF receptor family member Fn14 is highly expressed in recurrent glioblastoma and in GBM patient-derived xenografts with acquired temozolomide resistance. Neuro-Oncology 20 (10):1321–1330

88. Bouras A, Kaluzova M, Hadjipanayis CG (2015) Radiosensitivity enhancement of radioresistant glioblastoma by epidermal growth factor receptor antibody-conjugated iron-oxide nanoparticles. J Neuro-Oncol 124(1):13–22

89. Jiang X et al (2011) Self-aggregated pegylated poly (trimethylene carbonate) nanoparticles decorated with c(RGDyK) peptide for targeted paclitaxel delivery to integrin-rich tumors. Biomaterials 32(35):9457–9469

90. Jiang X et al (2013) Solid tumor penetration by integrin-mediated pegylated poly(trimethylene carbonate) nanoparticles loaded with paclitaxel. Biomaterials 34 (6):1739–1746

91. Wadajkar AS et al (2017) Decreased non-specific adhesivity, receptor targeted (DART) nanoparticles exhibit improved dispersion, cellular uptake, and tumor retention in invasive gliomas. J Control Release 267:144–153

92. Li Y et al (2012) A dual-targeting nanocarrier based on poly(amidoamine) dendrimers conjugated with transferrin and tamoxifen for treating brain gliomas. Biomaterials 33 (15):3899–3908

93. Porru M et al (2014) Medical treatment of orthotopic glioblastoma with transferrin-conjugated nanoparticles encapsulating zoledronic acid. Oncotarget 5(21):10446–10459

94. Kim SS et al (2014) The clinical potential of targeted nanomedicine: delivering to cancer stem-like cells. Mol Ther 22(2):278–291

95. Jain A et al (2015) Surface engineered polymeric nanocarriers mediate the delivery of transferrin-methotrexate conjugates for an improved understanding of brain cancer. Acta Biomater 24:140–151

96. Kim SS et al (2015) Encapsulation of temozolomide in a tumor-targeting nanocomplex enhances anti-cancer efficacy and reduces toxicity in a mouse model of glioblastoma. Cancer Lett 369(1):250–258

97. Xin H et al (2012) Anti-glioblastoma efficacy and safety of paclitaxel-loading Angiopep-conjugated dual targeting PEG-PCL nanoparticles. Biomaterials 33(32):8167–8176

98. Thorne RG, Nicholson C (2006) In vivo diffusion analysis with quantum dots and dextrans predicts the width of brain extracellular space. Proc Natl Acad Sci U S A 103 (14):5567–5572

99. Nance EA et al (2012) A dense poly(ethylene glycol) coating improves penetration of large polymeric nanoparticles within brain tissue. Sci Transl Med 4(149):149ra119

100. Dancy JG et al (2016) Non-specific binding and steric hindrance thresholds for penetration of particulate drug carriers within tumor tissue. J Control Release 238:139–148

101. Schneider CS et al (2015) Minimizing the non-specific binding of nanoparticles to the brain enables active targeting of Fn14-positive glioblastoma cells. Biomaterials 42:42–51

102. Cheng CJ et al (2015) A holistic approach to targeting disease with polymeric nanoparticles. Nat Rev Drug Discov 14(4):239–247

103. Lesniak MS, Brem H (2004) Targeted therapy for brain tumours. Nat Rev Drug Discov 3(6):499–508

104. Nance E et al (2014) Brain-penetrating nanoparticles improve paclitaxel efficacy in malignant glioma following local administration. ACS Nano 8(10):10655–10664

105. Zhang C et al (2017) Convection enhanced delivery of cisplatin-loaded brain penetrating nanoparticles cures malignant glioma in rats. J Control Release 263:112–119

106. Zhou J et al (2013) Highly penetrative, drug-loaded nanocarriers improve treatment of glioblastoma. Proc Natl Acad Sci U S A 110 (29):11751–11756

Chapter 6

Inorganic Nanostructures for Brain Tumor Management

Mohd Imran, Awais Ahmed Abrar Ahmed, Babak Kateb, and Ajeet Kaushik

Abstract

The nanoparticles have been widely investigated as therapeutic agents for cancer treatments in biomedical fields due to their unique physical/chemical properties, versatile synthetic strategies, easy surface functionalization, and excellent biocompatibility. Even though the advancement of certain treatment techniques is available for the diagnosis of the tumor, still the blood-brain barrier is the obstruction to the delivery of drug molecules to the tumor cells in the central nervous system (CNS) and brain parenchyma. Though nano enabled therapy make promise to deliver the anticancer drugs to cross the blood-brain barrier (BBB). This chapter focuses on the synthesis techniques and advanced characterization techniques adopted to design and develop inorganic nanostructures. Various inorganic nanostructure-based cancer therapeutic agents, including gold nanoparticles, magnetic nanoparticles, carbon nanotube, earth metal oxide nanoparticles, and other nanostructures, have also been discussed. Related challenges with this research area and future prospect are also discussed herein.

Key words Nano-biotechnology, Nano-therapeutics, Nanomedicine, Brain cancer, Drug delivery to the brain

1 Introduction

1.1 Brain Tumor and Its Occurrence

Brain tumor constitutes one of the most deadly and devastating diseases which if not detected in the early stages can't be treated. The rate of incidence for all primary malignant and nonmalignant brain and other CNS tumors is 22.64 cases per 100,000. Out of them 7.15 per 100,000 are malignant and 15.49 per 100,000 are nonmalignant, accounting for 379,848 cases each year. In between 2010 and 2014, approximately 75,271 people died because of primary malignant brain and other CNS tumors. Predicting further approximately 78,980 new cases of primary malignant and nonmalignant brain and other CNS tumors are anticipated to be found out in the United States in the year 2018; an estimated 16,616 deaths are predicted to occur because of primary malignant and other CNS tumors [1]. In UK, nearly 80,000 new cases of primary brain tumors are expected to be diagnosed in year 2018; approximately

Vivek Agrahari et al. (eds.), *Nanotherapy for Brain Tumor Drug Delivery*, Neuromethods, vol. 163, https://doi.org/10.1007/978-1-0716-1052-7_6, © Springer Science+Business Media, LLC, part of Springer Nature 2021

one third (32%) of these brain and CNS tumors are malignant [2]. Brain tumors are a diverse group of lesions that range from benign, slow-growing tumors to malignant, rapidly growing tumors that cause death within months [3]. In 1873, Russian scientist Gupta Longati was the professional who identified the brain cancer for the first time [4]. Surgeon William Macewen in 1879 successfully removed meningioma for the first time [5]. There are more than 100 histologically distinct types of primary brain and central nervous system tumors; this makes the diagnosis and treatment of these tumors very difficult [2]. Each different type requires multidisciplinary team and unique treatment protocol [3]. Brain tumors are usually classified and named according to their presumed cell lineage and progenitor cells (e.g., astrocytoma, glioblastoma, etc.) [6] (Fig. 1).

The common types of the primary brain tumors are gliomas (30–40%). It includes astrocytomas, oligodendrogliomas, and ependymomas. After glioma the next common primary brain tumors are meningiomas (32–35%) followed by vestibular schwannomas (10%) and pituitary gland tumors (8%) [8]. Secondary metastatic brain tumors that occur due to other organ cancers like lung cancer, etc., are also very common. Most metastases occur due to lung cancer followed by breast cancer, melanoma, renal cancer, and colorectal cancer [9].

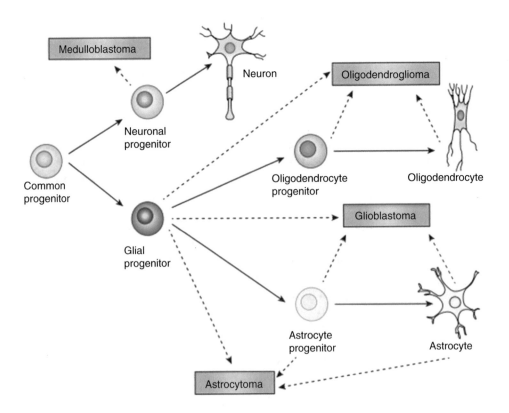

Fig. 1 The neuroglial lineage tree. (Source: Ref. 7)

1.2 Clinical Features

Brain tumors of any type may present with headache and slowly growing tumors present usually with convulsions. The first symptom noted usually is headache; it occurs because of high intracranial pressure due to vasogenic edema around the tumor or because of CSF blockage that leads into hydrocephalus [6]. These headaches may occur with or without nausea and vomiting, difficulty in understanding, personality changes, and gait disorders. Signs and symptoms of increased intracranial pressure, like severe headache, hypertension, bradycardia, vomiting, and irregular breathing, may also be found. In minority of the patients the classic headache of the brain tumor usually begins in the morning and improves during the day. Usually the headache is all over the head and sometimes on the same side of the location of the tumor [8]. There might be personality changes that look like depression; for example, the patient would not like being social or having apathy, etc. According to the site of tumor the focal or lateralizing signs, like hemiparesis, aphasia, and visual field defects, may occur. Focal or partial is also one of the common findings in the brain cancers; they are so mild sometimes that even the patient or the family may not know about the seizure. Depending on the location of the tumor, there may be loss of function in the body part or limb, which is called focal neurological deficit. Frontal lobe or temporal lobe lesions may present with dysphasia. Tumors that originate from pituitary gland may present with visual field defects, because of its proximity to optic chiasma. Patient may not know about these visual field defects; he may present with unusual consequences like having repeated driving accidents because of improper vision. Asymptomatic visual field defects may be accidentally found during the ophthalmological exam. Parinaud's syndrome that comprises of palsy of upward gaze, ptosis, and pupillary dilatation may also be present with hydrocephalus in the tumors that originate in the pineal region and posterior part of the third ventricle [6] (Fig. 2).

1.3 Diagnosis of the Brain Tumors

Mostly brain tumors are diagnosed using imaging techniques and biopsy of the resected tumor [3]. These techniques including CT and MRI provide noninvasive viewing of the structures. Using CT with contrast, diffusion-weighted MRI, and MR spectroscopy detailed information like functional, ultrastructural, and pathophysiological details may also be obtained. If the ring-enhancing lesion has been found by doing the CT scan with contrast, the possibilities may be glioblastoma, acute inflammatory demyelination, bacterial infections, toxoplasmosis, tuberculoma, and cysticercosis [11]. By using diffusion-weighted images of the brain MRI, we may differentiate a bacterial abscess and a brain tumor; usually the pus and hemorrhage show restricted diffusion while tumor cells show facilitated diffusion. Another technique that

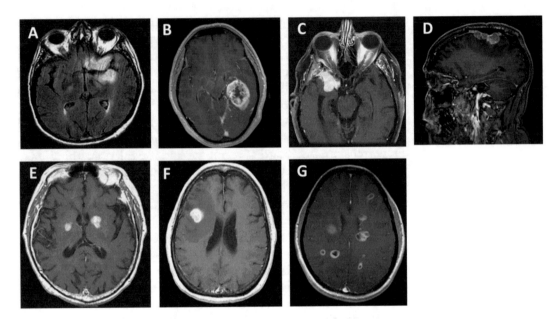

Fig. 2 (**a**) Low-grade glioma, (**b**) glioblastoma, (**c**) meningioma, (**d**) anaplastic meningioma, (**e**) central nervous system lymphoma, (**f**) single brain metastases, lung adenocarcinoma, and (**g**) multiple brain metastases, breast. (Source: Ref. 10)

can be useful is MR spectroscopy [12, 13]. It targets chemicals like choline, creatine *N*-acetyl aspartate, lipids, lactate, various amino acids, and myo-inositol in the area of the tumor. It may be used to target new metabolite that may be generated in the brain tissue due to mutation of genes like MR spectroscopic detection of 2-hydroxyglutarate (2-HG), an oncometabolite resulting from enzymatic activity of the mutated IDH gene product [6, 14].

1.4   Genetics of Brain Tumors

In 2008, the first time scientists noted change in isocitrate dehydrogenase genes (enzymes involved in citrate metabolism) in 12% of sequenced glioblastoma genome [15]. In further researches IDH mutation was also found in lower- and intermediate-grade and secondary glioblastoma [16]; out of them IDH1 (IDH1-R132H) is found to be the most common mutation [5]. In another study CIC gene mutation is also found to be one of the causes for oligodendroglioma [17]. In a study published in 2011 it shows characteristic losses in chromosome 1p and 19q in genetic analysis of 34 oligodendrogliomas. It revealed mutations in NOTCH1 in two tumors, PIK3CA in three tumors, CIC gene on the chromosome 19q in six cases, and mutation in FUBP1 gene on chromosome 1p in two cases; there were mutations in both genes in the remaining 27 oligodendrogliomas [18]. There is mutation in ATRX (alpha-thalassemia/mental retardation syndrome X-linked) found in 7% of glioblastomas out of total 363 glioblastomas [19].

1.5 Risk Factors

The cancer-associated risk factors are as follows:

1. Age: it can occur at any age, but it is found to be more common in children and older adults [20–22].

2. Gender: brain tumors are more common in men than women. Some specific tumors like meningioma are more common in women.

3. Family history: 5% of the tumors are because of genetic factors like Li-Fraumeni syndrome, neurofibromatosis, nevoid basal cell carcinoma syndrome, etc.

4. Exposure to infection: especially Epstein-Barr virus increases the risk of CNS lymphoma. Cytomegalovirus may also be the cause of the tumor.

5. Race and ethnicity: in some races the occurrence of some types of tumor is more common like in the USA; glioma is more common in white people than black, and similarly brain tumor is more common in people of Japan than the people of northern Europe.

6. Ionizing radiation: previous radiotherapy of the brain or head increases the risk of brain tumor [20, 21].

7. N-Nitroso compounds: some studies revealed N-nitroso compounds that are found in cured meats, cigarette smoke, and cosmetics may increase the risk of brain tumor [20].

1.6 Treatment Options for the Brain Tumor

Treatment options for the brain tumor range from conventional therapy that includes surgical removal, radiation therapy, and chemotherapy. In the initial stage and low-grade tumors surgery may provide complete cure [5]. The limitation of surgical removal is that it can be possible only if the tumor is localized and has not invaded the important structures of the brain. Radiotherapy using X-rays, gamma rays, and proton beams to kill the cancerous cells is also one of the treatment modalities. It involves two methods, delivering high amount of radiation as a single treatment that is called fractionated stereotactic radiosurgery and fractionated stereotactic radiotherapy in which a lower dose of radiation is delivered at each interval [23, 24]. Chemotherapy involves using the chemical agents to treat the brain tumor; these drugs usually act by disrupting the cell division or by stopping the blood supply to the carcinoma. The problem with this kind of treatment is along with cancerous cells these drugs may damage the normal cells as well, thus making the treatment more complicated [25]. Blood-brain barrier (BBB) is a barrier that prevents the harmful compounds to go inside the brain tissue and also prevents delivery of higher molecular weight and some small molecular weight therapeutic substances to enter into the tumor cells. The drugs that may cross the blood-brain barrier must have certain chemical properties; e.g., it must be lipid soluble,

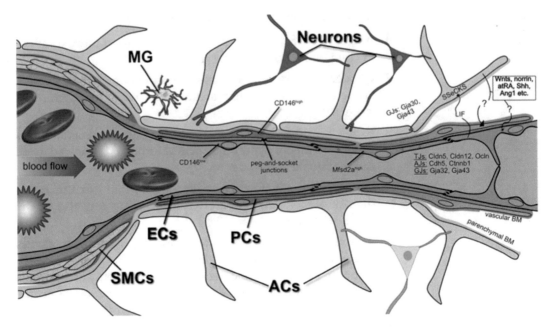

Fig. 3 Morphological and functional characteristics of the healthy NVU and BBB. A scheme of brain vessels ranging from an arteriole, via a capillary to a venule, representing the cellular and molecular composition of the NVU. (Source: Ref. 26)

have small molecular weight (500 Da), have weak bases, and must be electrically neutral molecule. Paul Ehrlich, Edwin Goldmann, and Lina Stern about a century ago formulated the concept of blood-brain barrier. In the late 1960s Reese and Karnovsky and Brightman and Reese identified the endothelial cells as the site of BBB. There are various components of neurovascular unit which are comprised of endothelial cells, pericytes, and astrocytic end-feet (Fig. 3) [26].

There is associated vasogenic edema with brain tumor; this is because of increased vascular permeability due to disruption in the blood-brain barrier; the endothelial cells in the blood-brain barrier make excellent barrier for water and water-soluble molecules [27]. This barrier as illustrated in Fig. 4 makes tight junction water-soluble compounds and toxic material inhibition across this junction. Along with it the progress of the glioma is highly affected by available circulation and blood vessel [28]. The major objective of the BBB is to guard neurons from systemically circulating potentially cytotoxic agents. The brain capillaries are different from capillaries in other organs as brain capillary has a very tight barrier (Fig. 4). The capillary endothelial cells constitute BBB which is surrounded by astrocytic perivascular end-feet and a basal membrane. In the lower part of the figure, enhanced illustration of a brain capillary endothelial cell illustrates the current view of the localization of drug efflux transporters at brain capillary endothelial

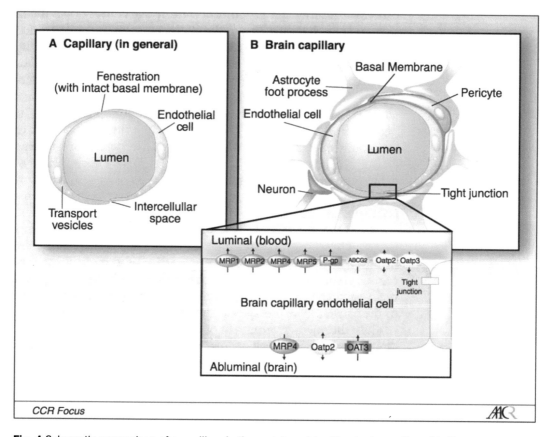

Fig. 4 Schematic comparison of a capillary in the periphery (**a**) with a brain capillary (**b**). (Source: Ref. 30)

cells that form the BBB. Because of this barrier it becomes very difficult for the drug to reach tumor cells. To overcome this problem the researchers are trying to either deliver the drug intracerebrally through arteries or by delivering the drug into a carrier that can cross the BBB [29].

1.7 Scope of Further Research in Brain Tumor Treatment

There has been continuous extensive research for the treatment of brain tumor. From chemotherapy, surgical removal and radiotherapy to new treatment modalities like use of electric field treatment using NovoTTF devices, immunotherapy, genetics, and nanoparticles have also been well established [5]. There is huge scope for the research for the activity of nanoparticles in the treatment of brain tumor. In this chapter we will include those nanoparticles, which have been proved to be effective in the treatment of brain tumor.

1.8 Known Hurdles in the Treatment of Brain Tumors

As already explained BBB is a physiological structure which restricts the entry of chemicals from the blood to the brain tumor cells or leads into decrease in amount of drug delivery into it. This is the reason that mostly chemical compounds which are considered anti-cancer efficiency in other organs have limited effect in the brain.

Moreover healthy cell may be affected by the conventional thera-pies, such as chemotherapy and radiotherapy. However researches are carrying to create an environment where drugs are getting hold of interaction by virtue of destruction of the tumor cells. Nano-technology enables the therapeutic assistances as it works at molec-ular level where engineered nanoparticles or functional nanoparticles accomplish the aims. Nanotechnology offers unique high surface area, excellent interacting surface, and confinement of electron within the system which are of great interest for the drug delivery. Nanotechnology is the only promising technology to deliver the drugs across the BBB. It has applications in neurosur-gery, and benefits in the treatments of glioblastoma multiforme (GBM), and other specifically therapeutics.

Glioblastoma (GBM) is a lethal human cancer which recurs followed by surgery; chemo and radiation therapy cause the demise of the patients. Gliomas are very aggressive in nature and surgery cannot eradicate the deeply infiltrated tumor cells in the adjacent brain parenchyma, and the residual glioma cells develop resistance to the conventional therapies such as chemo and radiation therapies [31]. In the microenvironment of GBM therefore evidence is found of regrowth of cancer cell because of production of cytokines which takes part in the growth of cancer cell and therefore therapeutic resistance occurs [32–34]. Obviously, innovative therapeutic approaches are desirable to inhibit tumor management resistance and regrowth. The engineering application of nanotechnology to the medicine has produced nanomedicine which enables to target the specific portion in brain tumor. The nanoparticles may accumu-late within the tumor and with imaging bulk modalities such as magnetic resonance imaging (MRI), ultrasound, and whole body fluorescence, which may target the specific cell and kill the cancer cell.

2   Potential and Advantages of Nanoformulations in Brain Tumor Therapy

2.1   Nanoparticle-Based System: A Promising Nanomedicine Approach for Brain Tumor

In recent years several cancer treatment techniques are accessible including surgery and radiotherapy followed by chemotherapy, photodynamic therapy, hyperthermia, and drug delivery. Before we discuss about the nanomedicine approach for the brain tumor treatment, we briefly elucidate the conventional chemotherapy and other treatments processes. Chemotherapy, in simple words, is a process of supply of specified drugs/chemical compound within the tumor after the radiation and surgery of the tumor. Nowadays advanced chemotherapy is in process with the specific targeted drugs to the cancer. However, the current technology of cancer therapy is limited due to the drug resistance and high degree of cancer clonal heterogeneity [35]. Hence more advanced treatment technique which can overcome such of these issues is the necessity with the use of combinatorial therapy which can end multiple targets and specified the accurate pathways. Therefore we need to discuss some of the advance currently used therapies which uses the certain new techniques with certain nanotechnologies. Firstly, we talk about the photodynamic therapy (PDT) which involves the process of interaction of photosensitizer with the light which produces the reactive oxygen species which intermingle with the surrounding of molecular oxygen and can kill the cancer cells [36]. The PDT technique is an effective treatment technique with less side effect and noninvasive as compared to conventional treatment techniques including chemotherapy. Certain limitation of the PDT is the systematic allocation of photosensitizer which leads to some unwanted side effects [37]. Another technique for cancer treatment is hyperthermia which leads to the apoptotic cell death caused by heat induced in cancer cells. Hyperthermia can also be used together with radiotherapy as well as with chemotherapy to increase the efficacy of the treatment [37]. When hyperthermia is generated with light irradiation then the therapy is also known as photothermal therapy. A light energy source such as laser is applied to the surface of metallic nanoparticles; the heat is produced as a result of conversion of nonionizing electromagnetic radiation energy. The temperature increases in the surrounding tissues which can lead to the cancer cell death. Hyperthermia technique is also relatively noninvasive with less side effect and easy to operate. Furthermore, an effective way of treatment is the delivery of certain drug molecules to the specific sites. The distribution of therapeutic agents into cancer cell and in microenvironment can be controlled by entrapping them in submicron colloidal systems which results in antitumor efficacy with little side effects. Though, there are certain limitations of the above-discussed cancer treatment techniques. Therefore introduction of nanoparticles into some of the above mentioned techniques is part of the treatment of brain tumor. With the application of engineered nanoparticles, a nanomedicinal approach can be found useful and may enhance curing capacity of drug delivery and other cancer treatment techniques. An emerging technique called theranostic nanomedicine or theranostic therapy is a combination of a therapy as well as diagnosis technique for cancer treatment. This technology has changed the way of treatment techniques by the development of molecular level specified targeted drugs. The clinical molecular imaging can provide the useful information of drug response, diagnosis, and prognoses of the disease. This approach has attracted attention of researchers as well as clinicians a lot as it provides the amazing platform of nano drug delivery system and imaging agent in vivo and in vitro process [38, 39].

2.2 Therapeutic Inorganic Nanostructures

Nanotechnology has provided an amazing platform for the scientists and researchers from diverse areas of science and technology to work together. Presently, almost no field remains untouched by nanotechnology where it has not reached either fully or partially. Nanotechnology is promising a number of applications in various fields of research and innovations because of unique, versatile, and remarkable properties associated with nanostructures. Inorganic nanostructures are essential and likely candidates which play a vital role in the numerous areas of science and technology comprising pharmaceutics, biomedicine, bioimaging, sensing and biosensing, advanced materials science, environmental sciences, physics, chemistry, agricultural sciences, electronics, information technology, catalysis, energy, and others [40]. Inorganic nanostructures have been extensively considered as therapeutic agents for cancer treatments in biomedical fields due to unique physical/chemical properties, small size, large surface/volume ratio, excellent biocompatibility, informal surface functionalization, and versatile preparation methods of nanostructures; they are extensively investigated and recognized as therapeutic agent for brain tumor treatment in the field of biomedical science. The small size of nanoparticles ranging from 1 nm to 100 nm is accessible through the body and compatible with the size of other structures such as cells (10–100 µm), viruses (20–450 nm), proteins (5–50 nm), and genes (2 nm wide by 10–100 nm long). The crucial nanostructures which are reliable and promising as therapeutic agent for brain tumor include metal nanoparticles (Fe, Cu, Ag, Au, Cr, Ni, Ti, Zn, etc.), metal oxide nanoparticles (Fe_3O_4, γ-Fe_2O_3, α-Fe_2O_3, CuO, ZnO, TiO_2, CeO_2 SiO_2, etc.), nano thin films, nanowires, nanorods, quantum dots, carbon nanotubes (CNT) (single wall and double wall nanotubes and multiwall nanotubes), graphene (single layered, double layered), graphene oxide, and core-shell nanoparticles (where core nanoparticles are encapsulated with inorganic shell such as carbon or silica and organic shell such as polymers, etc.). Inorganic nanoparticles can be of the same size of intracellular organelles depending on the size of the nanoparticles. In this way drug delivery through nanoparticles into the specific organelles can be possible and intracellular information can be extracted [41, 42]. Such information might be supportive for the recognition, diagnosis, and therapeutics of diseases [43].

3 Methodologies for Nanomaterials Synthesis and Characterization

3.1 Synthesis of Inorganic Nanostructures

The nanoparticles which are in the range of 1–100 nm of size can be synthesized in mainly two ways: one is "top-down" technology and the second is "bottom-up" approach. The former case is to cut down the bulk size materials into nanosize while the latter is to use small particles to make up nanoscale particles [44]. Only the fundamental similarities in both approach are to control the size of nanoparticles and the use of energies to synthesize the nanoparticles. The top-down technologies are derived from the mostly semiconductor industry used for the preparation of the chip which is generally known as the lithography. In the lithography an electron beam of light is used to remove selectively microstructures from precursor materials to obtain desired nanostructures. Lithography includes photolithography, scanning lithography, e-beam lithography, soft lithography, nano imprint lithography, nanosphere lithography, colloidal lithography, and scanning probe lithography [45–51]. The other top-down technology includes ball milling and sputtering techniques. The second method is the bottom-up approach where nanomaterials are prepared starting from atoms or molecules. In the bottom-up approach there are two main ways to produce nanoparticles: one is gas phase which includes plasma arcing, microwave-assisted plasma chemical vapor deposition (MWPCVD), and chemical vapor deposition (CVD), and the second way is liquid phase which includes coprecipitation method, sol-gel synthesis, hydrothermal method, microemulsion method, and reduction method. A short description of these synthesis techniques are described below.

| 3.1.1 Lithography | A conventional lithography is a process of transferring an image from a mask to a substrate. In this process a coating substrate, e.g., a glass or silicon wafer, is coated with a polymer layer which is called resist. This resist is exposed to the light, electrons, or beam of ions or X-rays. With the help of a developer which is a chemical, an image on a resist is developed which may be positive or negative depending on the types of resist used. Further the image is transferred from resist to the underlying substrate using chemical etching or dry plasma etching [45–51]. Another trend of miniaturization of products and consequently its components is done by micromachining. Micromachining is satisfactorily manufacturing techniques in which material is removed by geometrically determined cutting edges and referred to as mechanical micro cutting techniques [52–55]. All these methods have some advantages and disadvantages; they are mentioned below in the end of Subheading 3.1 in Table 1. |

3.1.1 Lithography

A conventional lithography is a process of transferring an image from a mask to a substrate. In this process a coating substrate, e.g., a glass or silicon wafer, is coated with a polymer layer which is called resist. This resist is exposed to the light, electrons, or beam of ions or X-rays. With the help of a developer which is a chemical, an image on a resist is developed which may be positive or negative depending on the types of resist used. Further the image is transferred from resist to the underlying substrate using chemical etching or dry plasma etching [45–51]. Another trend of miniaturization of products and consequently its components is done by micromachining. Micromachining is satisfactorily manufacturing techniques in which material is removed by geometrically determined cutting edges and referred to as mechanical micro cutting techniques [52–55]. All these methods have some advantages and disadvantages; they are mentioned below in the end of Subheading 3.1 in Table 1.

3.1.2 Ball Mill Method

Ball milling is the technique to reduce the size of the particles by the impact of the balls as they drop from near the top of the shell. This traditional technique is highly environmental friendly and an inexpensive technique for micro and nano size particle synthesis. It is a cost-effective and straight forward technique which can produce various types of nanostructures. A number of factors are important in ball milling such as frequency, number and size of milling balls, milling time, and the material used for milling balls [56–60].

Table 1
Various synthesis methods of nanoparticles with advantages and limitations of the respective methods

Synthesis methods	Platform	Advantages	Limitations	References
1. Top-down technology				
(i) Lithography				
(a) Photolithography	Silicon wafers, polymers, etc.	Micro and nanofabrication	Complex and expensive, clean room preparation, big tools required	[45]
(b) Electron beam lithography	Silicon wafers, polymers, etc.	Great resolutions, generate complex structures	Expensive and slow	[46, 47]
(c) X-ray lithograph	Silicon wafers, polymers, etc.	Ultrahigh resolution, high aspect ratio structures	Can damage the materials	[47–51]
(d) Micromachining	Any materials including plastics, polymers, silicon wafers, etc.	Small thickness, cost effective	Cleaning is critical at the end of the process	[52–55]
(ii) Ball milling	Metal, metal oxides, composites, etc.	Simple and safe, wide application, can mill highly abrasive materials	Limited size, contamination due to tear and wear, high noise	[56–60]
2. Bottom-up approach				
(a) Chemical vapor deposition	Metal, nanoparticles, metal oxides nanoparticles, carbon nanotubes, etc.	Films and substrates are free from plasma damage, films can be deposited on any sharp edge substances, large deposition area	Lifetime of metal catalyst	[61–64]
(b) Coprecipitation method	Metal, nanoparticles, metal oxide nanoparticles	Simple and rapid preparation, easy control of particle size and composition, modify the particle surface state and overall homogeneity, low temperature, energy efficient, does not involve use of organic solvent	Not applicable to uncharged species; trace impurities may also get precipitated with the product; batch-to-batch reproducibility problems	[93–95]

(continued)

Table 1
(continued)

Synthesis methods	Platform	Advantages	Limitations	References
(c) Sol-gel method	Metal oxides, ceramic and glass materials	Control the shape, morphology, and textual properties, control the size of particles and thickness of the film, high surface area, cost-effective synthesis, possibility of obtaining metastable materials, achieving superior purity and compositional homogeneity, moderate temperature synthesis	Limited precursor	[96–98]
(d) Hydrothermal synthesis	Metal nanoparticles, metal oxide nanoparticles	Control grain size, particle morphology, crystalline phase and surface chemistry, low cost	Expensive autoclave, high pressure required	[83–86, 99]
(e) Microemulsion method	Metal nanoparticles, metal oxide nanoparticles, polymer nanoparticles	Environmentally friendly, uniform distribution of size and shape, controlled crystallinity, reproducibility	Require large amount of surfactant, very sensitive to change	[87–90]
(f) Reduction method	Metal nanoparticles, metal oxide nanoparticles	Simple, low cost, dispersibility and stability	Agglomeration of nanoparticles, hazardous	[91, 92]

3.1.3 Chemical Vapor Deposition (CVD)

This technique provides the thin films of metals, metal oxides, and other mixed components. The crystalline, polycrystalline, and amorphous films can be produced using this technique. The compound deposited on a substrate is heated up and vapor of its component is deposited over a substrate. The heating may produce plasma arcing by laser and other heating techniques. One of the important techniques is microwave plasma-assisted CVD (MWPCVD). Certain power of microwave is set and therefore a temperature is generated which enables the component which has to be deposited into its vapor states and these compounds have deposited resulting into the thin film of metal or metal oxides. In the literatures, the thin film of metal oxides has deposited using

MWPCVD from its pure metallic thin film originally deposited by sputtered techniques (radio frequency (RF) sputtering, DC magnetron sputtering) [61–64].

3.1.4 Coprecipitation Method

In the chemical coprecipitation method aqueous salt of metal reacts with alkali solution resulting into a precipitate. This precipitate is kept at different pH and other conditions to control the shape and size of nanoparticles. It is a very simple and rapid technique for the synthesis of metal oxide nanoparticles. This method has some advantages and disadvantages as well and they are mentioned in Table 1. Generally metal oxide such as iron oxide, copper oxide, and other oxides are prepared by this method. The pH plays a very important role in the synthesis of nanoparticles and controls the shape and size of the particles [65–74].

3.1.5 Sol-Gel

Sol-gel synthesis technique is for inorganic and organic nanoparticles with extremely controlled nanostructures and tailored surface chemistry. A metal alkoxide or metal chloride is used as a precursor and an organic liquid such as tetraethyl orthosilicate (TEOS) (in case of silica nanoparticles) is used for hydrolysis. A sol is formed after hydrolysis due to the oligomerization through the condensation step. A gel is formed after the cross-linking of sol which results into a network [75–82].

3.1.6 Hydrothermal Synthesis

This technique involved a low-temperature and pressure process in which an autoclave is used. This technique is an ecofriendly technique which is carried out in a closed chamber where water is the reaction medium [65, 83]. Hydrothermal technique is generally used to control the size, shape, phase, and crystallinity of the particles by varying the conditions of hydrothermal. The surface modification or surface coating of inorganic particles such as carbon easily takes place by this technique [65]. This method provides smaller size, greater stability, and high specific area as compared to the others [84–86].

3.1.7 Microemulsion Method

This is a simple process in which water in oil or oil in water microemulsion is prepared using oil and aqueous solution of metal salts. Both the phases are stabilized by a surfactant (organic or inorganic). In the stabilized solution, an aqueous solution of alkali is added which results into the formation of precipitates. This precipitates are then washed by water and calcined in oven. This environmentally friendly technique provides uniform distribution of size and shape and controlled the crystallinity [87–90].

3.1.8 Reduction Method

Silver nitrate is reduced by ethylene glycol in the presence of poly (vinyl pyrrolidone) (PVP) resulting into the formation of monodispersed silver nanocubes in large scales. The shape and size of the nanoparticles can be controlled by various parameters. The properties of the nanoparticles are so important because they have strong correlation such as parameters [91, 92].

All of these methods have some advantages as well disadvantages. A summary of the abovementioned methods has been tabularized in Table 1. In Table 1, the synthesis methods have been provided with references, the name of materials synthesized or the techniques that work on the materials, and their advantages and disadvantages.

3.2 Characterization of Inorganic Nanostructures

Nanoparticles are characterized by various techniques. These techniques are essentials for the characterization of nanoparticles to analyze the surface, crystallinity, shape, size and the components of materials, etc., to be aware of their properties and applications. The techniques used to characterize the nanoparticles are X-ray diffraction (XRD), UV-visible spectroscopy, Fourier transform infrared spectroscopy (FT-IR), scanning electron microscopy (SEM), transmission electron microscopy (TEM), atomic force microscopy (AFM), Raman scattering spectroscopy, and Brunauer-Emmett-Teller (BET) surface area analysis. A short description of few important characterization techniques for materials and nanomaterials is described here. For example, XRD is the characterization technique which provides the information of crystal size, crystalline properties, and phase of the nanoparticles [65, 92–98]. The crystal size of the nanoparticles can be calculated by Scherrer formula in which the required value can be put which is obtained from the XRD of the samples (Fig. 5d). The other most important technique for the materials especially liquid materials is the UV-visible spectroscopy. UV-visible spectroscopy is also known as the absorption/reflectance spectroscopy where UV and visible range of light is passed through the samples. In this process light is absorbed or reflected where atoms or molecules get excited and provide the necessary information of the samples [65, 73, 74]. Generally dilute samples are characterized by this characterization technique. FTIR technique can be used for solid liquid and gas samples. This is a technique where spectrum is obtained for absorption or emission. Organic compounds, natural products, polymers, etc., can be analyzed by FTIR spectroscopy. The FTIR spectroscopy has significance over the conventional absorption/reflectance spectroscopy which measures the intensity over a narrow range of wavelengths at a time [65–74]. The morphology and surface characterization of the nanoparticles can be analyzed by scanning electron microscope. SEM also gives the texture topology and morphology of the nanoparticles [65, 73, 74]. SEM is for the study of the structural topography and compositions of metals, ceramics, polymers, composites, and biological material samples (Fig. 5a, b) [92]. TEM is one of the most important characterization techniques which are employed to get the information of the

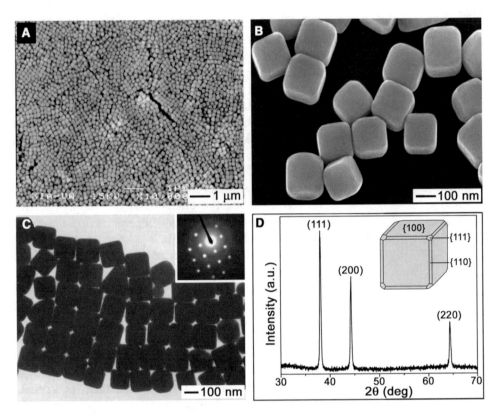

Fig. 5 (**a**) Low-magnification SEM images of somewhat reduced silver nanocubes synthesized with the polyol process and (**b**) high-magnification SEM images, taken at a tilting angle of 20°. (**c**) A TEM image of the same batch of silver nanocubes. The inset shows the diffraction pattern recorded by aligning the electron beam perpendicular to one of the square faces of an individual cube. (**d**) An XRD pattern of the same batch of sample, confirming the formation of pure fcc silver. a.u., arbitrary units. (Source: Ref. 92)

size and shape of the nanoparticles [73]. To study the local structures, morphology, and dispersion of multicomponent polymers, cross sections and crystallizations of metallic alloys, semiconductors, microstructure of composites, TEM is useful characterization technique (Fig. 5c).

The atomic force microscopy is a type of scanning probe microscopy (SPM) which provides the resolution of fractions of nanometer. It is a very high-resolution technique for the characterization of nanoparticles. The information can be collected by touching or feeling the substrate with a mechanical probe. The surface topography and mechanical properties such as stiffness, Young modulus, etc., can be measured by the atomic force microscopy. Furthermore AFM provides the information of atoms on the surface, information of the neighboring atoms, changes in the interaction between specific atoms, and changes in physical properties. Moreover the use of AFM can be used in molecular biology and cellular biology to image the surface and mechanical properties of protein complexes and assemblies. The Raman spectroscopy is useful for the nanoparticles especially inorganic nanoparticles which provide structural information by which atoms or molecules can be identified. Raman spectroscopy is mainly used in chemistry to identify the molecules and chemical bonding. Raman spectroscopy is used to observe vibrational, rotational, and other low-frequency modes in systems. The chemical bonding and symmetry in each molecule have its own vibrational frequency; therefore Raman spectroscopy provided the information about the identity of the molecules [62]. Last but not least, surface area analysis is characterized by BET model which explains the concept of calculation of specific surface area and pore size. This model requires the solid samples probed by gaseous molecules like N_2, argon, and other inert gases to measure the specific surface area using the gaseous adsorbate molecules in a solid sample [84].

4   Biomedical Application of Inorganic Nanostructures

4.1   Application of Nanoparticles as Nanocarrier in Drug Delivery

Carbon nanotube is the most versatile nanostructure which is widely used in the survey fields of science and technology because of its excellent physical and chemical properties. Carbon nanotube can cause oxidative stress which results into apoptosis of cancer cells. In a mouse ear circulating cancer cells could be detected magnetically by a gold-plated carbon nanotube which was conjugated with folic acid [100].

Among the rare earth metals, cerium oxide nanoparticles show the great affinity toward the cells and can protect pneumonitis which results after radiation therapy [101]. It can also act as a radical scavenger and protect healthy tissues which may get damage by the effect of radiation therapy [102]. Hence cerium oxide can be used to protect healthy tissues and regeneration of tissues from the effect of radiation therapy. Gold nanoparticles possess most of the biological applications as a passive agent. It can act as a vehicle of drug delivery and a probe for electron microscopy. Although in recent times Bhattacharya et al. verified that gold nanoparticles can be used as active agents to interfere directly with the cellular processes and possess antiangiogenic and antitumor properties [103, 104]. Gold nanoparticles are useful for the delivery of cytotoxic drugs with effective specific target. It can minimize toxicity of the drugs and minimize the risk of side effects. In an orthotopic model of pancreatic cancer, gold nanoparticles act as a drug delivery vehicle which results in the prevention of tumor [105–107]. The other application of gold nanoparticles is folate targeting which is a method used for drug delivery purpose. In various cancers such as breast, ovary, brain, and kidney, folate receptors (FRs) are known to overexpress in cancer cells [108, 109]. FRs become accessible for targeted drugs in blood circulation upon epithelial cell transformation and the cell polarity is lost [110]. FR expression is not destructive and necessary for healthy tissues as well [111]. In recent times, gold nanoparticles and polyethylene glycol (PEG) were synthesized to target the FR on a variety of cancer cells. The therapeutics of PEGylated molecules was accommodating to accomplish rapid renal clearance of nontargeted conjugates in a way to minimize the toxic effects toward healthy cells. The non-immunogenicity, minimal degradation, and stabilization during blood circulation is another advantage of conjugate PEGylated molecules of gold nanoparticles. One of the most versatile nanoparticles is the iron oxide nanoparticles or super paramagnetic iron oxide nanoparticles (SPION). SPION is considered MRI agent recently because of their unique magnetic property and controlled size. There is continuous progress in the development of inorganic nanoparticles in the biomedicine and other applications such as contrast enhancement agents for imaging, drug delivery vehicles, and, most recently, as a therapeutic component in initiating tumor cell death in magnetic and photonic ablation therapies [112]. Magnetic nanoparticles are gaining interest because of their tremendous properties such as sensitivity, biodegradable nature, and metabolic pathways of cellular iron [113]. Magnetic nanoparticles coated by dextran are also gaining interest as MRI contrast agent for clinical importance. The ferumoxides and ferumoxtran are approved by the Food and Drug Administration (FDA) in the application of detection of the solid tumors observed [114, 115]. The application of magnetic nanoparticles in vivo and vitro has also been used in detection of lymph node metastases in solid tumors. In the detection of lymph node metastases in solid tumors, imaging angiogenesis, neurodegeneration, macrophage-specific inflammatory pathologies, rheumatoid arthritis, and artheroscleroses, use of magnetic nanoparticles has been carried out by the researchers in vitro as well as in vivo [114–117]. Figure 6 illustrates the methods of drug delivery to brain cancer as intracerebral implants, intracerebroventricular infusion, and convection-enhanced diffusion (convection-enhanced diffusion in the brain of primate leads to astrogliosis).

4.2 Role of Nanoparticles in Brain Tumor Management

4.2.1 Therapeutic Cancer Vaccine

Nanoparticles are considering a carrier of antigen delivery in cancer vaccines. The biocompatibility, antigen protection from degeneration, and controlled antigen release make nanoparticles a promising candidate. Several types of nanoparticles (Fig. 7) like liposomes, polymeric nanoparticle micelles, carbon nanotubes, gold nanoparticles (AuNPs), mesoporous silica nanoparticles (MSNs), and virus nanoparticles can be used as delivery systems [119].

Recently a cancer therapeutic vaccine has attracted attention because this unique technique of killing cancer cells is due to the T-cells which are developed on the immune system. These T-cells can recognize and kill cancer cells, although it is difficult to produce large number of T-cells which can recognize the antigens on cancer

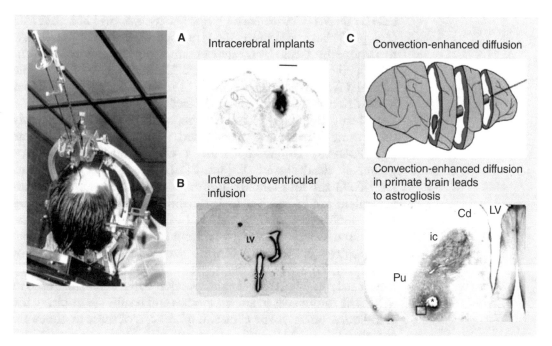

Fig. 6 Invasive methods of drug delivery for brain cancer therapy. Examples of invasive approaches to deliver drugs to the brain. (**a**) Intracerebral drug implants. (**b**) Intracerebroventricular infusion. (**c**) Convection-enhanced diffusion. The asterisk demarcates the hole left in the brain by the catheter, the insertion of which is shown in the top image of (**c**). *Cd* caudate, *ic* internal capsule, *LV* left ventricle, *Pu* putamen. (Source: Ref. 118)

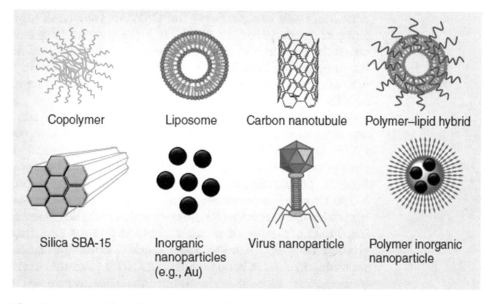

Fig. 7 Structural presentation of various nanoparticles for cancer vaccine carrier. (Source: Ref. 120)

cells by means of customary vaccine carrier systems [121, 122]. Li et al. [123] have used the α-Al$_2$O$_3$ nanoparticles for antigen carriers to activate the T-cells by means of minimizing the required amount of antigen carriers in vivo and in vitro. They have observed the role of α-Al$_2$O$_3$ nanoparticle in delivery of antigens to autophagosomes in dendritic cells which represent T-cells through autophagy.

Chiang C. S. et al. [124] have reported recently an inherently therapeutic fucoidan-dextran-based magnetic nanomedicine (IO@FuDex3) conjugated with a checkpoint inhibitor, programmed death-ligand 1 (anti-PD-L1), and T-cell activators (anti-CD3 and anti-CD28). Immunosuppressive tumor microenvironment can be repaired by this nanomedicine by reinvigorating tumor-infiltrating lymphocytes with the help of magnetic navigation to the localized tumor by limiting the off-target effects. Some nanoparticles coated with ligands, such as Herceptin, folate, or transferrin, improves the delivery and uptake of nanoparticles by tumor cells [125]. In recent study peptide-based ligands are used to target the tumor cells in animal models and results are inspiring the techniques; however the clinical implications of these methods are not clear [126, 127].

4.2.2 Imaging and Sensing Probes

Magnetic nanoparticles (MNPS) are of great interest in imaging and sensing probes because of their great magnetic properties. Functionalized magnetic nanoparticles such as dextran-coated MNPs are also gaining researcher's attention as MRI contrast agents [112]. There are two types of systems available as contrast imaging agents which contain magnetic nanoparticles coated with polymers or other nanoparticles known as SPION (superparamagnetic iron oxide nanoparticles) and USPION (ultrasmall superparamagnetic nanoparticles) or LCDIO (long circulating dextran-coated iron oxide) depending on the size of the nanoparticles. In the former system size of nanoparticles is 50 nm with coatings and in the latter, size of the whole nanosystems is less than 50 nm. The SPION which are available in the market are Lumirem® which contains silicon-coated iron oxide nanoparticles (300 nm size) and Endorem® containing dextran-coated iron oxide nanoparticles (150 nm size) [128]. The imaging function of Endorem® is the absorption of nanoparticles by Kupffer cells increased the contrast between healthy tissues and the tissues of tumor or metastases. USPION which are available in the market are the dextran-coated iron oxide nanoparticles (30 nm size) which are called Sinerem®. It helps in the imaging of tumor and blood pool by detecting and characterizing the lesions because of their vascular appearance. In a rat having malignant brain neoplasm, LCDIO were used in vivo for investigation. Human brain tumor was also investigated using Sinerem® [129]. Yamaguchi H. et al. [130] designed an imaging probes using silica nanoparticles coated by polyamidoamine (PAMAM) to target human epidermal growth factor receptor

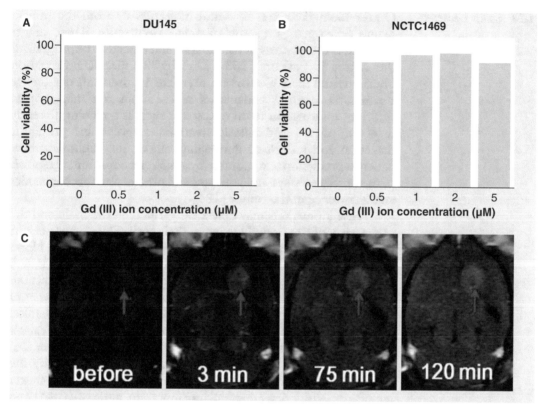

Fig. 8 In vitro cytotoxicity test for (**a**) DU145 and (**b**) NCTC1469 cell lines. (**c**) In vivo T1-weighted images of a rat brain with its tumor marked with arrows before and after injection of the ultrasmall gadolinium oxide nanoparticles. (Source: Ref. 131)

2 (HER2)-overexpressing breast cancer cells and achieve efficient target imaging of HER2-expressing tumors. Park J. Y. et al. [131] have demonstrated cytotoxicity test for human prostate cancer cell line (DU145) (Fig. 8a) and the mouse normal hepatocyte cell line (NCTC1469) (Fig. 8b) using ultrasmall gadolinium oxide (Gd III) nanoparticles coated with D-glucuronic acid in vitro. The nanoparticles show the nontoxicity to the applied concentration; however a little bit toxicity increased upon increasing the concentration. They have also performed in vivo T1 MRI images of a rat developing brain tumor. In Fig. 8c the T1 MRI images of brain tumor in a rat show the high contrast enhancement with time; however after 2 h contrast decreases slightly because of excretion of sample nanosystems.

In multiscale medical imaging, these nanoparticles have huge prospective. At multiple length scales and spatial resolutions these probes improve disease characterization by combining complimentary imaging capabilities. Such medical imaging capabilities not only facilitate disease study in laboratory but in clinical trials and assist in the planning of treatment at various stages [132].

4.2.3 Cancer Biomarker   Cancer biomarkers are biological molecules or from the human tissues developed as a result of cancer occurrence. These cancer biomarkers can be measured accurately and detected the progress of the cancer in patients [133]. Due to the specific information obtained from these biomarkers, there are requirements of effective treatment for cancer patients. A cancer biomarker may provide sufficient information to an oncologist such as screening for early detection of cancer, diagnosis, treatment response and prognosis detection and its risk of developing cancer, its diagnosis cancer patients, prediction and monitor treatment response and prognosis of the outcome of systemic therapy, and predict pharmacodynamics and recurrence of the tumor [134].

These biomarkers may detect early cancer in the brain and the other parts of the body because of their availability in the various parts of the body. The protein biomarkers which are found in blood serum can be useful in the early detection of the cancer. To detect the biomarker protein a sensor must be capable to detect an extremely lower range of conc. (say, one million times lower) of the protein as compared to other protein. A combined mechanical and optoplasmonic transduction has been used to detect cancer biomarkers in serum by using gold nanoparticles which is attached to an antibody which is present in the solution to identify the captured biomarker (Fig. 9). The colon cancer and prostate cancer are diagnosed by detecting carcinoembryonic antigen (CEA) and the prostate-specific antigen (PSA), respectively, in blood serum.

Gold nanoparticles act as a mass and plasmonic marker and these two signatures can be acknowledged by silicon cantilever which serves as a mechanical resonator to weigh the mass of captured nanoparticles and as an optical cavity which improve the plasmonic gesture from the nanoparticles [135]. Figure 9 illustrates the plasmonic detection of CEA protein biomarker on the cantilever optical microactivity. A detailed description of the working of the cantilever is given by Kosaka P. M. et al. [135]. Briefly, in Fig. 9 (a) the dark field optical image of cantilever in a control experiment and CEA detection assay (1 pg/ml in serum) is shown. Recognition of antibody by nanoparticles is also possible, a scattering signal is recorded which if found negligible in control experiment and in detection of CEA, a little scattering signal is usually observed in preclamping region while in cantilever region a bright signal appears. Figure 9(b) represents the marked area in Fig. 9(a) which shows the high-resolution dark field image in preclamping and cantilever region. The cantilever region shows the higher density of nanoparticles as compared to the preclamping region. In Fig. 9 (c) schematic pathways of generated dark field signal in the cantilever region via multiple internal reflection is shown. Figure 9(d) represents the scattering spectra in the preclamping and cantilever region which has the mean value of 40 µm in diameter of an optical signal.

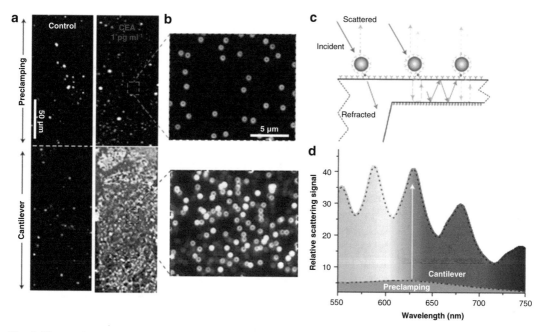

Fig. 9 Plasmonic detection of the CEA protein biomarker on the cantilever optical microcavity. (**a**) Dark-field optical images of the cantilever and preclamping region. (**b**) High-resolution dark-field images of the preclamping and cantilever regions marked in (**a**). (**c**) Schematics that illustrate the different pathways for the generation of the dark-field signal in the cantilever via multiple internal reflections. (**d**) Scattering spectra for CEA detection sandwich assay in the preclamping and cantilever regions. (Source: Ref. 135)

For cancer biomarker detection, nanoparticles are also attached to the peptides just like antibody and antigens. Due to their small size, low immunogenicity, and as being low-cost produced materials, peptides are attractive targeting molecules [136]. A tripeptide sequences containing arginylglycylaspartic acid are called RGD (arginine/glycine/asparagine) peptides broadly inspected for the detection of tumor lesions. Due to effective binding and recognizing properties of RGD peptides to cancerous angiogenesis and metastases surface marker integrin $\alpha\nu\beta3$, they are extensively used in the detection of cancerous cells and tumor lesions.

Aptamers are also used in the directing of cancer biomarkers, especially nucleic acid aptamers. These small nucleic acid ligands are low immunogenicity molecules containing 15–40 bases. Aptamers can be bonded to targets with high specificity because of their ability to fold into unique three-dimensional structures. Aptamers are easy to prepare on commercial scales without variation in any batch. The most common use of aptamers is against a prostate cancer cell biomarker, prostate-specific membrane antigen (PSMA) [137]. Another major breakthrough of detection of cancer biomarker is the sensing technology using conventional techniques such as mass spectrometry (MS), optical detection, and electrical and electrochemical detection.

4.2.4 *Drug Delivery Vehicles*

Nanoparticles play vital roles in drug delivery because of their selective specific targets. Nanoparticles with the drug molecules can improve the target delivery to get better biodistribution, increase circulation half-life, and protect drugs from the microenvironment. Nanoparticles with drug molecules hence increase the efficacy and minimize the side effects. Nanoparticles with drug molecule delivery to the specific target in the tumor have overcome the traditional chemotherapies in which the healthy cell also gets damage because of circulation of drugs into the whole body. Although organic nanoparticles like liposomes, micelles, and dendrimer-based drug delivery have been studied and found in most of the literature [106, 138–142], in recent studies the role of inorganic nanoparticles as nanocarriers has been approved by FDA for clinical trials. The coating of some organic molecules over the inorganic nanoparticles has solved the problem of drug binding and it became a hot spot in the research of nanomedicine because of its unique physical properties which offer extraordinary imaging and controlled capabilities. The three important areas in drug delivery are image-guided drug delivery, magnetic drug targeting, and combined chemotherapy and thermal therapy [131].

A class of nanoparticles where molecules get attached and are delivered to the specific part are collected in Table 2. This table comprises the types of nanoparticles, molecules which are attached on the surface of nanoparticles, and the name of treatment techniques where the combination has used. Nanoparticles which are actively involved in nanotherapy for brain tumor management are collected together in Table 2.

5 Challenges and Future Prospects

The brain tumor has challenged the diagnostic care and treatment techniques because of its complicated nature. The tumor has complex structure which involves the mixtures of stem cells, malignant cells, nonmalignant cells, and progenitor cells [153]. Numerous cancer cells are present and each is different in nature within a tumor microenvironment. These cancer cells are not similar in genetic mutations and its expression, hence this complexity of tumor makes the treatment and diagnosis very difficult. Although there are certain treatment techniques available but those are not enough to provide effective treatment and reduce the risk of life and complete curing from the disaster. To overcome the limitations which occur due to the present treatment techniques, current therapies, and surgery of the tumor, the use of inorganic NPs could be fruitful to target the tumor cells in the specific tumor microenvironment. This specific target therapy may be developed in more effective way after getting the interaction between tumor

Table 2
Various nanostructures with their functions against cancer/tumor

Nanoparticles (NPs)	Functionalization	Functions	Targets/disease	References
CNT	–	Cause oxidative stress within the cell	Cancer	[103]
Gold plated CNT	Folic acid	Detection of cancer cells	Tumor in ear cells	[100]
Gold NPs	Gemcitabine anticancer drug using cetuximab (C225) as targeting agent	Inhibit cell proliferation	Pancreatic cancer	[107]
Gold NPs	EGFR antibody	SPR scattering	Cervical cancer	[143, 144]
PEGylated gold NPs	Folic acid	Folate receptor overexpression	Renal clearance	[145]
Iron oxide NPs	Dextran as coating	MRI contrast materials	Solid tumors	[114, 115]
	Silica NPs as coating	Hyperthermia (induced heat)	Cancer cell	[146, 147]
	–	T1 and T2 MRI contrast agent	Clinical tumor detection	[148]
Starch-coated SPION	Mitoxantrone (MTX)	–	Hind limb tumor model of rabbit	[149]
Silica NPs	Organic fluorophores	High-resolution image	Visualize lymphatic drainage in peritumoral region	[150]
Silver NPs	Starch as coating	Cell cycle arrest, damage organelles	GBM	[151]
TiO$_2$ NPs	Alcohol, amine carboxylic acid	Membrane disruption	Lung carcinoma, prostate cancer, melanoma	[152]

cells and stromal cells. A leaky vasculature is found in microenvironment of tumor cells and high permeability which is the result of angiogenesis triggered by different cytokines [105]. Because of enhanced permeability and retention (EPR) effect, the inorganic NPs may be used to target passively within this leaky vasculature in tumor [105, 154]. The role of NPs may be ideal because the conjugated NPs and anticancer agents such as liposomes and macromolecules may accumulate together within the tumor lesions, although the interstitial pressure unfortunately is a constant threat for the successful delivery of nanoconjugates because of the EPR effect. However antiangiogenic therapy and accurate administration of nanoconjugates can help in the normalization of tumor vasculature and can lead to better therapeutic efficacy [155]. Targeted delivery using inorganic NPs may reduce the non-specificity of the drugs [156], although inorganic nanostructures have provided many advantages and benefits for the cancer treatment using nanotherapy in near future if handled some of its challenges carefully. The functionalization of small size NPs, uniform distribution, and compatibility with drugs must be focused in more elaborated way. The nanosystems with drug molecules may be homogenized and movement of these nanosystems into blood stream may be in a controlled way. In vivo and in vitro process the method of working of these nanosystems must be explored after they get attached with the surface of tumor cells; for example, how do they penetrate into the tumor cells and how do they kill the cancer cells [157]? However, it is more complicated to have bypassed the blood-brain barrier for these nanosystems [158]. To understand the behavior in vivo, the toxicity, biodistribution, and destiny [159] is necessity of the research before applying these nanosystems into clinical trials. The effective nanoconjugates are facing another issue whether they are biodegradables or nonbiodegradables. For early diagnosis of cancer in the brain the detection of cancer biomarkers using such nanosystems is also challenging. One of the biggest challenges of the NP-enabled therapy to brain tumor is to provide general public safety and their interest and to prove the use of NPs into the effective treatment of brain tumor. Will the NP therapy be helpful in preventing the growth of tumor after surgery or treatment? On focusing such challenges in near future NP-based therapy can boom the new era of research and development in the field of brain tumor treatment and diagnosis [160]. The effective methodologies of combining NPs to their counterpart coating materials which will result into successful delivery of drugs to the tumor lesions can overcome the challenges. These include biodegradability, non-cytotoxicity, highly sensitivity to protein biomarkers, high stability within tumor lesions, and surface interacting properties to the tumor tissues.

The future of NPs in the therapeutics and diagnosis are very interesting as they can simultaneously deliver the drugs to the specific target and work as an imaging agents. The unique properties of NPs are making them effective contrast agents which can help a surgeon in dissection of tumor completely without harming healthy tissues. Certain NPs are effectively working in the detection of protein biomarkers at very low concentration which can early diagnose the brain tumor or cancer. The advantages of NPs are that their modifiable size, shape, and different combination with multi-drug molecules and their delivery to the specific targets make them materials of future in the research topic of brain tumor management [161]. Various efforts have been made to develop effective drug delivery systems to deliver nano-enabled therapeutics across the BBB to manage CNS diseases [162–169] and significant effort must be made to strategically address the present challenges for the complete eradication and management of brain cancer.

6 Conclusion

The complex physiological structures of the brain surrounded by BBB prohibit or make a little entry of most of the chemical compounds from blood to the brain. The effective anticancer drugs which are used for other organ treatments have limited effect in the brain tumor. The therapies such as radiotherapy, chemotherapy, and other diagnostic techniques have produced harmful effect for the healthy cells as well. Therefore the concept of nano-enabled therapy has come into existence where the drug molecules are attached with the NPs which act as nanocarriers to treat the tumor cells. The NPs have similarity with the organelles and have excellent nano properties such as surface to volume ratio, biocompatibility, magnetic properties (iron oxides), electrical properties, mechanical properties, optical properties, and other surface interacting properties. Therefore use of NPs in cancer therapeutics has opened a new chapter for the effective drug delivery.

The references of uses of NPs into therapeutic cancer vaccine, contrast imaging, cancer biomarker, and drug delivery are collected together in this chapter. Functional NPs, such as carbon nanotube, earth metal NPs, gold NPs, SPION, and other NPs, have been used in the drug delivery with anticancer drugs to minimize the risk of side effect and to enhance the effective diagnosis and treatment techniques.

Acknowledgments

Authors do acknowledge respective department and institution for providing resources and facilities.

References

1. "CBTRUS—2016 CBTRUS Fact Sheet." Accessed March 9, 2018. ("Explore Brain Cancer Treatment Options & Advanced Therapies," 2018)
2. "Brain, Other CNS and Intracranial Tumours Statistics." Cancer Research UK, May 14, 2015. http://www.cancerresearchuk.org/health-professional/cancer-statistics/statistics-by-cancer-type/brain-other-cns-and-intracranial-tumours
3. Weathers S-P, O'Brien B, de Groot J (2016) Commentary: Anita Mahajan, and Commentary: Sujit S. Prabhu. "Tumors of the central nervous system". In: Kantarjian HM, Wolff RA (eds) The MD Anderson manual of medical oncology, 3rd edn. McGraw-Hill Medical, New York, NY. http://accessmedicine.mhmedical.com/content.aspx?aid=1126744985
4. "Gupta Longati And Brain Cancer." prezi.com. Accessed March 29, 2018. https://prezi.com/erleq_tkx5if/guptalongati-and-brain-cancer/
5. Shah V, Kochar P (2018) Brain cancer: implication to disease, therapeutic strategies and tumor targeted drug delivery approaches.

Recent Pat Anticancer Drug Discov 13 (1):70–85. http://www.eurekaselect.com/157831/article

6. "Craniospinal Malignancies—Oxford Medicine. Accessed March 29, 2018. http://oxfordmedicine.com/view/10.1093/med/9780199656103.001.0001/med-9780199656103-chapter-56

7. Huse JT, Holland EC (2010) Targeting brain Cancer: advances in the molecular pathology of malignant glioma and Medulloblastoma. Nat Rev Cancer 10(5):319–331. https://doi.org/10.1038/nrc2818

8. DeAngelis LM, Wen PY (2015) Primary and metastatic tumors of the nervous system. In: Kasper D, Fauci A, Hauser S, Dan L, Larry Jameson J, Loscalzo J (eds) Harrison's principles of internal medicine, 19th edn. McGraw-Hill Education, New York, NY. http://accessmedicine.mhmedical.com/content.aspx?aid=1129103876

9. Saha A, Ghosh SK, Roy C, Choudhury KB, Chakrabarty B, Sarkar R (2013) Demographic and clinical profile of patients with brain metastases: a retrospective study. Asian J Neurosurg 8(3):157–161. https://doi.org/10.4103/1793-5482.121688

10. Kantarijan HM, Wolff RA (2016) Tumors of the central nervous system. In: The MD Anderson manual of medical oncology, 3rd edn. McGraw-Hill Medical, New York

11. Grant R (2004) Overview: brain tumour diagnosis and management/Royal College of Physicians guidelines. J Neurol Neurosurg Psychiatry 75:ii18–ii23. https://doi.org/10.1136/jnnp.2004.040360

12. Bozgeyik Z, Onur MR, Poyraz AK (2013) The role of diffusion weighted magnetic resonance imaging in oncologic settings. Quant Imaging Med Surg 3:269–278. https://doi.org/10.3978/j.issn.2223-4292.2013.10.07

13. Kono K, Inoue Y, Nakayama K, Shakudo M, Morino M, Ohata K, Wakasa K, Yamada R (2001) The role of diffusion-weighted imaging in patients with brain tumors. Am J Neuroradiol 22:1081–1088

14. Maier SE, Sun Y, Mulkern RV (2010) Diffusion imaging of brain tumors. NMR Biomed 23:849–864. https://doi.org/10.1002/nbm.1544

15. Williams PD, Jones S, Zhang X, Lin JC-H, Leary RJ, Angenendt P, Mankoo P et al (2008) An integrated genomic analysis of human glioblastoma multiforme. Science 321:5897. https://doi.org/10.1126/science.1164382

16. Hartmann C, Meyer J, Balss J, Capper D, Mueller W, Christians A, Felsberg J et al (2009) Type and frequency of IDH1 and IDH2 mutations are related to astrocytic and oligodendroglial differentiation and age: a study of 1,010 diffuse gliomas. Acta Neuropathol 118(4):469–474

17. Yip S, Butterfield YS, Morozova O, Chittaranjan S, Blough MD, An J, Birol I et al (2012) Concurrent CIC mutations, IDH mutations and 1p/19q loss distinguish oligodendrogliomas from other cancers. J Pathol 226(1):7–16. https://doi.org/10.1002/path.2995

18. Bettegowda C, Agrawal N, Jiao Y, Sausen M, Wood LD, Hruban RH, Rodriguez FJ et al (2011) Mutations in CIC and FUBP1 Contribute to Human Oligodendroglioma. Science 333(6048):1453–1455. https://doi.org/10.1126/science.1210557

19. Heaphy CM, de Wilde RF, Jiao Y, Klein AP, Edil BH, Shi C, Bettegowda C et al (2011) Altered Telomeres in Tumors with ATRX and DAXX Mutations. Science 333(6041):425. https://doi.org/10.1126/science.1207313

20. Brain Tumor: Risk Factors. Cancer.Net, June 25, 2012. https://www.cancer.net/cancer-types/brain-tumor/risk-factors

21. Brain Tumors & Brain Cancer - ABC2 [WWW Document], n.d. https://abc2.org/guidance/brain-cancer-facts/risk-factors. Accessed May 27 2019

22. Identify Top Brain Cancer Causes & Factors That Put You at Risk [WWW Document], 2018. Cancer Treat. Cent. Am. https://www.cancercenter.com/cancer-types/brain-cancer/risk-factors. Accessed May 27 2019

23. Radiology (ACR), R.S. of N.A. (RSNA) and A.C. of, n.d. Brain tumor treatment [WWW Document]. https://www.radiologyinfo.org/en/info.cfm?pg=thera-brain. Accessed May 27 2019

24. Radiotherapy treatment | Brain tumour (primary) | Cancer Research UK [WWW Document], n.d. https://www.cancerresearchuk.org/about-cancer/brain-tumours/treatment/radiotherapy/radiotherapy-treatment. Accessed May 27 2019

25. Explore Brain Cancer Treatment Options & Advanced Therapies [WWW Document], 2018. Cancer treat. Cent. Am. https://www.cancercenter.com/cancer-types/brain-cancer/treatments. Accessed May 27 2019

26. Liebner S, Dijkhuizen RM, Reiss Y, Plate KH, Agalliu D, Constantin G (March 2018) Functional morphology of the blood–brain barrier

in health and disease. Acta Neuropathol 135 (3):311–336. https://doi.org/10.1007/s00401-018-1815-1

27. Engelhardt B, Liebner S (March 2014) Novel insights into the development and maintenance of the blood–brain barrier. Cell Tissue Res 355(3):687–699. https://doi.org/10.1007/s00441-014-1811-2

28. Quail DF, Joyce JA (2017) The microenvironmental landscape of brain tumors. Cancer Cell 31(3):326–341. https://doi.org/10.1016/j.ccell.2017.02.009

29. Bhadoriya, Santosh Singh, Ankita Thakur, Major Hurdles for Brain Tumour Therapy and the Ways to Overcome Them: A Review. https://www.academia.edu/4974978/6_Major_hurdles_for_brain_tumour_therapy_and_the_ways_to_overcome_them. Accessed May 15, 2018

30. Deeken JF, Löscher W (2007) The blood-brain barrier and Cancer: transporters, treatment, and Trojan horses. Clin Cancer Res 13 (6):1663–1674. https://doi.org/10.1158/1078-0432.CCR-06-2854

31. Van Meir EG et al (2010) Exciting new advances in neuro-oncology; the avenue to cure for malignant glioma. CA Cancer J Clin 60:166–193

32. Van Meir EG, Sawamura Y, Diserens AC, Hamou MF, de Tribolet N (1990) Human glioblastoma cells release interleukin 6 in vivo and in vitro. Cancer Res 50:6683–6688

33. Desbaillets I, Diserens A-C, de Tribolet N, Hamou M-F, Van Meir EGJ (1997) Upregulation of interleukin 8 by oxygen deprived cells in glioblastoma suggests a role in leukocyte activation, chemotaxis and angiogenesis. Exp Med 186:1201–1212

34. Osuka S, Van Meir EG (2017) Overcoming therapeutic resistance in glioblastoma: the way forward. J Clin Invest 127:415–426

35. Tomitaka A, Kaushik A, Kevadiya B, Mukadam I, Gendelman HE, Khalili K, Liu G, Nair M (2019) Surface-engineered multimodal magnetic nanoparticles to manage CNS diseases. Drug Discov Today 24:873–882

36. Juarranz Á, Jaén P, Sanz-Rodríguez F, Cuevas J, González S (2008) Photodynamic therapy of cancer: basic principles and applications. Clin Transl Oncol 10:148–154

37. Ericson MB, Wennberg A-M, Larkö O (2008) Review of photodynamic therapy in actinic keratosis and basal cell carcinoma. Ther Clin Risk Manag 4:1–9

38. Jokerst JV, Gambhir SS (2011) Molecular imaging with theranostic nanoparticles. Acc Chem Res 44:1050–1060

39. Thakor AS, Gambhir SS (2013) Nanooncology: "The future of cancer diagnosis and therapy.". CA Cancer J Clin 63:395–418

40. Chatterjee K, Sarkar S, Jagajjanani Rao K, Paria S (2014) Core/shell nanoparticles in biomedical applications. Adv Colloid Interf Sci 209:8–39

41. Thrall JH (2004) Nanotechnology and medicine. Radiology 230:315–318

42. Jiang W, KimBetty YS, Rutka JT, Warren CCW (2008) Nanoparticle-mediated cellular response is sizedependent. Nat Nanotechnol 3:145–150

43. Peng G, Tisch U, Adams O, Hakim M, Shehada N, Broza YY, Billan S, Abdah-Bortnyak R, Kuten A, Haick H (2009) Diagnosing lung cancer in exhaled breath using gold nanoparticles. Nat Nanotechnol 4:669–673

44. Arole VM, Munde SV (2014) Fabrication of nanomaterials by top-down and bottom-up approaches-an overview. J Mater Sci 1:89–93

45. Sha D, Hsu S, Che Z, Chen C (2006) A dispatching rule for photolithography scheduling with an on-line rework strategy. Comput Indust Eng 50(3):233–247. https://doi.org/10.1016/j.cie.2006.04.002

46. Yang J, Berggren K (2007) Using high contrast salty development of hydrogen silsequioxane for sub-10-nm half pitch lithography. J Vac Sci Technol A 25(6):2025

47. Maldonado JR, Peckerar M (2016) X-ray lithography: some history, current status and future prospects. Microelectron Eng 161:87–93

48. Vladimirsky Y, Bourdillon A, Vladimirsky O, Jiang W, Leonard Q (1999) Demagnification in proximity X-ray lithography and extensibility to 25 nm by optimizing Fresnel diffraction. J Phys D Appl Phys 32(22):114

49. Vladimirsky Y (2003) X-ray lithography towards 15 nm, Jefferson Laboratory Technical Note 03-016

50. Cerrina F (2000) X-ray imaging: applications to patterning and lithography. J Phys D Appl Phys 33:R103

51. Maldonado JR, Acosta RE, Angelopoulos M, Doany FE, Narayan C, Chandrasekhar C, Pomerene ATS, Shaw JM, Kimmel K (1998), X-ray mask pellicle, US Patent 5,793,836

52. Petar P, Keran Z, Math M (2014) Micromachining–review of literature from 1980 to 2010. Interdisciplinary Description of Complex Systems: INDECS 12, no. 1: 1–27

53. Gietzelt T, Eichhorn L (2012) Mechanical Machining by Drilling, Milling and Slotting. In: Kahrizi M (ed) Micromachining Techniques for Fabrication of Micro and Nano Structures. InTech, London, pp 159–182

54. Chae J, Park SS, Freiheit T (2006) Investigation of micro-cutting operations. Int J Mach Tool Manu 46(3–4):313–332. https://doi.org/10.1016/j.ijmachtools.2005.05.015

55. Dornfeld D, Min S, Takeuchi Y (2006) Recent advances in mechanical micromachining. CIRP Ann Manuf Technol 55 (2):745–768. https://doi.org/10.1016/j.cirp.2006.10.006

56. Sina B, Arsalani N, Khataee A, Tabrizi AG (2018) Comparison of ball milling-hydrothermal and hydrothermal methods for synthesis of ZnO nanostructures and evaluation of their photocatalytic performance. J Ind Eng Chem 62:265–272

57. Xing T, Sunarso J, Yang W, Yin Y, Glushenkov AM, Li LH, Howlett PC, Chen Y (2013) Ball milling: a green mechanochemical approach for synthesis of nitrogen doped carbon nanoparticles. Nanoscale 5:7970

58. Amirkhanlou S, Ketabchi M, Parvin N (2012) Nanocrystalline/nanoparticle ZnO synthesized by high energy ball milling process. Mater Lett 86:122

59. Stolle A, Ranu B (2014) Ball Milling Towards Green Synthesis: Applications, Projects, Challenges. Royal Society of Chemistry, UK

60. Abdel-Magid AF, Caron S (2006) Fundamentals of early clinical drug development: from synthesis design to formulation. John Wiley and Sons, Inc., Hoboken, New Jersey

61. Ansari AR, Imran M, Yahia IS, Abdel-Wahab MS, Alshahrie A, Khan AH, Sharma C (2018) Effect of microwave power on morphology of AgO thin film grown using microwave plasma CVD. Int J Surf Sci Eng 12(1):1–2

62. Ansari AR, Hussain S, Imran M, Abdel-wahab MS, Alshahrie A (2018) Synthesis, characterization and oxidation of metallic cobalt (Co) thin film into semiconducting cobalt oxide (Co3O4) thin film using microwave plasma CVD. Mater Res Exp 5(6):065003

63. Ozaydin-Ince G, Coclite AM, Gleason KK (2012) CVD of polymetric thin films:applications in sensors, biotechnology, microelectronics/organic electronics, microfluidics, MEMS, composites and membranes. Rep Prog Phys 75:016501

64. Matsumura H (2019) Current status of catalytic chemical vapor deposition technology—history of research and current status of industrial implementation. Thin Solid Films 679:42–48

65. Sharma R, Agrawal VV, Srivastava AK, Govind, Nain L, Imran M, Kabi SR, Sinha RK, Malhotra BD (2013) Phase control of nanostructured iron oxide for application to biosensor. J Mater Chem B 1:464

66. Kaushik A, Khan R, Solanki PR, Pandey P, Alam J, Ahmad S, Malhotra BD (2008) Iron oxide Nanoparticles- Chitosan composite based glucose biosensor. Biosens Bioelectron 24:676–683

67. Javed M, Shaik AH, Khan TA, Imran M, Aziz A, Ansari AR, Chandan MR (2018) Synthesis of stable waste palm oil based CuO nanofluid for heat transfer applications. Heat Mass Transf 54:1–7

68. Kaushik A, Solanki PR, Ansari AA, Sumana G, Ahmad S, Malhotra BD (2009) Iron oxide-chitosan nanobiocomposite for urea sensor. Sensors Actuators B Chem 138(2):572–580

69. Kaushik A, Solanki PR, Ansari AA, Ahmad S, Malhotra BD (2008) Chitosan–iron oxide nanobiocomposite based immunosensor for ochratoxin-A. Electrochem Commun 10 (9):1364–1368

70. Solanki PR, Kaushik A, Ansari AA, Sumana G, Malhotra BD (2008) Zinc oxide-chitosan nanobiocomposite for urea sensor. Appl Phys Lett 93(16):163903

71. Kaushik A, Solanki PR, Ansari AA, Malhotra BD, Ahmad S (2009) Iron oxide-chitosan hybrid nanobiocomposite based nucleic acid sensor for pyrethroid detection. Biochem Eng J 46(2):132–140

72. Kaushik A, Solanki PR, Kaneto K, Kim CG, Ahmad S, Malhotra BD (2010) Nanostructured iron oxide platform for impedimetric cholesterol detection. Electroanalysis 22 (10):1045–1055

73. Imran M, Ansari AR, Shaik AH, Hussain S, Khan A, Chandan MR (2018) Ferrofluid synthesis using oleic acid coated Fe_3O_4 nanoparticles dispersed in mineral oil for heat transfer applications. Mater Res Exp 5(3):036108

74. Imran M, Shaik AH, Ansari AR, Aziz A, Hussain S, Abouatiaa AF, Khan A, Chandan MR (2018) Synthesis of highly stable γ-Fe_2O_3 ferrofluid dispersed in liquid paraffin, motor oil and sunflower oil for heat transfer applications. RSC Adv 8(25):13970–13975

75. Riaz S, Naseem S (2015) Controlled nanostructuring of TiO_2 nanoparticles: a sol-gel approach. J Sol-Gel Sci Technol 74:299–309

76. Solanki PR, Kaushik A, Ansari AA, Tiwari A, Malhotra BD (2009) Multi-walled carbon nanotubes/sol–gel-derived silica/chitosan nanobiocomposite for total cholesterol sensor. Sensors Actuators B Chem 137 (2):727–735

77. Solanki PR, Kaushik A, Ansari AA, Malhotra BD (2009) Nanostructured zinc oxide platform for cholesterol sensor. Appl Phys Lett 94 (14):143901

78. Kaushik A, Solanki PR, Ansari AA, Ahmad S, Malhotra BD (2009) A nanostructured cerium oxide film-based immunosensor for mycotoxin detection. Nanotechnology 20 (5):055105

79. Ansari AA, Kaushik A, Solanki PR, Malhotra BD (2010) Nanostructured zinc oxide platform for mycotoxin detection. Bioelectrochemistry 77(2):75–81

80. Ansari AA, Kaushik A, Solanki PR, Malhotra BE (2009) Electrochemical cholesterol sensor based on tin oxide-chitosan nanobiocomposite film. Electroanalysis 21(8):965–972

81. Solanki PR, Dhand C, Kaushik A, Ansari AA, Sood KN, Malhotra BD (2009) Nanostructured cerium oxide film for triglyceride sensor. Sensors Actuators B Chem 141 (2):551–556

82. Solanki PR, Kaushik A, Chavhan PM, Maheshwari SN, Malhotra BD (2009) Nanostructured zirconium oxide based genosensor for Escherichia coli detection. Electrochem Commun 11(12):2272–2277

83. Sayilkan F, Erdemoglu S, Asilturk M, Akarsu M, Sener S, Sayilkan H, Erdemogl M, Mater EA (2006) Photocatalytic performance of pure anatase nanocrystallite TiO2 sythesized under low temperature hydrothermal conditions. Res Bull 41:2276–2285

84. Yu JG, Wang GH, Cheng B, Zhou MH (2007) Effects of hydrothermal temperature and time on the photocatalytic and microstructures of bimodal mesoporous TiO2 powders. Appl Catal B 69:171–180

85. Rooymans JM (1972) In: Hagenmuller P (ed) Preparative methods in solid state chemistry. Academic Press, New York

86. Asiltürk M, Sayilkan F, Erdemoglu S, Akarsu M, Sayilkan H, Erdemoglu M, Arpac E (2006) J Hazard Mater 129:164–170

87. Pineda-Reyes AM, M de la Olvera L (2018) Synthesis of ZnO nanoparticles from water-in-oil (w/o) microemulsions. Mater Chem Phys 203:141–147

88. Yalçınöz Ş, Erçelebi E (2018) Potential applications of nano-emulsions in the food systems: an update. Mater Res Exp 5(6):062001

89. Asgari S, Saberi AH, McClements DJ, Lin M (2019) Microemulsions as nanoreactors for synthesis of biopolymer nanoparticles. Trends Food Sci Technol 86:118–130

90. Ahmad T, Wani IA, Al-Hartomy OA, Al-Shihri AS, Kalam A (2015) Low temperature chemical synthesis and comparative studies of silver oxide nanoparticles. J Mol Struct 1084:9–15

91. Suriati G, Mariatti M, Azizan A (2014) Synthesis of silver nanoparticles by chemical reduction method: effect of reducing agent and surfactant concentration. Int J Automot Mech Eng 10:1920

92. Sun Y, Xia Y (2002) Shape-controlled synthesis of gold and silver nanoparticles. Science 298(5601):2176–2179

93. Petcharoen K, Sirivat A (2012) Synthesis and characterization of magnetite nanoparticles via the chemical co-precipitation method. Mater Sci Eng B 177(5):421–427

94. Liu X, Kaminski MD, Guan Y, Chen H, Lui HAJ (2006) Preparation and characterization of hydrophobic superparamagnetic magnetic gel. J Magn Magn Mater 306:248–253

95. Sun S, Zeng H, J. (2002) Size controlled synthesis of magnetite nanoparticles. Am Chem Soc 124:8204–8205

96. Unal B, Durmus Z, Kavas H, Baykal A, Toprak MS (2010) Synthesis, conductivity and dielectric characterization of salicylic acid- Fe3O4 nanocomposite. Mater Chem Phys 123:184–190

97. Uchiyama H, Bando T, Kozuka H (2019) Effect of the amount of H2O and HNO3 in Ti(OC3H7i)(4) solutions on the crystallization of sol-gel-derived TiO2 films. Thin Solid Films 669:157–161

98. Lopez AJ, Urena A, Rams J (2011) Wear resistant coatings; Silica sol-gel reinforced with carbon nanotubes. Thin Solid Films 519:7904–7910

99. Poienar M, Martin C, Lebedev OI, Maignan A (2018) Advantage of low-temperature hydrothermal synthesis to grow stoichiometric crednerite crystals. Solid State Sci 80:39–45

100. Galanzha EI, Shashkov EV, Kelly T, Kim JW, Yang L, Zharov VP (2009) In vivo magnetic enrichment and multiplex photoacoustic detection of circulating tumour cells. Nat Nanotechnol 4:855–860

101. Colon J, Herrera L, Smith J, Patil S, Komanski C, Kupelian P, Seal S, Jenkins DW, Baker CH (2009) Protection from radiation-induced pneumonitis using cerium oxide nanoparticles. Nanomedicine 5:225–231

102. Tarnuzzer RW, Colon J, Patil S, Seal S (2005) Vacancy engineered ceria nanostructures for protection from radiation-induced cellular damage. Nano Lett 5:2573–2577

103. Bhattacharyya S, Kudgus RA, Bhattacharya R, Mukherjee P (2011) Inorganic nanoparticles in cancer therapy. Pharm Res 28(2):237–259

104. Mukherjee P, Bhattacharya R, Wang P, Wang L, Basu S, Nagy JA, Atala A, Mukhopadhyay D, Soker S (2005) Antiangiogenic properties of gold nanoparticles. Clin Cancer Res 11:3530–3534

105. Patra CR, Bhattacharya R, Mukhopadhyay D, Mukherjee P (2008) Application of gold nanoparticles for targeted therapy in cancer. J Biomed Nanotechnol 4:99–132

106. Peer D, Karp JM, Hong S, Farokhzad OC, Margalit R, Langer R (2007) Nanocarriers as an emerging platform for cancer therapy. Nat Nanotechnol 2:751–760

107. Patra CR, Bhattacharya R, Wang E, Katarya A, Lau JS, Dutta S, Muders M, Wang S, Buhrow SA, Safgren SL, Yaszemski MJ, Reid JM, Ames MM, Mukherjee P, Mukhopadhyay D (2008) Targeted delivery of gemcitabine to pancreatic adenocarcinoma using cetuximab as a targeting agent. Cancer Res 68:1970–1978

108. Toffoli G, Cernigoi C, Russo A, Gallo A, Bagnoli M, Boiocchi M (1997) Overexpression of folate binding protein in ovarian cancers. Int J Cancer 74:193–198

109. Coney LR, Tomassetti A, Carayannopoulos L, Frasca V, Kamen BA, Colnaghi MI, Zurawski VR Jr (1991) Cloning of a tumor-associated antigen: MOv18 and MOv19 antibodies recognize a folate binding protein. Cancer Res 51:6125–6132

110. Elnakatand M, Ratnam H (2004) Distribution, functionality and gene regulation of folate receptor isoforms: implications in targeted therapy. Adv Drug Deliv Rev 56:1067–1084

111. Kamenand AK, Smith BA (2004) A review of folate receptor alpha cycling and 5-methyltetrahydrofolate accumulation with an emphasis on cell models in vitro. Adv Drug Deliv Rev 56:1085–1097

112. Revia RA, Zhang M (2016) Magnetite nanoparticles for cancer diagnosis, treatment, and treatment monitoring: Recent advances. Mater Today 19:157–168

113. Kaushik A, Jayant RD, Sagar V, Nair M (2014) The potential of magneto-electric nanocarriers for drug delivery. Expert Opin Drug Deliv 11:1635–1646

114. Shubayev VI, Pisanic TR 2nd, Jin S (2009) Magnetic nanoparticles for theragnostics. Adv Drug Deliv Rev 61:467–477

115. Arbab AS, Liu W, Frank JA (2006) Cellular magnetic resonance imaging: current status and future prospects. Expert Rev Med Devices 3:427–439

116. McCarthy JR, Kelly KA, Sun EY, Weissleder R (2007) Targeted delivery of multifunctional magnetic nanoparticles. Nanomedicine 2:153–167

117. Gao X, Cui Y, Levenson RM, Chung LWK, Nie S (2004) In vivo cancer targeting and imaging with semiconductor quantum dots. Nat Biotechnol 22:969–976

118. Pardridge WM (2005) The blood-brain barrier: bottleneck in brain drug development. NeuroRx 2(1):3–14

119. Wen R, Banik B, Pathak RK, Kumar A, Kolishetti N, Dhar S (2016) Nanotechnology-inspired tools for mitochondrial dysfunction related diseases. Adv Drug Deliv Rev 99:52–69

120. Wen R, Umeano AC, Kou Y, Xu J, Farooqi AA (2019) Nanoparticle systems for cancer vaccine. Nanomedicine 14(5):627–648

121. Pardoll DM (1998) Cancer vaccines. Nat Med 4:525–531

122. Finn OJ (2003) Cancer vaccines: between the idea and the reality. Nat Rev Immunol 3:630–641

123. Li H, Li Y, Jiao J, Hu HM (2011) Alpha-alumina nanoparticles induce efficient autophagy-dependent cross-presentation and potent antitumour response. Nat Nanotechnol 6(10):645

124. Chiang C-S, Lin Y-J, Lee R, Lai Y-H, Cheng H-W, Hsieh C-H, Shyu W-C, Chen S-Y (2018) Combination of fucoidan-based magnetic nanoparticles and immunomodulators enhances tumour-localized immunotherapy. Nat Nanotechnol 13(8):746

125. Shao K, Singha S, Clemente-Casares X, Tsai S, Yang Y, Santamaria P (2015) Nanoparticle based immunotherapy for cancer. ACS Nano 9(1):16–30. https://doi.org/10.1021/nn5062029

126. Gao H, Yang Z, Zhang S, Cao S, Shen S, Pang Z, Jiang X (2013) Ligand modified nanoparticles increases cell uptake, alters endocytosis and elevates glioma distribution and internalization. Sci Rep 3:2534

127. Kuang Y, An S, Guo Y, Huang S, Shao K, Yang L, Li J, Ma H, Jiang C (2013) T7 peptide-functionalized nanoparticles utilizing

RNA interference for glioma dual targeting. Int J Pharm 454(1):11–20

128. Bonnemain B (1998) Superparamagnetic agents in magnetic resonance imaging: physicochemical characteristics and clinical applications—a review. J Drug Target 6:167–174

129. Enochs WS, Harsh G, Hochberg F, Weissleder R (1999) Improved delineation of human brain tumors on MR images using long-circulating, superparamagnetic iron oxide agent. J Magn Reson Imaging 9:228–232

130. Yamaguchi H, Tsuchimochi M, Hayama K, Kawase T, Tsubokawa N (2016) Dual-labeled near-infrared/99mTc imaging probes using PAMAM-coated silica nanoparticles for the imaging of HER2-expressing cancer cells. Int J Mol Sci 17(7):1086

131. Park JY, Baek MJ, Choi ES et al (2009) Paramagnetic ultrasmall gadolinium oxide nanoparticles as advanced T1 MRI contrast agent: account for large longitudinal relaxivity, optimal particle diameter, and in vivo T1 MR images. ACS Nano 3(11):3663–3669

132. Bao G, Mitragotri S, Tong S (2013) Multifunctional nanoparticles for drug delivery and molecular imaging. Annu Rev Biomed Eng 15:253–282

133. Füzéry AK, Levin J, Chan MM, Chan DW (2013) Translation of proteomic biomarkers into FDA approved cancer diagnostics: issues and challenges. Clin Proteomics 10:13

134. Ye F, Zhao Y, El-Sayed R, Muhammed M, Hassan M (2018) Advances in nanotechnology for cancer biomarkers. Nano Today 18:103–123

135. Kosaka PM, Pini V, Ruz JJ, da Silva RA, González MU, Ramos D, Calleja M, Tamayo J (2014) Detection of cancer biomarkers in serum using a hybrid mechanical and optoplasmonic nanosensor. Nat Nanotechnol 9:1047–1053

136. Yu MK, Park J, Jon S (2012) Targeting strategies for multifunctional nanoparticles in cancer. Theranostics 2:3–44

137. Chinen AB, Guan CM, Ferrer JR, Barnaby SN, Merkel TJ, Mirkin CA (2015) Nanoparticle probes for the detection of cancer biomarkers, cells, and tissues by fluorescence. Chem Rev 115:10530–10574

138. Chen G, Roy I, Yang C, Prasad PN (2016) Nanochemistry and nanomedicine for nanoparticle-based diagnostics and therapy. Chem Rev 116:2826–2885

139. Kaasgaard T, Andresen TL (2010) Liposomal cancer therapy: exploiting tumor characteristics. Expert Opin Drug Deliv 7:225–243

140. Astruc D, Boisselier E, Ornelas C (2010) Dendrimers designed for functions: from physical, photophysical, and supramolecular properties to applications in sensing, catalysis, molecular electronics, photonics, and nanomedicine. Chem Rev 110:1857–1959

141. Dreaden EC, Mackey MA, Huang X, Kang B, El-Sayed MA (2011) Beating cancer in multiple ways using nanogold. Chem Soc Rev 40:3391–3404

142. Duncan R (2006) Polymer conjugates as anticancer nanomedicines. Nat Rev Cancer 6:688–701

143. Sokolov K, Follen M, Aaron J, Pavlova I, Malpica A, Lotan R, Richards-Kortum R (2003) Real-time vital optical imaging of Precancer using anti-epidermal growth factor receptor antibodies conjugated to gold nanoparticles. Cancer Res 63:1999–2004

144. Maruo T, Yamasaki M, Ladines-Llave CA, Mochizuki M (1992) Immunohistochemical demonstration of elevated expression of epidermal growth factor receptor in the neoplastic changes of cervical squamous epithelium. Cancer 69:1182–1187

145. Dixit V, Van Den Bossche J, Sherman DM, Thompson DH, Andres RP (2006) Synthesis and grafting of thioctic acid-PEG-folate conjugates onto au nanoparticles for selective targeting of folate receptor-positive tumor cells. Bioconjug Chem 17:603–609

146. Gupta AK, Naregalkar RR, Vaidya VD, Gupta M (2007) Recent advances on surface engineering of magnetic iron oxide nanoparticles and their biomedical applications. Nanomedicine 2:23–39

147. Lu AH, Salabas EL, Schuth F (2007) Magnetic nanoparticles: synthesis, protection, functionalization, and application. Angew Chem Int Ed 46:1222–1244

148. Clement O, Siauve N, Cuenod CA, Frija G (1998) Liver imaging with ferumoxides (Feridex): fundamentals, controversies, and practical aspects. Top Magn Reson Imaging 9:167–182

149. Alexiou C, Arnold W, Klein RJ, Parak FG, Hulin P et al (2000) Locoregional cancer treatment with magnetic drug targeting. Cancer Res 60:6641–6648

150. Larson DR, Ow H, Vishwasrao HD, Heikal AA, Wiesner U, Webb WW (2008) Silica nanoparticles architecture determines radiative properties of encapsulated fluorophores. Chem Mater 20:2677–2684

151. AshaRani PV, Mun GLK, Hande MP, Valiyaveettil S (2009) Cytotoxicity and genotoxicity

of silver nanoparticles in human cells. ACS Nano 3:279–290

152. Thevenot P, Cho J, Wavhal D, Timmons RB, Tang L (2008) Surface chemistry influences cancer killing effect of TiO2 nanoparticles. Nanomedicine 4:226–236

153. Heppner GH (1984) Tumor heterogeneity. Cancer Res 44:2259–2265

154. Jain RK (1999) Transport of molecules, particles, and cells in solid tumors. Annu Rev Biomed Eng 1:241–263

155. Jainand TP, Padera RK (2003) Development. Lymphatics make the break. Science 299:209–210

156. Koo OM, Rubinstein I, Onyuksel H (2005) Role of nanotechnology in targeted drug delivery and imaging: a concise review. Nanomedicine 1:193–212

157. Davis ME, Chen Z, Shin DM (2008) Nanoparticle therapeutics: An emerging treatment modality for cancer. Nat Rev Drug Discov 7:771–782

158. Saraiva C, Praça C, Ferreira R, Santos T, Ferreira L, Bernardino L (2016) Nanoparticle-mediated brain drug delivery: overcoming blood–brain barrier to treat neurodegenerative diseases. J Control Release 235:34–47

159. Arami H, Khandhar A, Liggitt D, Krishnan KM (2015) In vivo delivery, pharmacokinetics, biodistribution and toxicity of iron oxide nanoparticles. Chem Soc Rev 44:8576–8607

160. Wang F, Li C, Cheng J, Yuan Z (2016) Recent advances on inorganic nanoparticle-based cancer therapeutic agents. Int J Environ Res Public Health 13(12):1182

161. Meyers JD, Doane T, Burda C, Basilion JP (2013) Nanoparticles for imaging and treating brain cancer. Nanomedicine 8 (1):123–143

162. Kaushik A, Jayant RD, Bhardwaj V, Nair M (2018) Personalized nanomedicine for CNS diseases. Drug Discov Today 23:1007–1015

163. Vashist A, Kaushik A, Vashist A, Sagar V, Ghosal A, Gupta YK, Ahmad S, Nair M (2018) Advances in carbon nanotubes–hydrogel hybrids in nanomedicine for therapeutics. Adv Healthc Mater 7:1701213

164. Vashist A, Kaushik A, Ghosal A, Bala J, Nikkhah-Moshaie R, Wani WA, Manickam P, Nair M (2018) Nanocomposite hydrogels: advances in nanofillers used for nanomedicine. Gels 4:75

165. Kaushik A, Jayant RD, Nair M (2016) Advancements in nano-enabled therapeutics for neuroHIV management. Int J Nanomedicine 11:4317

166. Nair M, Jayant RD, Kaushik A, Sagar V (2016) Getting into the brain: potential of nanotechnology in the management of NeuroAIDS. Adv Drug Deliv Rev 103:202–217

167. Kaushik A, Jayant RD, Nair M (2017) Advances in personalized Nanotherapeutics. Springer International Publishing, New York

168. Vashist A, Kaushik A, Vashist A, Bala J, Nikkhah-Moshaie R, Sagar V, Nair M (2018) Nanogels as potential drug nanocarriers for CNS drug delivery. Drug Discov Today 23:1436–1443

169. Kaushik A, Jayant RD, Nair M (2018) Nanomedicine for neuroHIV/AIDS management. Nanomedicine 13:669–673

Chapter 7

Strategies to Enhance the Distribution of Therapeutic Nanoparticles in the Brain by Convection Enhanced Delivery

Karina Negron, Namir Khalasawi, and Jung Soo Suk

Abstract

Convection enhanced delivery (CED) is an attractive method to bypass the blood-brain barrier for therapeutic delivery to the brain as well as to facilitate widespread therapeutic distribution within brain parenchyma by creating a continuous pressure-driven bulk flow. However, rapid removal of therapeutics from the brain by the physiological clearance mechanism remains a critical challenge to achieving widespread therapeutic delivery by CED. Nanoparticle (NP)-based delivery systems that stay longer in the brain while convoying a high concentration of payloads can potentially provide widespread and efficient therapeutic delivery to the highly disseminated disease areas within the brain. In particular, we have recently demonstrated that CED of NPs designed to efficiently penetrate the highly adhesive and nanoporous brain extracellular matrix synergistically enhances the payload distribution within healthy and tumor-bearing brains. In this book chapter, we first briefly overview the mechanism and current limitations of CED, as well as strategies to maximize the CED-mediated therapeutic and/or NP delivery to the brain. We then describe a detailed methodology for preclinical CED experiments, including device/animal setup, NP preparation, tissue processing, and image/data analysis. Finally, we conclude the chapter with a few troubleshooting tips.

Key words Intracranial drug delivery, Brain parenchyma, Brain-penetrating nanoparticles, Volume of distribution, Infusion parameters

---

1 Introduction

1.1 Key Challenge to Therapeutic Delivery to the Brain: Limited Therapeutic Distribution

The blood-brain barrier (BBB) is a primary impediment to the delivery of systemically administered therapeutic agents to the brain [1]. Strategies to overcome the BBB have been widely explored and preclinically validated, including transient BBB opening via chemical [2] or physical [3–5] methods and circumvention of the barrier by local infusion [6, 7]. The conventional strategy to bypass the BBB involves direct intracerebroventricular or intrathecal injection of therapeutics into lateral ventricles of the brain or into the lumbar spine where the cerebrospinal fluid (CSF) circulates, respectively [8]. However, an additional challenge, regardless of the route of administration, is to achieve widespread therapeutic

Vivek Agrahari et al. (eds.), *Nanotherapy for Brain Tumor Drug Delivery*, Neuromethods, vol. 163,
https://doi.org/10.1007/978-1-0716-1052-7_7, © Springer Science+Business Media, LLC, part of Springer Nature 2021

distribution in the brain [9]. It is now well established that simple diffusion-mediated dispersion of conventional therapeutics, including small molecule- and protein-based drugs, does not provide widespread coverage throughout highly disseminated disease areas within the brain parenchyma [10, 11]. In addition to the slow diffusion rates, the rapid drug clearance by convective flow of the CSF, from brain tissue to bloodstream, remains a contributor to the suboptimal drug distribution and their local retention [12]. The human CSF is produced at a rate of 550 mL/day, with ~100–140 mL of CSF turning over every 4–5 h [13]. Thus, injected drugs carried by the interstitial fluid bulk flow are continuously destine to the CSF and subsequently eliminated from the brain parenchyma by the physiological CSF turnover [9].

Due to the combined effect of slow diffusion and rapid clearance, the concentrations of locally injected drugs have been estimated to decrease by ~10-folds for every millimeter of the traveled distance from the site of administration [10, 14, 15]. This decrease in drug concentrations augments with the increase in the molecular weight or size of the molecule [16], which is primarily attributed to the reduction in the diffusion rates, as expected from the Stokes-Einstein diffusion equation. The bis-chloroethylnitrosourea (BCNU) wafer, Gliadel®, developed and clinically used to provide sustained drug concentrations within the brain also suffers from a similar reality; it has been reported that the BCNU concentration reduces by ~10-folds every 500 μm of distance away from the wafer [17]. Thus, the amount of drug necessary to achieve therapeutic concentration only a few millimeters away from the site of administration would be several orders of magnitude greater than its effective therapeutic dose. The required drug dose can be even greater, noting that the overall distance that needs to be traveled by injected drugs can be as far as ~50 mm or more depending on the administration site [10]. Such drug doses will most likely lead to significant toxicity, thereby demanding a safer method to achieve therapeutic concentrations throughout the target diseased areas within the brain.

1.2 Convection Enhanced Delivery as a Strategy to Enhance the Therapeutic Distribution in the Brain

Convection enhanced delivery (CED) was initially introduced and implemented as a means to bypass the BBB in the early 1990s [18], but has later become appreciated as a strategy to facilitate widespread distribution of locally infused therapeutics [19–22]. The CED involves continuous infusion of a solution into the brain at a predetermined rate [23]. The procedure creates a pressure-driven bulk flow that pushes the infusate away from the administration site into the extracellular space of the brain parenchyma [23, 24]. Thus, CED provides a markedly enhanced volumetric drug distribution

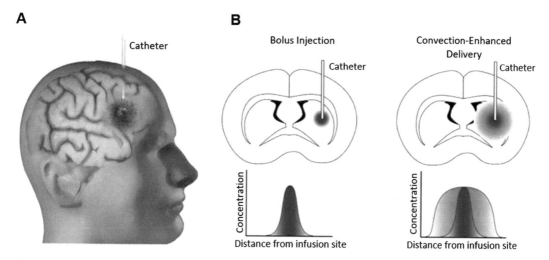

Fig. 1 Volume of distribution achieved by bolus injection versus convection-enhanced delivery (CEO). (**a**) A catheter is inserted into the brain tumor (red) or a cavity created after tumor resection, and therapeutic agents (blue) are continuously infused. Schematics (upper panel) and diffusion profiles (i.e., concentration gradients away from the administration site; lower panel) of therapeutic agents (blue) administered by (**b**) bolus injection and (**c**) CED. CED provides a marked greater therapeutic distribution compared to bolus injection that counts solely on simple diffusion. (CED Reproduced from Seo, Y.E. et al. 2017 with permission from Elsevier)

within the brain as it is primarily mediated by convection, unlike the bolus injection that entirely counts on nondirectional random diffusion (Fig. 1) [25, 26].

In addition, instantaneous pressure buildup is avoided by the CED due to the slow infusion, which reduces reflux that inevitably occurs during bolus injection [23]. Further, the gradual infusion of CED can potentially mitigate systemic toxicity of therapeutic agents by reducing the amount of drugs exposed to the systemic circulation at a given time [21, 27]. These unique advantages offered by CED have prompted the use of the technique for localized chemotherapeutic delivery to treat malignant gliomas, primarily glioblastoma multiforme [28, 29] and diffuse intrinsic pontine glioma [24, 30], of which recurrence generally occurs as far as centimeters away from the primary tumor site [31]. The CED has been also explored in various preclinical and/or clinical settings for treatment or diagnosis of other brain diseases beyond tumors, including neurodegenerative diseases, epilepsy, stroke, traumatic injuries, and brainstem lesions [32–37]. Specifically, the utility of CED has been expanded to achieve widespread delivery of a wide array of therapeutic or diagnostic agents, including monoclonal antibodies, therapeutic toxins, proteins, imaging tracers, as well as drug and gene delivery systems [7, 18, 21, 38–40]. Table 1 lists general advantages and disadvantages around the use of CED in comparison to other local or systemic administration modalities commonly applied to brain delivery.

Table 1
Advantages and disadvantages of CED

Advantages	Disadvantages
• Bypasses BBB • Reduces risk of local and systemic toxicity • Widespread and uniform delivery of agents by pressure-driven continuous bulk flow • Distribution minimally affected by molecular weights of agents • Less prone to human errors • Reduces backflow	• Relatively invasive procedure • Prolonged infusion times • Potential buildup of intracranial pressure • Requires refilling for a long-term infusion

1.3 Challenges for Therapeutic Distribution in the Brain by CED

While the CED provides a method to enhance drug distribution within the brain, multiple studies have revealed that the benefit is rapidly lost due to the physiological fluid clearance mechanism described above [23, 25, 41]. It has been reported that the peak volume of drug distribution precipitously reduces as early as 30 min after CED and drugs are often found completely removed from the brain within a few hours [42]. Of note, the clearance is even greater in the brains of patients with diseases characterized by increased vascular permeability and interstitial pressure, such as tumors [43–45].

One of the methods to circumvent the clearance problem involves prolonging the infusion of therapeutic agents over days using an implantable catheter, which has shown proof-of-principle benefits in treating brain tumors [46–51]. However, widespread clinical use of this long-term infusion modality is limited by potential risk of infection and mechanical injury, need for constant monitoring of drug toxicity, and complex and high-cost catheter design. As an alternative approach, nanoparticle (NP)-based delivery systems have been widely explored in preclinical settings to provide sustained therapeutic concentrations in the brain. Due to their larger sizes, therapeutic NPs carrying a high concentration of payloads delivered into the brain parenchyma stay longer without being rapidly cleared unlike drugs administered "naked" [52, 53]. NPs can carry and deliver controlled release of various types of payloads, ranging from small molecule drugs to biological macromolecules [39, 53, 54]. Further, NPs can be designed to facilitate specific cell targeting and subsequent uptake [54, 55], providing another mechanism by which rapid drug clearance from the brain is avoided.

While NPs can potentially delay the drug clearance, released payloads spread in the brain via slow diffusion remain identical to directly injected naked drugs. Thus, NPs must broadly disperse throughout the brain parenchyma via CED, followed by timely release of payloads, to provide widespread therapeutic coverage within the brain. However, widespread NP distribution is not readily achieved due to the presence of highly adhesive and nano-porous brain extracellular matrix (ECM) that fills the brain extra-cellular space through which NPs spread in the brain [56, 57]. The brain ECM is composed of negatively charged or hydrophobic macromolecules, including hyaluronic acids, proteoglycans, glyco-saminoglycans, and fibrous proteins [58–60], which hampers NP transport in the brain via multivalent adhesive interactions such as electrostatic and hydrophobic interactions [56, 61]. In addition, these macromolecules form a dense meshwork, rendering the ECM a steric barrier to NP diffusion within the brain [39, 54, 56]. The pore sizes of the brain ECM have been previously estimated to be 38–64 nm [62], but we have recently re-evaluated them using nonadhesive probes to be as large as or potentially greater than 110 nm [56]. Physicochemical barrier properties of the ECM within tumor tissues are most likely distinct from those of normal ECM due to significant changes in cellularity and macromolecular compositions that are highly varied depending on the type of tumor [63–65]. There have been several reports suggesting that the pore sizes of tumor ECM may be smaller than 100 nm [62, 66]. Of note, interstitial pressure buildup in tumor tissues [44, 67] often pro-motes outward brain tumor cell migration and consequently, greater volume of therapeutic distribution is likely needed [39, 67–69].

1.4 Strategies to Enhance NP Distribution in the Brain Following CED

As described above, brain ECM is an adhesive and steric barrier to NP diffusion. Thus, NPs possessing small particle diameters to fit through the ECM pores as well as nonadhesive surface coatings that resist adhesive interactions with ECM components would be able to penetrate the brain tissue. Indeed, we have previously demonstrated that NPs as large as ~110 nm efficiently penetrate rodent and human brain parenchyma ex vivo and in vivo, but only if the particle surfaces are densely passivated with hydrophilic and neutrally charged polyethylene glycol (PEG) (Fig. 2) [56]. Specifi-cally, NPs possessing diameters of ~100 nm were shown to effi-ciently diffuse through normal brain ECM when their surfaces were coated with PEG at the surface densities of ≥ 8 PEG molecules per 100 nm^2 particle surface area [56]. However, minimal surface PEG density cutoff may vary depending on particle size due to the difference in surface curvature. We also note that smaller particle diameters may be needed for efficient NP penetration through tumor ECM meshes. More recently, we found that CED of this "brain-penetrating" NP (BPN) resulted in widespread distribution in healthy or tumor-bearing brain striatum; in contrast, the distri-bution of similarly sized uncoated or conventionally PEGylated NPs was confined to the site of administration despite the pressure-driven convective flow provided by the CED [38, 65, 70–74]. Importantly, NPs must retain the physicochemical

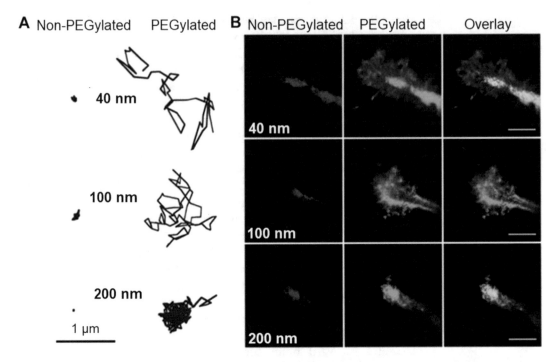

Fig. 2 Ex vivo diffusion and in vivo distribution of nanoparticles (NPs) in human and rodent brain tissues, respectively. (**a**) Representative particle trajectories for non-PEGylated (i.e., adhesive) and PEGylated (i.e., nonadhesive) NPs of various sizes in brain tissues freshly harvested from humans. (**b**) Direct comparisons of the distribution of non-PEGylated (red) and PEGylated (green) NPs of various sizes following intracranial co-injection into mouse brains. Scale bar = 50 μm. (Reproduced from Nance, E. et at. 2012 with permission from The American Association for the Advancement of Science)

properties required for efficient brain penetration in the physiological brain environment (i.e., brain interstitial fluid or CSF).

The physicochemical properties and stability that allow efficient brain penetration are likely applicable commonly to nanoscale particulate matters regardless of type of core materials, including polymers, lipids, and inorganic compounds. We have previously demonstrated that BPNs formulated with chemically distinct core polymers equally provide widespread distribution in rodent brain tissue in vivo following CED [38, 65, 70–72, 74]. Other than PEG, various coating materials have been used to endow NP-based delivery systems with colloidal stability; those include but are not limited to sugars, pluronics, polyglycerols, polyacrylic, and polyvinyl polymers [75, 76]. However, these materials, often possessing chemical properties similar to PEG (i.e., hydrophilic and neutrally charged), can potentially be utilized to engineer BPNs although it is yet to be experimentally determined. We note that while not necessarily geared specifically toward creating NPs capable of penetrating brain ECM, a variety of NPs have been preclinically investigated in conjunction with CED. Interested readers are referred to a recent review article [54].

Adhesive NPs small enough to avoid steric obstruction imposed by the brain ECM can widely spread throughout the brain parenchyma following CED if NPs are administered at a very high concentration [38]. Specifically, CED of uncoated NPs administered at 25 mg/mL exhibited the volumetric distribution approaching that achieved by identically administered BPN at 0.1 mg/mL [38]. This phenomenon is likely due to potential masking of the adhesive moieties of ECM by an excess of adhesive NPs, thereby enabling the residual NPs to pass through the ECM pores without being associated with ECM meshes. However, it should be noted that this approach is likely prone to dose-limiting toxicity, particularly when payloads possess intrinsic toxicity such as chemotherapeutic agents.

Alternatively, NP distribution in the brain following CED can be enhanced by manipulating the barrier properties of ECM. In particular, the pore sizes can be transiently modulated by chemical or physical methods. As an example, we have recently reported that distribution of BPNs in rodent brain following CED significantly increases when administered in a hyperosmotic solution [38]. This is most likely attributed to the mechanism described in a prior study that hyperosmolar saline administered in brain tissue leads to enlargement of the ECM mesh spacings as water is drawn out of cells into the extracellular space via an osmotic gradient established by the hyperosmolar saline [77, 78]. Likewise, it has been recently demonstrated that focused ultrasound, by transiently enlarging the extracellular space, can enhance NP dispersion in the brain [79].

1.5 CED in Clinical Trials

Previous and ongoing clinical trials involving the implementation of CED for treating diseases affecting the central nervous system (CNS) are listed in Table 2. Presumably due to the somewhat invasive nature of the procedure, most (i.e., 37 out of 41) of trials were conducted in patients with malignant gliomas who are already receiving surgery. The rest covers neurodegenerative disorders, including Parkinson's disease (NCT01621581, NCT00921128), aromatic L-amino acid decarboxylase (AADC) deficiency (NCT02852213) [80, 81], and Gaucher disease [82]. A majority of trials to date involve delivery of naked therapeutic agents for cancer therapy [21], but the number of trials that include delivery systems is increasing, particularly regarding gene therapy applications. Specifically, viruses (NCT02852213, NCT01621581) and liposomes [83–86] have been or are currently being explored in clinical settings to achieve widespread therapeutic transgene expression following CED.

Table 2
Clinical trials conducted to evaluate CED for treating CNS diseases as of early 2018

Target disease	Therapeutic agent	Year	Trial phase	Current status	NCT#	Reference
Malignant gliomas	124I-8H9	2011–present	I	Recruiting	NCT01502917	–
	Carboplatin	2015		Withdrawn	NCT01317212	[87]
		2012–2017		Recruiting	NCT01644955	–
	Cotara	2005	I/II	Completed	–	[88]
	CpG-28	2005–2008	I	Recruiting	NCT00190424	[89, 90]
	D2C7-IT	2015–present	I	Recruiting	NCT02303678	[91]
	Delta-24-rgd	2010–2014	I/II	Completed	NCT01582516	–
	GRm13Z40-2 CTL	2010–2013	I		NCT01082926	[92]
	hrBMP4	2017–present		Recruiting	NCT02869243	–
	HSV-1-tk[a] + ganciclovir	2003	I/II	Completed	–	[83]
	IL13-PE38QQR	2000–2007			NCT00024570	[93]
		2001–2006	I		NCT00024557	[94]
		2004–2007	III		NCT00076986	[95, 96]
		2009–2015	I	Terminated	NCT00880061	[97]
	IL13-PE38QQR + temozolomide	2004–2007		Completed	NCT00089427	[98]
	LSFV-IL12[a]	2003	I/II		–	[84]
	mAb 425	1997	I		–	[99]
	MDNA55	2016–present	II	Recruiting	NCT02858895	–
	MRI-1KDEL	2006–2012	I	Terminated	NCT01009866	–
	Irinotecan[a]	2014–present		Enrolling by invitation	NCT02022644	[86]
		2017–present		Recruiting	NCT03086616	–
	NBI-3001	2003		Completed	NCT00014677	[100]
	Nimustine hydrochloride	2011		–	–	[101]
	Paclitaxel	2001, 2004	I/II	Completed	–	[102, 103]
	PRX321	2009–2010	II	Withdrawn	NCT00797940	–
	PVS-RIPO	2012–2017	I	Active, not recruiting	NCT01491893	[104]
		2017–present		Recruiting	NCT03043391	–
	TF-CRM107	1997		Completed	–	[105]
		2003	I/II		–	[106]

Disease	Agent	Years	Phase	Status	NCT number	References
	Topotecan	2004–2011	I	Recruiting	NCT00308165	[30, 48, 50]
		2018–present			NCT03154996	–
		2014–2015		Completed	NCT02278510	[107]
		2015–present		Recruiting	NCT02500459	–
		2017–present			NCT03193463	–
	TP-38	2004–2007	I/II	Completed	NCT00104091	[108]
	Trabedersen	2008–2012	III	Terminated	NCT00761280	–
		2003–2009	II	Completed	NCT00431561	[109]
PD	AAV2-GDNF[b]	2012–present	I	Active, not recruiting	NCT01621581	–
	Muscimol	2009		Withdrawn	NCT00921128	–
Type 2 Gaucher disease	Glucocerebrosidase	2005–2006	I	Completed	NCT00244582	[82]
AADC deficiency	AAV2-hAADC[b]	2016–present	I	Recruiting	NCT02852213	–

[a]Liposomal formulations
[b]Adeno-associated virus (AAV)

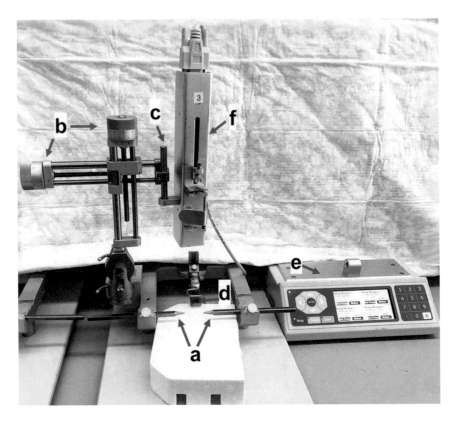

Fig. 3 CED apparatus. The CED device is composed of two distinct functional compartments, including structural (i.e., stereotaxic frame) and injection (injection module) elements. The primary components of the former include (**a**) non-rupture ear bars, (**b**) 100 μm three-axes manipulator arm, (**c**) a corner clamp probe holder, and (**d**) an adaptor. The latter is composed of (**e**) nanojet control box and (**f**) nanojet syringe header or injector. Detailed functions of individual components are described in Subheading 2.1

2   Materials

2.1   CED Apparatus

The CED apparatus is composed of two main components, including the ultraprecise rodent "U"-stereotaxic frame (Stoelting, Wood Dale, IL) and the Chemyx NanoJet injector module (Chemyx, Stafford, TX) (Fig. 3). The primary components of the stereotaxic frame are (a) non-rupture ear bars, which help place rodent head flat and parallel to the table of operation; (b) 100 micron three-axes manipulator arm, which allows movement of a probe holder in x-, y-, and z-axis directions (measured in mm); (c) a corner clamp probe holder that grips the syringe header; and (d) an adaptor where the rodent head is placed. The second component of the CED apparatus, injection module, consists of (e) NanoJet control box where infusion parameters, including infusion rate, volume, and type of syringe, are selected and (f) NanoJet syringe header, which is connected to the control box and holds the syringe in place. We also refer this part to "injector," as it controls the infusion rates as programmed by the control box.

2.2 Syringe and Cleaning Solutions

The infusate solution carrying therapeutics or NPs is loaded into a 5 or 50 µL Hamilton Neuros syringe (Hamilton, Reno, NV) connected to a 33-gauge needle for treating mice or rats, respectively. This syringe is designed to endow an enhanced needle rigidity specifically for applications to brains, which allows infusion of a micro-volume of fluid precisely into a desired location within the brain while minimizing the risk of injection site damage [110]. There are five different types of needle catheters used in CED, including end-port catheter, multi-port catheter, porous-tipped catheter, balloon-tip catheter, and stepped-down catheter [111] (*see* [21] for details). We routinely use the stepped-down catheter designed to allow feasible adjustment of a Hamilton Neuros syringe. This catheter is composed of a needle sleeve with or without a blind stop and a needle that passes through the sleeve hollow. The distance between the sleeve end and needle tip or the length of the needle exposed from the sleeve end is defined as the step distance, which can be manually adjusted to range 1–20 mm [112]. For the former setup (with a blind stop), the step distance is predetermined and maintained by the sleeve-end blind stop placed on top of the skull during the administration. In contrast, needle sleeve of the latter passes through the burr hole and the step distance can be subsequently adjusted for desired stereotaxic applications. The term "stepped" originated from a sharp dimensional transition from the wider needle sleeve to a narrower needle [112]. Of note, a 1-mm step distance has been previously optimized to minimize the possibility of reflux and leakage in rodent brain (Fig. 4) [112] and thus most widely employed for CED application.

A fresh solvent used to carry and administer therapeutics and/or NPs (e.g., ultrapure water, medical-grade saline, etc.) is

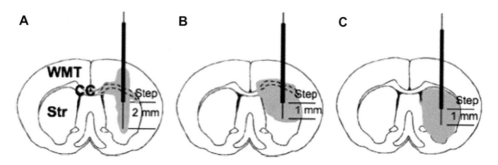

Fig. 4 Adjustment of stepped-down catheter for optimal infusion into the striatum. (**a**) Either a long (i.e., 2 mm) step distance or (**b**) an optimal 1-mm step distance with a short cortical distance (i.e., distance between cerebral cortex and catheter step or the end of needle sleeve) increases the probability of reflux and/or leak, leading to a suboptimal infusate distribution (gray shade). (**c**) A 1-mm step distance with a longer cortical distance minimizes the probability of reflux and/or leakage, thereby leading to a greater infusate distribution. (Reproduced from Yin, D. et at. 2010 with permission from Elsevier)

utilized to rinse the syringe between sequential infusions when multiple CED experiments are involved. A biodegradable and non-detergent-based cleaning agent such as the Cleaning Solution Concentrate (Hamilton, Reno, NV) is used to rigorously wash the syringe and needle upon the completion of CED experiments in order to avoid potential clogging and/or contamination. Ultrapure water in conjunction with ethanol or acetone is used for final wash.

2.3 Preparation of NPs and Infusate Solutions

Based on aforementioned physicochemical design criteria required for efficient brain tissue penetration (*see* Subheading 1.4; [54, 56]), we have engineered numerous BPN formulations for delivery of a wide array of payloads, ranging from small molecule chemotherapeutics [63, 74] to reporter or therapeutic nucleic acids [65, 71, 72]. For the purpose of this chapter, however, we use commercially available latex beads possessing highly controlled particle diameters and defined surface chemistries as a model therapeutic delivery NP. In particularly, we utilize fluorescently labeled carboxylated polystyrene beads (PS-COOH) (Life Technologies, Grand Island, NY) to engineer model BPNs (i.e., PS-PEG) [56], while the unmodified PS-COOH NPs serve as a conventional control that readily interacts with brain ECM components and thus cannot efficiently penetrate the brain parenchyma [56]. Methoxy-PEG-NH_2 (5 kDa; Creative PEGWorks, Durham, NC), *N*-hydroxysulfosuccinimide (NHS; Sigma, St. Louis, MO), 1-ethyl-3-(3-dimethylaminopropyl) carbodiimide (EDC; Invitrogen, Carlsbad, CA), and 200 mM sterile sodium borate buffer pH 8.2 (Growcells, Irvine, CA) are used to synthesize model BPNs possessing dense surface PEG coatings. All NPs are washed in Amicon Ultra 0.5 mL 100 K MW filters (Millipore Sigma, Burlington, MA) and characterized with a Zetasizer Nano ZS (Malvern Instruments, Columbia, MD) and ^{1}H NMR (400 MHz; REM400; Bruker, Billerica, MA).

Saline-based infusate solutions possessing different osmolality are prepared in ultrapure DNase/RNase-free water (Thermo Fisher Scientific, Waltham, MA) and filtered through a 0.20 μm sterile syringe filter (Corning Incorporated, Corning, NY). Specifically, we use 0.9% (~300 mOsm/kg) or 3% (~1000 mOsm/kg) NaCl solution as an isotonic or hypertonic infusate solution, respectively. Osmolality is measured using the Vapro® Vapor Pressure Osmometer (EliTech Wescor, Logan, UT).

2.4 Animal Setup

While technically any type of rodents can be used, we routinely conduct CED experiments with two different strains each for mice (20–30 g), including CF-1 (Jackson Laboratory, Bar Harbor, ME) and C57BL/6J mice (Charles River, Wilmington, MA), and rats (120–200 g), including Fischer 344 and Sprague Dawley rats (Harlan Laboratories, Frederick, MD). Of note, inbred rodents are often preferred when an identical genetic background is desired, such as in case of gene therapy applications [113]. Animals are housed in a standard animal husbandry facility and are given a free access to food and water. All animals are treated in accordance with the policies and guidelines of the Johns Hopkins University Animal Care and Use Committee.

A mixture of ketamine and xylazine is used to anesthetize rodents as routinely conducted [114]. Optixcare eye lubricant (CLC Medica, Ontario, Canada) and Betadine® antiseptic solution (Purdue Pharma L.P., Stamford, CT) are used for lubricating eyes and sterilizing skin, respectively, prior to surgery. A scalpel (Aspen Surgical, Caledonia, MI) equipped with a sterile surgical blade (Cincinnati Surgical, Cincinnati, OH) is used to create a midline scalp incision, and a micro-drill (Harvard Apparatus, Holliston, MA) is utilized to make a small 1.0-mm burr hole through the rodent skull. We use biodegradable sutures (Polysorb Braided Absorbable Sutures 5-0; Medtronic, Minneapolis, MN) to seal the incision and bacitracin as a topical antibiotic, following CED experiments.

2.5 Tissue Processing and Confocal Microscopy

For brain tissue fixation and cryoprotection, we use a ready-to-use working solution composed of 4% formaldehyde (i.e., equivalent to 10% formalin) dissolved in a phosphate buffer (Sigma-Aldrich, St. Luis, MO) and gradient sucrose (Sigma-Aldrich, St. Luis, MO) solutions prepared by dissolving sucrose in Dulbecco's phosphate-buffered saline (DPBS, 1×) (Corning™, Manassas, VA) at 10%, 20%, and 30% w/v, respectively. Tissue-Tek optimal cutting temperature (OCT) compound (Sakura Finetek, Torrance, CA) is used to prepare frozen tissue samples for subsequent cryo-section with a research cryostat (model CM3050 S, Leica Biosystems, Buffalo Grove, IL). Dako fluorescence mounting medium (Dako, Carpinteria, CA) is utilized to mount cryosectioned brain tissue slices. Confocal microscopy of the tissue slices is conducted by using a Zeiss LSM 710 confocal laser scanning microscope (Carl Zeiss Microscopy, Jena, Germany) equipped with a 5×/0.25 M27 air objective (Carl Zeiss Microscopy, Jena, Germany). We use Metamorph® image analysis software (Metamorph, Sunnyvale, CA), ImageJ (NIH, Bethesda, MD), and Imaris (Bitplane, South Windsor, CT) for image analysis and/or reconstruction.

3 Methods

The outcomes of CED applications are contingent to the selection of different variables, including device arrangement, infusion parameters, brain coordinates, and NP design. We here describe how we adjust these variables to achieve widespread distribution of model NPs following CED in rodent brains.

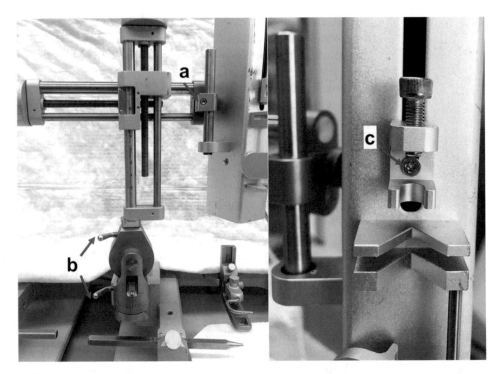

Fig. 5 CED device arrangement. All components should be fully stabilized prior to CED experiments. Specifically, it is critical to securely fasten (**a**) the corner clamp probe holder to firmly hold the syringe header during the procedure, (**b**) the base of three-axes manipulator arm to avoid its rotation, and (**c**) the screw on the syringe header must be fastened to ensure that a syringe is tightly associated with the header during the infusion

3.1 CED Device Arrangement

To ensure consistent and stable infusions, each component of the CED apparatus must be assembled and adjusted properly following the manufacturer's manual. While the syringe header (i.e., injector) presses the syringe plunger for infusion as programmed by the control box, all other components should be fully stabilized. Specifically, it is critical to securely fasten the base of the three-axes manipulator arm to avoid its rotation and the corner clamp probe holder to firmly hold the syringe header during the procedure (Fig. 5). In addition, the screw on the syringe header must be fastened to ensure that a syringe is tightly associated with the header during the infusion. The stereotaxic frame must be placed on a flat surface on which a rodent is laid facedown for a subsequent CED experiment.

3.2 Determination of Infusion Parameters

Infusion parameters are generally perceived as important factors that dictate the final outcomes of CED. However, reports that have systemically evaluated the impacts of infusion parameters are rare. It is conceivable that increase in the infusion rate, by potentially increasing the pressure gradient, may improve the distribution of therapeutics or NPs within the brain tissue following CED.

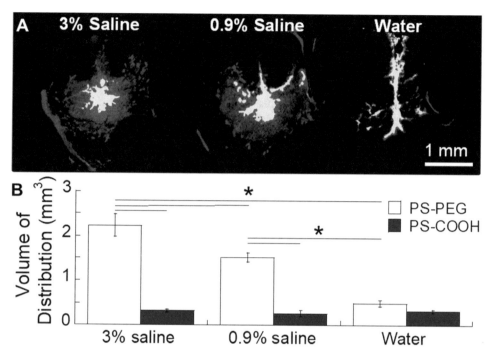

Fig. 6 Effect of infusate osmolality on in vivo distribution of NPs in mouse brains following CED. (**a**) Representative tissue sections depicting the coronal plane within the mouse striatum where PS-PEG (red) and PS-COOH (green) NPs were infused via CED. Yellow fluorescence represents overlay of two different types of NPs. Scale bar = 1 mm. (**b**) Volumetric distribution of PS-PEG and PS-COOH NPs quantified by an image-based MATLAB analysis described in Subheading 3.5. *$p < 0.05$ denotes a statistically significant difference. (Reproduced from Zhang, C. et al. 2017 with permission from Elsevier)

However, greater infusion rates have been shown to enhance reflux, thereby offsetting their otherwise positive impacts on the distribution [115]. For example, it has been demonstrated that the difference in distribution of soluble proteins (i.e., bovine serum albumin; 69 kDa) in rat brains was negligible regardless of infusion rates ranging from 0.1 to 5.0 μL/min due to the greater refluxes at higher rates [25]. However, it has been also shown that the reflux can be minimized by a careful optimization of the step distance where no significant reflux was observed up to the infusion rate of 10 μL/min when 1-mm step distance was employed [116]. We note that infusion rates of 0.1 to 1 μL/min and 0.33 to 5 μL/min are commonly used for mice [85, 117–119] and rats [38, 116, 120–122], respectively. Based on these previously reported ranges and our in-house optimization, we infuse overall volumes of 2 and 20 μL at rates of 0.2 and 0.33 μL/min for mice and rats, respectively [38, 72, 73].

As described above, hypertonic infusate solution can enhance volume of NP distribution by osmotically increasing the brain ECM pore sizes [77, 78] when NP diffusion in the brain tissue is primarily hindered by steric hindrance imposed by the ECM mesh (e.g., BPN or PS-PEG NP) (Fig. 6) [38]. However, conventional NPs (e.g., PS-COOH NP) that readily interact with ECM components are unable to fully exploit this mechanism due to the contribution of adhesion on their diffusion in brain tissue [38]. It should be also noted that the colloidal stability of NPs must be retained in a hypertonic infusate solution as the increase in NP sizes in an infusate solution can offset the effect of infusate-mediated enlargement of the ECM pores on steric hindrance to NP diffusion [38].

3.3 NP Formulation and Characterization

Model BPNs are engineered by amidation reaction between carboxyl groups on the surface of fluorescently labeled PS-COOH NPs and methoxy-PEG-NH$_2$, following a previously reported protocol [123, 124] with some modification. Briefly, PS-COOH NPs are diluted in ultrapure water to the NP concentration of 1.25% w/v (i.e., fourfold dilution from the stock concentration), followed by an addition of methoxy-PEG-NH$_2$ at two-molar equivalents of the surface carboxyl groups. Subsequently, 10-molar equivalents (to methoxy-PEG-NH$_2$) of NHS is added to the mixture and diluted with a 200 mM borate buffer to the final NP concentration of 0.375% w/v. Finally, 1-molar equivalent (to methoxy-PEG-NH$_2$) of EDC is added and mixed until all reagents are fully dissolved. The particle suspension is incubated at 25 °C on a rotary incubator overnight, washed in Amicon Ultra 0.5 mL 100 K MW filters by extensive centrifugation to remove unreacted species as well as to recover the starting NP concentration (i.e., 5% w/v), and then stored at 4 °C until use.

NPs are suspended in 10 mM NaCl for fundamental physicochemical characterization via a Zetasizer Nano ZS. Specifically, hydrodynamic diameter and polydispersity index are measured by dynamic light scattering (DLS) with 90° scattering optics. The ζ-potential, an indicative of NP surface charge, is determined by laser Doppler anemometry, also known as laser Doppler velocimetry. Table 3 demonstrates physicochemical properties of PS-COOH and PS-PEG (i.e., BPN) NPs described above. The surface PEG density is quantified by ^{1}H NMR as previously reported [56]. Briefly, PS-PEG NPs are fully dissolved in a mixture

Table 3
Physicochemical properties of PS NPs

Particle types	Hydrodynamic diameter ± SEM (nm)[a]	Polydispersity index (PDI)[a]	ζ-potential ± SEM (mV)[b]
PS-COOH	42 ± 10	0.05	−39 ± 3
PS-PEG	69 ± 8	0.04	−3 ± 1

[a]Hydrodynamic diameter and polydispersity index (PDI) were measured by dynamic light scattering (DLS) in 10 mM NaCl at pH 7.0. Mean ± SEM ($n \geq 3$)
[b]ζ-potential was measured by laser Doppler anemometry in 10 mM NaCl at pH 7.0. Mean ± SEM ($n \geq 3$)

of deuterated chloroform and trifluoroacetic acid containing 0.5% w/v BTSB and subjected to ^{1}H NMR measurement. Total amount of PEG molecules on NP surfaces is then determined using a standard curve established with ^{1}H NMR integrals measured at various concentrations of 5 kDa methoxy-PEG-NH$_2$ (3.6 ppm) dissolved in the same deuterated solvent mixture. In parallel, NP surface area is estimated by the hydrodynamic diameter of NP measured by DLS and the density of PS provided by the manufacturer (i.e., 1.055 g/mL). The surface PEG density is then calculated by dividing the total amount of PEG molecules with NP surface area. The colloidal stability of NPs in a physiologically relevant brain microenvironment is assessed by incubation of NPs in artificial CSF at 37 °C, followed by DLS measurement at different time points post-incubation. Likewise, the colloidal stability of NPs in a predetermined infusate solution over time is assessed by DLS. Fresh NPs are store at 4 °C and diluted to 1 mg/mL in an infusate solution prior to CED experiments.

3.4 Animal Preparation and Execution of CED

Rodents are anesthetized by an intraperitoneal administration of ketamine and xylazine at 75 and 7.5 mg/kg, respectively. We then lubricate eyes, shave the hair on the head, and sterilize the skin with an antiseptic solution. Subsequently, a midline scalp incision is made to expose the coronal and sagittal sutures of the skull and a small 1-mm burr hole is made by carefully drilling through the skull. It is critical not to damage the brain tissue (e.g., hemorrhage) during this procedure by perforating beyond the skull.

We sequentially wash the Hamilton Neuros syringe with ethanol or acetone and with ultrapure water or an infusate solution prior to the loading of the syringe with a NP-suspended infusate solution. The syringe filled with a NP suspension is then vertically mounted onto the NanoJet syringe header. Of note, the syringe is subjected to the aforementioned two-step wash between consecutive injections to avoid potential NP contamination when multiple CED experiments with different NP types are involved.

We most routinely conduct CED experiments targeting the striatum where several preclinical neurological disease as well as orthotopic brain tumor models are often established [125], while several other anatomical locations can potentially be target sites (e.g., hippocampus for Alzheimer's disease). The syringe is lowered to a depth of 2.5 mm from the dura at the coordinate of 2 mm lateral to and 0.5 mm anterior to bregma for mice; and to a depth of 3.5 mm from the dura at the coordinate of 3 mm lateral to and 0.5 mm posterior to bregma for rats (Fig. 7) [126]. The pre-infusion sealing time of 5–10 min is applied after inserting the catheter into the brain in order to minimize tissue damage and to provide tissue-syringe equilibration (*see* Fig. 8 for animal setup) [25], and then infusion is commenced as programmed by the control box. The syringe should be withdrawn slowly (i.e., at a

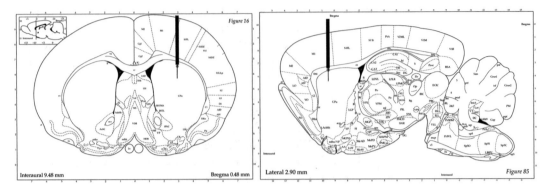

Fig. 7 Coordinates to infusing therapeutics or NPs into rodent striatum. A stepped-down catheter with a 1-mm step distance points: a rat coordinate of medial-lateral (ML) 3 mm, anterior-posterior (AP) 0.5 mm, and dorsal-ventral (DV) 3.5 mm

Fig. 8 Final CED setup with rodents mounted onto stereotaxic frames. Rats are under anesthesia, and stepped-down catheters are inserted into rat brains targeting striatum (i.e., a coordinate in Fig. 7b)

rate of 1 mm/min) 5 min after the completion of a CED experiment to minimize the reflux. Subsequently, animals are placed on a heating pad for quick recovery while the skin opening is washed with a sterile medical-grade saline and sutured with biodegradable sutures, followed by topical application of an antiseptic.

3.5 Tissue Processing and Imaging Analysis

Using the model BPNs, we have confirmed the previous finding that the volume of NP distribution remains consistent from immediately up to 24 h after the completion of CED [41, 122]. We thus routinely harvest brain tissues immediately after the infusion and fix the tissue with 4% paraformaldehyde for 24 h. Subsequently, the brain tissues are immersed in cryoprotective gradient sucrose solutions and then frozen at -80 °C. For the cryosection, frozen brain tissues are first equilibrated in a cryostat for more than 30 min at a chamber and an operating temperature of -24 and -22 °C, respectively. The brain tissues are then mounted on the tissue stage, firmly secured with the OCT compound and sectioned at ± 2 mm of the coronal infusion plate with a section thickness of 50 and 100 μm for mice and rats, respectively. Slides with tissues sections are subsequently fixed with the Dako fluorescence mounting medium and stored at 4 °C until confocal microscopy. While fixed tissue samples are generally considered amenable to a long-term storage at 4 °C [127, 128], it has been also reported that the autofluorescence often observed with fixed samples increases with the storage period [129]. We thus conduct confocal microscopy of tissue sections no later than a few days of CED experiments.

We acquire 2D images of fluorescent NP distribution within the striatal tissue sections captured through the $5\times/0.25$ M27 air objective using a Zeiss LSM 710 confocal laser scanning microscope. We start from the tissue section of the coronal infusion plate and take images of all sections exhibiting NP fluorescence beyond background. The laser power and the master gain are carefully adjusted using the range indicator to prevent saturation of fluorescence. Other software settings, including but not limited to pinhole size, scan zoom, and tile numbers, are also adjusted depending on type of fluorophore, region of interest, thickness of tissue sections, etc.

It is critical to properly subtract background fluorescence to accurately quantify the NP fluorescence signal and its coverage area. We use a custom-made MATLAB script featuring a binarization algorithm named Otsu's thresholding that distinguishes the pixels with positive (i.e., foreground) fluorescence signals from those with background signals [55]. The code then quantifies background-subtracted NP-positive area in each tissue section and the volume of NP distribution is finally calculated by multiplying the measured areas with the thickness between individual sections. Images of individual sections are stacked and aligned by Metamorph® image analysis software and StackReg plugin functionality of ImageJ, respectively. Finally, 3D-rendered volume of NP distribution is created by Imaris (Bitplane, South Windsor, CT) at 10% maximum fluorescent intensity (Fig. 9).

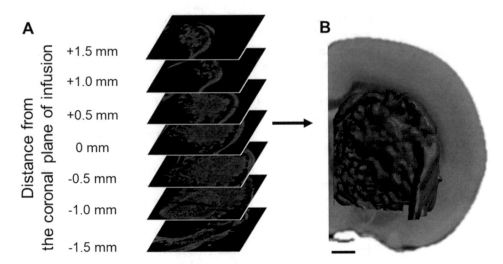

Fig. 9 Workflow for image-based analysis of NP distribution in the brain. (a) Representative 2D images of consecutive rat brain tissue sections away from the coronal plane of infusion captured by confocal microscopy. Red fluorescence indicates model BPNs (i.e., PS-PEG NPs) administered into the rat striatum via CED. (b) Representative 3D-rendered volumetric distribution of BPNs created by stacking and aligning the series of 2D images. Scale bar = 1 mm. The detailed procedure is described in Subheading 3.5

4   Notes for Troubleshooting

4.1   Cleaning of Needle Sleeve

Use of the stepped-down catheter without a blind stop involves insertion of the sleeve-end into the brain tissue. During the catheter insertion and subsequent CED experiments, brain tissue residues and/or blood often squeezes into the sleeve hollow, which cannot be cleared by the standard cleaning procedure described in an earlier section. We thus recommend that the sleeve should be slid toward the syringe to expose the needle beyond the step distance and completely remove the remnant biological specimen by alcohol swipe. It should be noted that residual tissues and/or blood clots can potentially contaminate brain tissues of rodents undergoing future CED experiments.

4.2   Removal of Air Bubbles

The syringe plunger should be pulled up slowly to minimize the aspiration of air into the syringe during the sample loading. Nevertheless, we occasionally encounter air bubble accumulation within needle and/or syringe, which often promotes pressure spikes and reflux during the CED experiments, thereby leading to suboptimal distribution of therapeutics or NPs in the brain [7, 130]. We also note that air bubbles in the infusate solution would be included in a preset volume of infusion and thus a fraction of sample would be left within the syringe without being administered (i.e., infusion of a lower dose than planned). We thus recommend that the syringe is loaded with an excess of sample volume and purged at a high infusion rate (i.e., no greater than 16.7 μL/min) for a few minutes to remove air bubbles prior to the catheter insertion into the rodent brain [39, 131].

5 Conclusion

In this book chapter, we have introduced a marriage of CED and BPN as an attractive strategy to achieve widespread distribution of therapeutic or diagnostic payloads in the brain following local administration. We have walked through rationales, materials, methods, and a few troubleshooting remarks based on our expertise and experience as well as findings from relevant studies reported by excellent scientists and clinicians devoted to the field. The consolidated efforts have led to multiple clinical trials as described earlier and the safety of CED procedure is now well established in humans. However, later stage trials have revealed technical shortcomings that need to be addressed to fulfil its therapeutic potential, which include design and refinement of CED compartments and parameters [21]. In addition, combined approach of CED and NP-based delivery systems, including BPNs, is yet at its infancy with most of the studies at their preclinical stages. Encouragingly, the field has recently experienced expansion regarding simultaneous optimization of CED procedure and NP engineering, and lessons learned will pave a way for its clinical development.

References

1. Wohlfart S, Gelperina S, Kreuter J (2012) Transport of drugs across the blood-brain barrier by nanoparticles. J Control Release 161(2):264–273

2. Kroll RA, Neuwelt EA (1998) Outwitting the blood-brain barrier for therapeutic purposes: osmotic opening and other means. Neurosurgery 42(5):1083–1099; discussion 1099-100

3. Vykhodtseva N, McDannold N, Hynynen K (2008) Progress and problems in the application of focused ultrasound for blood-brain barrier disruption. Ultrasonics 48(4):279–296

4. Kennedy JE (2005) High-intensity focused ultrasound in the treatment of solid tumours. Nat Rev Cancer 5(4):321–327

5. Curley CT et al (2017) Focused ultrasound immunotherapy for central nervous system pathologies: challenges and opportunities. Theranostics 7(15):3608–3623

6. Patel T et al (2012) Polymeric nanoparticles for drug delivery to the central nervous system. Adv Drug Deliv Rev 64(7):701–705

7. Mehta AM, Sonabend AM, Bruce JN (2017) Convection-enhanced delivery. Neurotherapeutics 14(2):358–371

8. Calias P et al (2014) Intrathecal delivery of protein therapeutics to the brain: a critical reassessment. Pharmacol Ther 144(2):114–122

9. Hammarlund-Udenaes M et al (2008) On the rate and extent of drug delivery to the brain. Pharm Res 25(8):1737–1750

10. Pardridge WM (2012) Drug transport across the blood-brain barrier. J Cereb Blood Flow Metab 32(11):1959–1972

11. Jain RK (1989) Delivery of novel therapeutic agents in tumors: physiological barriers and strategies. J Natl Cancer Inst 81(8):570–576

12. Oldendorf WH (1972) Cerebrospinal fluid formation and circulation. Prog Nucl Med 1:336–358

13. Cutler RW et al (1968) Formation and absorption of cerebrospinal fluid in man. Brain 91(4):707–720

14. Yan Q et al (1994) Distribution of intracerebral ventricularly administered neurotrophins in rat brain and its correlation with trk receptor expression. Exp Neurol 127(1):23–36

15. Fung LK et al (1998) Pharmacokinetics of interstitial delivery of carmustine, 4-hydroperoxycyclophosphamide, and paclitaxel from a biodegradable polymer implant in the monkey brain. Cancer Res 58 (4):672–684

16. Wolak DJ, Thorne RG (2013) Diffusion of macromolecules in the brain: implications for drug delivery. Mol Pharm 10(5):1492–1504

17. Fung LK et al (1996) Chemotherapeutic drugs released from polymers: distribution of 1,3-bis(2-chloroethyl)-l-nitrosourea in the rat brain. Pharm Res 13(5):671–682

18. Bobo RH et al (1994) Convection-enhanced delivery of macromolecules in the brain. Proc Natl Acad Sci U S A 91(6):2076–2080

19. Bruce JN et al (2011) Regression of recurrent malignant gliomas with convection-enhanced delivery of topotecan. Neurosurgery 69 (6):1272–1279; discussion 1279-80

20. Degen JW et al (2003) Safety and efficacy of convection-enhanced delivery of gemcitabine or carboplatin in a malignant glioma model in rats. J Neurosurg 99(5):893–898

21. Jahangiri A et al (2017) Convection-enhanced delivery in glioblastoma: a review of preclinical and clinical studies. J Neurosurg 126(1):191–200

22. Lieberman DM et al (1995) Convection-enhanced distribution of large molecules in gray matter during interstitial drug infusion. J Neurosurg 82(6):1021–1029

23. Raghavan R et al (2006) Convection-enhanced delivery of therapeutics for brain disease, and its optimization. Neurosurg Focus 20(4):E12

24. Zhou Z, Singh R, Souweidane MM (2017) Convection-enhanced delivery for diffuse intrinsic pontine glioma treatment. Curr Neuropharmacol 15(1):116–128

25. Chen MY et al (1999) Variables affecting convection-enhanced delivery to the striatum: a systematic examination of rate of infusion, cannula size, infusate concentration, and tissue—cannula sealing time. J Neurosurg 90 (2):315–320

26. Kawakami K et al (2004) Distribution of bolus or convection-enhanced delivery of IL-13 receptor-directed cytotoxin to intracranial brain tumors in a mouse model. Cancer Res 64(7 Supplement):1246–1247

27. Lonser RR et al (2015) Convection-enhanced delivery to the central nervous system. J Neurosurg 122(3):697–706

28. Vogelbaum MA, Aghi MK (2015) Convection-enhanced delivery for the treatment of glioblastoma. Neuro Oncol 17(Suppl 2):ii3–ii8

29. Ung TH et al (2015) Convection-enhanced delivery for glioblastoma: targeted delivery of antitumor therapeutics. CNS Oncol 4 (4):225–234

30. Anderson RC et al (2013) Convection-enhanced delivery of topotecan into diffuse intrinsic brainstem tumors in children. J Neurosurg Pediatr 11(3):289–295

31. Barker FG 2nd et al (1998) Survival and functional status after resection of recurrent glioblastoma multiforme. Neurosurgery 42 (4):709–720; discussion 720-3

32. Miranpuri GS et al (2012) Gene-based therapy of Parkinson's disease: translation from animal model to human clinical trial employing convection enhanced delivery. Ann Neurosci 19(3):133–146

33. Song DK, Lonser RR (2008) Convection-enhanced delivery for the treatment of pediatric neurologic disorders. J Child Neurol 23 (10):1231–1237

34. Lonser RR et al (2007) Real-time image-guided direct convective perfusion of intrinsic brainstem lesions. Technical note. J Neurosurg 107(1):190–197

35. Martin Bauknight W et al (2012) Convection enhanced drug delivery of BDNF through a microcannula in a rodent model to strengthen connectivity of a peripheral motor nerve bridge model to bypass spinal cord injury. J Clin Neurosci 19(4):563–569

36. Rogawski MA (2009) Convection-enhanced delivery in the treatment of epilepsy. Neurotherapeutics 6(2):344–351

37. Haar PJ et al (2010) Quantification of convection-enhanced delivery to the ischemic brain. Physiol Meas 31(9):1075–1089

38. Zhang C et al (2017) Strategies to enhance the distribution of nanotherapeutics in the brain. J Control Release 267:232–239

39. Allard E, Passirani C, Benoit JP (2009) Convection-enhanced delivery of nanocarriers for the treatment of brain tumors. Biomaterials 30(12):2302–2318

40. Morrison PF et al (1994) High-flow microinfusion: tissue penetration and pharmacodynamics. Am J Phys 266(1 Pt 2):R292–R305

41. Singh R et al (2017) Volume of distribution and clearance of peptide-based nanofiber after convection-enhanced delivery. J Neurosurg 129(1):10–18

42. Asthagiri AR et al (2011) Effect of concentration on the accuracy of convective imaging

distribution of a gadolinium-based surrogate tracer. J Neurosurg 115(3):467–473

43. Blakeley J (2008) Drug delivery to brain tumors. Curr Neurol Neurosci Rep 8 (3):235–241

44. Heldin CH et al (2004) High interstitial fluid pressure - an obstacle in cancer therapy. Nat Rev Cancer 4(10):806–813

45. Teo CS et al (2005) Transient interstitial fluid flow in brain tumors: effect on drug delivery. Chem Eng Sci 60(17):4803–4821

46. Barua NU et al (2016) A novel implantable catheter system with transcutaneous port for intermittent convection-enhanced delivery of carboplatin for recurrent glioblastoma. Drug Deliv 23(1):167–173

47. Boiardi A et al (2008) Treatment of recurrent glioblastoma: can local delivery of mitoxantrone improve survival? J Neuro-Oncol 88 (1):105–113

48. Lopez KA et al (2011) Convection-enhanced delivery of topotecan into a PDGF-driven model of glioblastoma prolongs survival and ablates both tumor-initiating cells and recruited glial progenitors. Cancer Res 71 (11):3963–3971

49. Patchell RA et al (2002) A phase I trial of continuously infused intratumoral bleomycin for the treatment of recurrent glioblastoma multiforme. J Neuro-Oncol 60(1):37–42

50. Sonabend AM et al (2011) Prolonged intracerebral convection-enhanced delivery of topotecan with a subcutaneously implantable infusion pump. Neuro-Oncology 13 (8):886–893

51. Bienemann A et al (2012) The development of an implantable catheter system for chronic or intermittent convection-enhanced delivery. J Neurosci Methods 203(2):284–291

52. Gelperina S et al (2005) The potential advantages of nanoparticle drug delivery systems in chemotherapy of tuberculosis. Am J Respir Crit Care Med 172(12):1487–1490

53. Mudshinge SR et al (2011) Nanoparticles: emerging carriers for drug delivery. Saudi Pharm J 19(3):129–141

54. Seo YE, Bu T, Saltzman WM (2017) Nanomaterials for convection-enhanced delivery of agents to treat brain tumors. Curr Opin Biomed Eng 4:1–12

55. Song E et al (2017) Surface chemistry governs cellular tropism of nanoparticles in the brain. Nat Commun 8:15322

56. Nance EA et al (2012) A dense poly(ethylene glycol) coating improves penetration of large polymeric nanoparticles within brain tissue. Sci Transl Med 4(149):149ra119

57. Sykova E, Nicholson C (2008) Diffusion in brain extracellular space. Physiol Rev 88 (4):1277–1340

58. Zimmermann DR, Dours-Zimmermann MT (2008) Extracellular matrix of the central nervous system: from neglect to challenge. Histochem Cell Biol 130(4):635–653

59. George N, Geller HM (2018) Extracellular matrix and traumatic brain injury. J Neurosci Res 96(4):573–588

60. Frantz C, Stewart KM, Weaver VM (2010) The extracellular matrix at a glance. J Cell Sci 123:4195–4200

61. Suk JS et al (2016) PEGylation as a strategy for improving nanoparticle-based drug and gene delivery. Adv Drug Deliv Rev 99 (Pt A):28–51

62. Thorne RG, Nicholson C (2006) In vivo diffusion analysis with quantum dots and dextrans predicts the width of brain extracellular space. Proc Natl Acad Sci U S A 103 (14):5567–5572

63. Nance E et al (2014) Brain-penetrating nanoparticles improve paclitaxel efficacy in malignant glioma following local administration. ACS Nano 8(10):10655–10664

64. Xu Q et al (2015) Impact of surface polyethylene glycol (PEG) density on biodegradable nanoparticle transport in mucus ex vivo and distribution in vivo. ACS Nano 9 (9):9217–9227

65. Mastorakos P et al (2017) Biodegradable brain-penetrating DNA nanocomplexes and their use to treat malignant brain tumors. J Control Release 262:37–46

66. Cragg B (1980) Preservation of extracellular space during fixation of the brain for electron microscopy. Tissue Cell 12(1):63–72

67. Ferrer VP, Moura Neto V, Mentlein R (2018) Glioma infiltration and extracellular matrix: key players and modulators. Glia 66 (8):1542–1565

68. Naumann U et al (2013) Glioma cell migration and invasion as potential target for novel treatment strategies. Transl Neurosci 4 (3):314–329

69. Serwer LP, James CD (2012) Challenges in drug delivery to tumors of the central nervous system: an overview of pharmacological and surgical considerations. Adv Drug Deliv Rev 64(7):590–597

70. Berry S et al (2016) Enhancing intracranial delivery of clinically relevant non-viral gene vectors. RSC Adv 48(6):41665–41674

71. Mastorakos P et al (2015) Highly PEGylated DNA nanoparticles provide uniform and

widespread gene transfer in the brain. Adv Healthc Mater 4(7):1023–1033

72. Mastorakos P et al (2016) Biodegradable DNA nanoparticles that provide widespread gene delivery in the brain. Small 12 (5):678–685

73. Schneider CS et al (2015) Minimizing the non-specific binding of nanoparticles to the brain enables active targeting of Fn14-positive glioblastoma cells. Biomaterials 42:42–51

74. Zhang C et al (2017) Convection enhanced delivery of cisplatin-loaded brain penetrating nanoparticles cures malignant glioma in rats. J Control Release 263:112–119

75. Salmaso S, Caliceti P (2013) Stealth properties to improve therapeutic efficacy of drug nanocarriers. J Drug Deliv 2013:374252

76. Mogosanu GD et al (2016) Polymeric protective agents for nanoparticles in drug delivery and targeting. Int J Pharm 510(2):419–429

77. Kume-Kick J et al (2002) Independence of extracellular tortuosity and volume fraction during osmotic challenge in rat neocortex. J Physiol 542(Pt 2):515–527

78. Chen KC, Nicholson C (2000) Changes in brain cell shape create residual extracellular space volume and explain tortuosity behavior during osmotic challenge. Proc Natl Acad Sci U S A 97(15):8306–8311

79. Hersh DS et al (2018) MR-guided transcranial focused ultrasound safely enhances interstitial dispersion of large polymeric nanoparticles in the living brain. PLoS One 13(2):e0192240

80. Christine CW et al (2009) Safety and tolerability of putaminal AADC gene therapy for Parkinson disease. Neurology 73 (20):1662–1669

81. Fiandaca MS et al (2009) Real-time MR imaging of adeno-associated viral vector delivery to the primate brain. NeuroImage 47 (Suppl 2):T27–T35

82. Barton NW et al (1991) Replacement therapy for inherited enzyme deficiency--macrophage-targeted glucocerebrosidase for Gaucher's disease. N Engl J Med 324 (21):1464–1470

83. Voges J et al (2003) Imaging-guided convection-enhanced delivery and gene therapy of glioblastoma. Ann Neurol 54(4):479–487

84. Ren H et al (2003) Immunogene therapy of recurrent glioblastoma multiforme with a liposomally encapsulated replication-incompetent Semliki forest virus vector carrying the human interleukin-12 gene--a phase I/II clinical protocol. J Neuro-Oncol 64 (1-2):147–154

85. Chen PY et al (2013) Comparing routes of delivery for nanoliposomal irinotecan shows superior anti-tumor activity of local administration in treating intracranial glioblastoma xenografts. Neuro-Oncology 15(2):189–197

86. Butowski N et al (2014) A phase i study of convection-enhanced delivery of liposomal-irinotecan using real-time imaging with gadolinium in patients with recurrent high grade glioma. Neuro Oncol 16(Suppl 3):iii13

87. White E et al (2012) A phase I trial of carboplatin administered by convection-enhanced delivery to patients with recurrent/progressive glioblastoma multiforme. Contemp Clin Trials 33(2):320–331

88. Patel SJ et al (2005) Safety and feasibility of convection-enhanced delivery of Cotara for the treatment of malignant glioma: initial experience in 51 patients. Neurosurgery 56 (6):1243–1252; discussion 1252-3

89. Carpentier A et al (2010) Intracerebral administration of CpG oligonucleotide for patients with recurrent glioblastoma: a phase II study. Neuro-Oncology 12(4):401–408

90. Carpentier A et al (2006) Phase 1 trial of a CpG oligodeoxynucleotide for patients with recurrent glioblastoma. Neuro-Oncology 8 (1):60–66

91. Randazzo D et al (2017) Phase 1 single-center, dose escalation study of D2C7-IT administered intratumorally via convection-enhanced delivery for adult patients with recurrent malignant glioma. J Clin Oncol 35 (15_suppl):e13532

92. Keu KV et al (2017) Reporter gene imaging of targeted T cell immunotherapy in recurrent glioma. Sci Transl Med 9:373

93. Kunwar S (2003) Convection enhanced delivery of IL13-PE38QQR for treatment of recurrent malignant glioma: presentation of interim findings from ongoing phase 1 studies. Acta Neurochir Suppl 88:105–111

94. Kunwar S et al (2007) Direct intracerebral delivery of cintredekin besudotox (IL13-PE38QQR) in recurrent malignant glioma: a report by the Cintredekin Besudotox Intraparenchymal study group. J Clin Oncol 25 (7):837–844

95. Kunwar S et al (2010) Phase III randomized trial of CED of IL13-PE38QQR vs Gliadel wafers for recurrent glioblastoma. Neuro-Oncology 12(8):871–881

96. Mueller S et al (2011) Effect of imaging and catheter characteristics on clinical outcome for patients in the PRECISE study. J Neuro-Oncol 101(2):267–277

97. Chittiboina P et al (2014) Magnetic resonance imaging properties of convective delivery in diffuse intrinsic pontine gliomas. J Neurosurg Pediatr 13(3):276–282

98. Vogelbaum MA et al (2007) Convection-enhanced delivery of cintredekin besudotox (interleukin-13-PE38QQR) followed by radiation therapy with and without temozolomide in newly diagnosed malignant gliomas: phase 1 study of final safety results. Neurosurgery 61(5):1031–1037; discussion 1037–8

99. Wersall P et al (1997) Intratumoral infusion of the monoclonal antibody, mAb 425, against the epidermal-growth-factor receptor in patients with advanced malignant glioma. Cancer Immunol Immunother 44(3):157–164

100. Weber FW et al (2003) Local convection enhanced delivery of IL4-pseudomonas exotoxin (NBI-3001) for treatment of patients with recurrent malignant glioma. Acta Neurochir Suppl 88:93–103

101. Saito R et al (2011) Regression of recurrent glioblastoma infiltrating the brainstem after convection-enhanced delivery of nimustine hydrochloride. J Neurosurg Pediatr 7(5):522–526

102. Lidar Z et al (2004) Convection-enhanced delivery of paclitaxel for the treatment of recurrent malignant glioma: a phase I/II clinical study. J Neurosurg 100(3):472–479

103. Mardor Y et al (2001) Monitoring response to convection-enhanced taxol delivery in brain tumor patients using diffusion-weighted magnetic resonance imaging. Cancer Res 61(13):4971–4973

104. Desjardins A et al (2014) Oncolytic polio/rhinovirus recombinant (PVSRIPO) in recurrent glioblastoma (GBM): first phase i clinical trial evaluating the intratumoral administration. Neuro Oncol 16(Suppl 3):iii43

105. Laske DW, Youle RJ, Oldfield EH (1997) Tumor regression with regional distribution of the targeted toxin TF-CRM107 in patients with malignant brain tumors. Nat Med 3(12):1362–1368

106. Weaver M, Laske DW (2003) Transferrin receptor ligand-targeted toxin conjugate (Tf-CRM107) for therapy of malignant gliomas. J Neuro-Oncol 65(1):3–13

107. Vogelbaum M et al (2017) Surg-10. Convection enhanced delivery of topotecan and gadolinium for recurrent GBM via the cleveland multiport catheter: a first in human study. Neuro Oncol 19(suppl_6):vi237

108. Sampson JH et al (2008) Intracerebral infusion of an EGFR-targeted toxin in recurrent malignant brain tumors. Neuro Oncol 10(3):320–329

109. Bogdahn U et al (2011) Targeted therapy for high-grade glioma with the TGF-beta2 inhibitor trabedersen: results of a randomized and controlled phase IIb study. Neuro Oncol 13(1):132–142

110. Company H. Syringe Care and Use. https://websites.pmc.ucsc.edu/~silab/pdf/Hamilton_Syringe%20Care%20and%20Use%20Guide.pdf

111. Debinski W, Tatter SB (2009) Convection-enhanced delivery for the treatment of brain tumors. Expert Rev Neurother 9(10):1519–1527

112. Yin D, Forsayeth J, Bankiewicz KS (2010) Optimized cannula design and placement for convection-enhanced delivery in rat striatum. J Neurosci Methods 187(1):46–51

113. Liu Y et al (2002) Strain-based genetic differences regulate the efficiency of systemic gene delivery as well as expression. J Biol Chem 277(7):4966–4972

114. Recinos VR et al (2010) Combination of intracranial temozolomide with intracranial carmustine improves survival when compared with either treatment alone in a rodent glioma model. Neurosurgery 66(3):530–537; discussion 537

115. Lewis O et al (2016) Chronic, intermittent convection-enhanced delivery devices. J Neurosci Methods 259:47–56

116. Krauze MT et al (2005) Reflux-free cannula for convection-enhanced high-speed delivery of therapeutic agents. J Neurosurg 103(5):923–929

117. Wang M et al (2017) A murine model for quantitative, real-time evaluation of convection-enhanced delivery (RT-CED) using an (18)[F]-positron emitting, fluorescent derivative of Dasatinib. Mol Cancer Ther 16(12):2902–2912

118. Sehedic D et al (2017) Locoregional confinement and major clinical benefit of (188)reloaded CXCR4-targeted Nanocarriers in an Orthotopic human to mouse model of glioblastoma. Theranostics 7(18):4517–4536

119. Lin CY et al (2018) Controlled release of liposome-encapsulated temozolomide for brain tumour treatment by convection-enhanced delivery. J Drug Target 26(4):325–332

120. Singleton WGB et al (2018) The distribution, clearance, and brainstem toxicity of panobinostat administered by convection-enhanced delivery. J Neurosurg Pediatr 22:288–296

121. Nordling-David MM et al (2017) Liposomal temozolomide drug delivery using convection enhanced delivery. J Control Release 261:138–146

122. Singleton WG et al (2017) Convection enhanced delivery of panobinostat (LBH589)-loaded pluronic nano-micelles prolongs survival in the F98 rat glioma model. Int J Nanomedicine 12:1385–1399

123. Lai SK et al (2007) Rapid transport of large polymeric nanoparticles in fresh undiluted human mucus. Proc Natl Acad Sci U S A 104(5):1482–1487

124. Popielarski SR, Pun SH, Davis ME (2005) A nanoparticle-based model delivery system to guide the rational design of gene delivery to the liver. 1. Synthesis and characterization. Bioconjug Chem 16(5):1063–1070

125. Huszthy PC et al (2012) In vivo models of primary brain tumors: pitfalls and perspectives. Neuro-Oncology 14(8):979–993

126. Paxinos G, Watson C (2018). The rat and mouse brain in stereotaxic coordinates. 2001–2006.   http://labs.gaidi.ca/rat-brain-atlas/

127. Grillo F et al (2015) Factors affecting immunoreactivity in long-term storage of formalin-fixed paraffin-embedded tissue sections. Histochem Cell Biol 144(1):93–99

128. Ono Y et al (2018) Quality assessment of long-term stored formalin-fixed paraffin embedded tissues for histopathological evaluation. J Toxicol Pathol 31(1):61–64

129. Davis AS et al (2014) Characterizing and diminishing autofluorescence in formalin-fixed paraffin-embedded human respiratory tissue. J Histochem Cytochem 62(6):405–423

130. Casanova F, Carney PR, Sarntinoranont M (2014) Effect of needle insertion speed on tissue injury, stress, and backflow distribution for convection-enhanced delivery in the rat brain. PLoS One 9(4):e94919

131. Sillay KA et al (2014) Image-guided convection-enhanced delivery into agarose gel models of the brain. J Vis Exp 14(87):51466

Chapter 8

Focused Ultrasound-Mediated Blood-Brain Barrier Disruption for Enhanced Drug Delivery to Brain Tumors

Pavlos Anastasiadis, Jeffrey A. Winkles, Anthony J. Kim, and Graeme F. Woodworth

Abstract

Transcranial focused ultrasound (FUS) in conjunction with intravenous circulating ultrasound contrast agents (UCAs) or microbubbles (MBs) can promote a transient and spatiotemporally selective increase in blood-brain barrier (BBB) permeability. The coupling of FUS with image guidance, such as magnetic resonance imaging (MRI)-guided FUS (MRgFUS), can further facilitate the visualization of targeted, localized therapeutic delivery to the central nervous system (CNS), potentially revolutionizing current treatment approaches for a spectrum of neurological diseases and conditions. Targeted and effective treatments of numerous CNS diseases are hindered by the BBB, a neurovascular structural and biological interface impeding the transport of molecules, particles, and cells between the systemic circulation and the underlying brain tissues. The functionality of FUS in conjunction with UCAs or MBs to safely and reversibly disrupt the BBB has become an area of intense research efforts both preclinically and clinically. Clinical efforts related to this approach are currently underway in patients with brain tumors, early Alzheimer's disease, and amyotrophic lateral sclerosis (ALS). Critical ongoing issues include maintaining and monitoring safe levels of ultrasound exposure within different brain regions and tissues. Ongoing work related to real-time monitoring of the levels of acoustic energy deposition within brain regions warrants further investigation and integration into preclinical and clinical FUS systems. Despite the technological advancements in the field and the great promise of this new approach, there is a relative void in the knowledge pertaining to the underlying biological and molecular alterations induced by FUS in the brain. In this chapter, we review basic concepts of the BBB, the cellular and molecular composition of the BBB, the acoustic emissions and bio-effects associated with FUS-mediated activation of UCAs, and provide a summary of recent advances in strategies for detecting, controlling, and mapping FUS activity in the brain.

Key words Focused ultrasound (FUS), Glioblastoma (GBM), Magnetic resonance imaging-guided focused ultrasound (MRgFUS), Targeted therapy, Microbubbles (MBs), Ultrasound contrast agents (UCAs), Blood-brain barrier (BBB), Neuroimmunology, Neurotherapeutics, Magnetic resonance imaging (MRI), Brain tumors, Central nervous system (CNS)

Vivek Agrahari et al. (eds.), *Nanotherapy for Brain Tumor Drug Delivery*, Neuromethods, vol. 163,
https://doi.org/10.1007/978-1-0716-1052-7_8, © Springer Science+Business Media, LLC, part of Springer Nature 2021

1   Blood-Brain Barrier and the Central Nervous System

The brain is one of the most perfused organs in the body. It is, therefore, not surprising that it comprises a vast network of specialized blood vessels. Among the most prominent vessels are cerebral arteries, which branch into progressively smaller arteries succeeded by arterioles, both of which are positioned along the surface of the brain and termed pial arteries [1]. Pial arteries consist of an endothelial monolayer, a smooth muscle layer, and an outer layer of leptomeningeal cells separated from the brain by the Virchow-Robin space. As the arterioles form deeper networks into the brain, the Virchow-Robin space is progressively replaced by the basal vascular membrane, which is shared with the astrocytic endfeet [2, 3].

Cerebral endothelial cells possess unique characteristics in that they are not fenestrated and form connections with neighboring endothelial cells by focal adhesions, termed tight junctions [4]. Smooth muscle cells, together with pericytes, exert control over hemodynamic patterns by regulating vasodilation and vasoconstriction and by converting biochemical signals from the associated endothelium, neurons, and astrocytes [4, 5].

Contrary to previous assumptions that the endothelium was a passive organ, vital for the conduit of blood flow, it has become increasingly evident that neurons, glia, and the cerebral microvasculature together form a modular neurovascular entity (Fig. 1) that, among other vital functions, plays a significant role in the regulation of cerebral hemodynamic responses to brain activity [4]. Another essential function of the neurovascular unit is to preserve the homeostasis in the cerebral microenvironment and maintain the BBB [6, 7]. While further modular elements exist within this unit (e.g., gliovascular unit), they go beyond the scope of this chapter and will not be discussed here. The interested reader is encouraged to refer to the review articles by Abbot et al. and Arvanitis et al. that discuss in detail all elements of the neurovascular unit [8, 9].

The BBB is a highly complex structure performing tasks that are critically important for maintaining cerebral homeostasis. The evolutionary conservation of the CNS's functional role across different species stands to witness the importance it plays [9]. Three main barriers in the CNS impede molecular exchanges at the interface between the blood and the underlying parenchyma or the fluid spaces. The BBB forms between the blood and the brain interstitial fluid, the choroid plexus epithelium is located between the blood and the ventricular cerebrospinal fluid (CSF), and, finally, the arachnoid epithelium forms between the blood and the subarachnoid CSF. This chapter is limited to the BBB, the role it plays in the

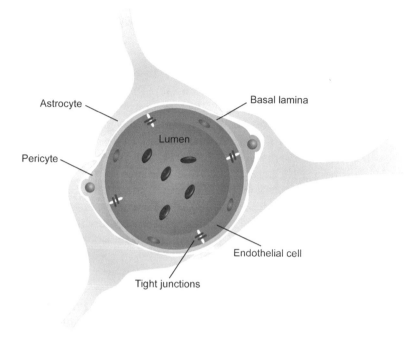

Fig. 1 Structure of the BBB. Depicted is the cross section of a cerebral capillary in the central nervous system (CNS) with surrounding astrocytic endfeet. Capillaries in the CNS share a common basal membrane with pericytes. More than 99% of the brain surface of the capillary basal lamina is engulfed by astrocytic endfeet with occasional neuronal innervation of the capillary (not shown). Vascular permeability of the BBB is predominantly regulated by endothelial cells, astrocytes, and intercellular tight junctions [124, 125]. The neurovascular unit is the chief regulator of transvascular transport in the CNS

pathophysiology of brain tumors, and how FUS can modulate it for achieving enhanced, temporary, and safe drug delivery [8].

In brain tumor pathologies, the BBB's function is modulated by endothelial-glial interactions resulting in enhanced capillary leakiness of some glial tumors compared to healthy brain tissue [10, 11]. It does not come as a surprise, therefore, that claudin-1 and claudin-5 expression have been reported to be altered in GBM along with tight junction morphologies [12]. The increase in cerebral microvascular permeability in human gliomas points to the dysregulation of junctional proteins [12, 13]. Moreover, there is an intricate relationship between the extracellular matrix (ECM) and the loss of astrocytic polarity due to glioma cell-mediated matrix metalloproteinase-3 (MMP-3) activity followed by the degradation of agrin and the redistribution of astrocytic aquaporin 4 (AQP4) [8, 11]. This cascade of events often leads to cerebral edema and brain swelling, as well as increases in intracerebral and interstitial pressures [11].

2 Blood-Brain Barrier and the Immunological Landscape of the Central Nervous System

The notion that the CNS is relatively immune-privileged has been based on early studies conducted by Peter Medawar over five decades ago, showing that implanting cells into the rodent brain led to their successful engraftment and growth. However, the same cells failed to engraft when injected into peripheral tissues outside the CNS [14–16]. Moreover, the perceived lack of an obvious lymphatic network and the presence of the BBB was seen as an additional barrier preventing peripheral immune cells from gaining entry into the brain [17, 18]. Recent findings have advanced our understanding of immunological mechanisms in the CNS, pointing to the presence of active immunosurveillance, which can be responsible for robust immune responses to specific antigens and pathogens [19–25]. While the CNS is compartmentalized including the brain parenchyma, the ventricles including the choroid plexus and the CSF, the meningeal layers enveloping the parenchyma, the BBB, blood-CSF, and the blood-leptomeningeal barrier, it has become progressively clear that the extent of immune-privilege varies strongly between its various compartments [21, 26]. Outside the brain parenchyma, immune surveillance is facilitated by bone marrow-derived dendritic cells (DCs) and macrophages lining the meninges, choroid plexus, and perivascular spaces [27–29]. The parenchyma appears to be the most immune-privileged compartment within the CNS, with microglia—specialized tissue macrophages—being the only resident leukocytes [20, 30, 31].

Neuroinflammatory conditions modulate the immunological landscape of the CNS, where resident immune cells are activated, and the brain parenchyma can be infiltrated by peripheral leukocytes [26, 32]. The underlying mechanisms regulating the transport of immune constituents into and out of the CNS have been reassessed after the discovery of a functional lymphatic system lining the dural sinuses capable of carrying both fluid and immune cells from the CSF [33]. While immune cells do not typically cross the BBB, the meninges surrounding the CNS constantly remain under immunosurveillance [34]. The meninges house diverse populations of both innate and adaptive immune cells capable of producing cytokines than can cross the BBB [33–35]. Besides, the meninges contain an intricate network of lymphatic vessels capable of draining soluble molecules originating from the brain parenchyma and the CSF [34, 36–39].

Cytokines, chemokines, and immune cells can modulate the BBB in many different ways [40]. Several reports have pointed to a delicate balance of homeostatic support by adaptive immune activity inside the CNS, demonstrating the importance of meningeal immune cells along with their derived cytokines in brain physiology [34, 41].

3   Glioblastoma

Gliomas are intrinsic brain tumors likely derived from neuroglial progenitors based on their anatomical location, the morphological similarities to nonmalignant neuroglial cells, and the generation of preclinical glioma models by targeting neural progenitor cells [42–44]. However, there is also evidence that GBM may arise from neural cells with more robust stem cell-like properties [45]. Brain tumors encompass a diverse range of CNS cancers, including astrocytomas, oligodendrogliomas, and ependymomas. Some of these tumors typically occur in childhood (e.g., pilocytic astrocytoma), while others, including GBM, have a peak incidence at an older age with a male and Caucasian predominance [24, 46]. GBM constitutes the most common primary malignant brain tumor accounting for 16% of the entirety of primary brain and CNS neoplasms [47]. The average age-adjusted incidence rate is 3.2 per 100,000 population [24, 48]. GBMs manifest almost exclusively in the brain but can also appear in the brain stem and the spinal cord. Among all primary gliomas, 61% occur in the four lobes of the brain, frontal (25%), temporal (20%), parietal (13%), and occipital (3%) [49]. While GBM tumors rarely metastasize to extracranial sites, circulating tumor cells and exosomes have been detected in patient blood [50–57]. The prevailing assumption is that glioma cells either lack survival mechanisms outside the brain tumor microenvironment or, due to the rapid progression of the disease, cells do not have adequate time to establish secondary colonies at distant extracranial sites [58].

Extensive research has been conducted to shed light on the molecular pathogenesis of GBM, including analyses for gene copy number variations, chromosomal alterations, and transcriptional signatures. The plethora of studies has reported dysregulation in at least three major signaling pathways: (1) the receptor tyrosine kinase and phosphatidylinositol 3-kinase (PI3K)-AKT axis, (2) the p53 tumor suppressor pathway, and (3) the retinoblastoma pathway involved in cell cycle progression [59–61]. Genetic alterations that have been recognized and are typical of primary GBM are (1) epidermal growth factor receptor (EGFR) overexpression, (2) mutations in the phosphate and tensin homolog (PTEN) gene, and (3) loss of chromosome 10q [62–64]. For a more in-depth review of GBM molecular aberrations, the interested reader is encouraged to refer to Brennan et al., Verhaak et al., and Wang et al. [65–67].

The standard of care treatment approach for primary GBM includes maximal safe tumor resection followed by concurrent chemoradiation therapy. Complete surgical resection is not feasible as tumor cells are generally invasive and are often embedded in normal brain parenchyma. Therefore, surgical resection is not curative, and tumor recurrence is nearly universal due to persistent heterogeneous, highly plastic populations of invading and stemlike tumor cells [63, 68–70]. Commonly used adjuvant chemotherapy drugs are carmustine (i.e., BCNU), PCV (i.e., procarbazine, lomustine, and vincristine), and temozolomide, an oral DNA alkylating agent [71–73]. These chemotherapeutic drugs prolong survival by only a few months and can be accompanied by significant systemic toxicities [74]. Radiotherapy typically consists of fractionated regional irradiation at a dose of 2 Gy per fraction delivered to the gross tumor volume at a margin of 2 cm to 3 cm around the resection cavity over a period of 6 weeks amassing to a total dose of 60 Gy [71]. New techniques that will allow cancer drugs to reach the CNS at therapeutically effective concentrations while minimizing off-target effects and systemic toxicities are needed.

The integrity of the glioma BBB is often heterogeneously compromised, rendering varying degrees of permeability within the same tumor. The BBB permeability in GBM — as is the case in all other brain tumors—peaks in the area of the tumor core and decreases at the tumor periphery [75–78]. At the same time, brain tumor cells can break away from the tumor core and migrate into healthy brain tissue far from the contrast-enhancing regions [79, 80]. As tumor-infiltrating regions retain their BBB functionality, they impede chemotherapeutics from crossing, thereby detrimentally affecting treatment efficacy and patient outcomes. The tumor-infiltrated margin regions around the resection cavity are also the sites where the almost unavoidable GBM recurrence commonly occurs. Therefore, targeting these brain regions could translate into improved treatment efficacy and prolonged survival [81].

Several approaches for the delivery of chemotherapeutic agents for GBM treatment have been developed over the years. Intravenously administered chemotherapeutics reach the tumor regions in the brain by crossing either the BBB or the blood-CSF barrier. Interstitial delivery involves either convection-enhanced delivery or the direct implantation of biodegradable wafer-containing drugs inside the resection cavity in order to evade the BBB [82–86]. Yet another approach consists of intra-arterial routes using osmotic or hypertonic solutions [87–89]. Each one of these techniques has its unique strengths, but they are all plagued with a series of severe complications, including lack of spatial specificity, neurological deficits, CSF leakage, seizures, new tumor nodule formation, intracranial hypertension, and brain edema [90–93].

4   Focused Ultrasound-Induced BBB Disruption for Enhanced Drug Delivery

Pertinent to the concept of improving therapeutic delivery within the CNS is MRgFUS coupled with intravenously administered UCAs [94] (Fig. 2). This relatively recent technology has emerged

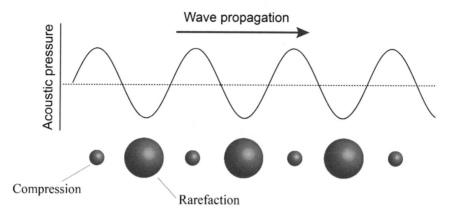

Fig. 2 Ultrasound contrast agents. Ultrasound contrast agents (UCAs) interact with focused ultrasound when they come into the vicinity of the acoustic field. Acoustic waves follow a sinusoidal pattern of compression and rarefaction. When acoustic pressures alternate between their peak minimum and peak maximum, they cause UCAs to expand and compress, respectively. The ensuing oscillation of UCAs induces mechanical stress that disrupts endothelial cell tight junctions, thereby mediating BBB disruption

as an attractive noninvasive alternative for the safe and transient disruption of the BBB in various pathologies, including brain tumors [75, 95, 96]. The frequencies used for the FUS-mediated BBBD are approximately threefold lower compared to the ablative frequencies used for the treatment of medically refractory essential tremor [75, 76]. UCAs in the vicinity of the FUS acoustic field (Fig. 3) cause a transient mechanical disruption of the BBB in the absence of thermal injury [70, 77, 78].

UCAs were originally developed for enhancing tissue contrast in ultrasound imaging and assist as diagnostic agents in the clinic. They are microspheres consisting of gas bubbles typically stabilized by a shell of albumin, lipid, or polymer. The coating material serves the purpose of reducing surface tension and prevents diffusion of the encapsulated gas in the core from escaping, thereby increasing the half-life of UCAs while in circulation. The most widely used UCAs are available from Definity, Imagent, Levovist, SonoVue (lipid coating), Optison (albumin coating), Imagify, and Sonovist (polymer coating) [79].

Among several technical challenges associated with transcranial FUS, a major obstacle is the presence of the cranium itself. The skull, with its considerably higher acoustic impedance compared to the soft brain tissue, reflects overwhelmingly the acoustic energy of the ultrasound beam. The amount of transmitted energy depends on the orientation of the transducer and the incident angle between the transducer and the target region [97]. Furthermore, the geometry and heterogeneity of the human calvarium present a challenge due to the attenuating and aberrating effects on ultrasound propagation [98]. The calvarium not only attenuates the ultrasound beam but also distorts it, thereby negatively influencing the

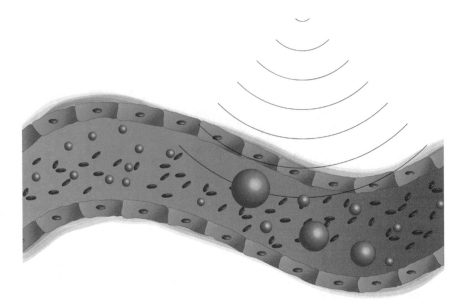

Fig. 3 Ultrasound contrast agents for BBB disruption. The expansion and compression of ultrasound contrast agents (UCAs) within an acoustic field contribute to the BBB disruption at lower and safer acoustic pressures than would be feasible without their presence. Higher acoustic pressures, while technically feasible, can induce significant tissue structural damage and cause microhemorrhages. The prime feature of UCAs in diagnostic ultrasound (e.g., echogenicity) allows them to be used clinically for optimal tissue contrast. In therapeutic ultrasound (e.g., BBB disruption, gene delivery) UCAs translate this oscillatory pattern into mechanical stress that is exerted onto endothelial cells, which is thought to lead to the mechanical disruption of tight junctions

acoustic energy deposited at the target location within the brain. The development of multielement hemispherical large-aperture ultrasound arrays consisting of hundreds of individually controlled transducers is one solution to this issue, which maximizes the sonication surface area [99, 100]. The individual components of these phased arrays can be rapidly steered electronically to multiple targets, while different elements can be independently inactivated or deactivated to reduce internal reflections [101].

Currently, ultrasonic beam aberrations in clinical MRgFUS systems are corrected based on previously acquired computed tomography (CT) scans of the patient's head. The CT scanning data are imported into the MRgFUS interface software during treatment planning and aligned with MR images obtained in situ to match the skull's position. Following this alignment, the software performs numerical simulations of acoustic wave propagation to determine the exact phase and amplitude corrections for each transducer element. The simulations are based on a three-layer model including the coupling water, the cranial bone, and the soft brain tissue [102, 103] or a heterogeneous model of the internal structure of the skull [104–106]. MRI is not commonly used for skull imaging due to the low MR signal density in cortical bone. Therefore, CT is the imaging modality of choice for the generation of a skull map as it provided excellent resolution with impeccable bone contrast [107].

Accurate targeting is of critical importance for disrupting the BBB at discretely defined locations. The magnitude of the interruption depends not only on the activity of UCAs but also on the degree of vascularity in the target brain region. Pre-treatment vascular imaging to delineate vascularity and perfusion rate might offer valuable information for treatment planning [108, 109].

For real-time monitoring and control, passive cavitation detection has been employed, allowing to record the diverging acoustic pressure waves emitted by oscillating UCAs during FUS-mediated BBB disruption [110–114]. Inertial cavitation appears in the frequency domain of the acoustic emissions as a broadband signal and is associated with vascular damage. Harmonic, subharmonic, or ultraharmonic acoustic emissions in the absence of a broadband signal are signifying stable volumetric oscillation that falls into the safe regime of BBB disruption [110–112, 115].

5 Focused Ultrasound-Induced BBB Disruption for Enhanced Drug Delivery in Preclinical Studies

In one study, FUS-mediated BBB disruption was used to enhance doxorubicin delivery in a rat 9L gliosarcoma cell model of intracranial tumor growth. FUS treatment reduced tumor growth that was accompanied by a modest median survival increase [116]. Puri et al. investigated the FUS-mediated pharmacokinetics of doxorubicin delivery in NOD-scid mice using a patient-derived GBM cell model. Animals receiving doxorubicin in combination with FUS treatment exhibited enhanced accumulation of the drug in the brain tumor region [117]. Another study investigated the potential for FUS-mediated gene therapy in the CNS. In this work, liposomes were used to carry plasmid DNA containing a luciferase reporter gene. The liposomal-plasmid DNA (LpDNA) complex exhibited improved cell transduction compared to plasmid DNA alone. Expression of the luciferase reporter gene in vivo was evaluated via an In Vivo Imaging System (IVIS) and quantified by immunoblotting and immunohistology. The study demonstrated improved FUS-mediated LpDNA delivery into the CNS that peaked at 2 days with a fivefold increase when compared to the untreated brain hemisphere [118]. Following up on the successful FUS-mediated delivery of LpDNA, the same group went further to conjugate LpDNA with UCAs prior to FUS exposures. This led to improvements in the delivery of the glial cell line-derived neurotrophic factor (GDNF) gene indicating yet another FUS-assisted vehicle system for enhanced delivery to the brain [119].

More recently, the use of FUS and liposomal-gene combination approach was used with the adoption of a phospholipid complex that was optimized in terms of gene cargo loading and peptide-mediated targeting to delay glioma growth. UCAs were conjugated with Birc5 small interfering RNA prior to intravenous injection following FUS-aided BBB disruption. This led to a significantly elevated therapeutic effect compared to the experimental groups where liposomal particles, siRNA, or UCAs were infused separately from one another demonstrating FUS' promising role in RNA interference approaches for the treatment of glioma [120]. These studies collectively demonstrate the utility of FUS-mediated BBB disruption for greatly enhancing brain therapeutic delivery.

As gene therapy approaches are gaining traction as a treatment modality for a wide spectrum of indications, it is not surprising that efforts have been underway to tackle brain disorders including brain tumors. In a study conducted by Mead et al., systemically administered DNA-bearing nanoparticles were delivered across the BBB after disrupting it by means of FUS and UCAs leading to a dose-dependent transgene expression only in the FUS-treated hemisphere. More importantly, this was achieved without astrocyte activation and with no indication of toxicity [121].

The ability of FUS to selectively deliver neurotherapeutic agents to brain regions at impressively high spatiotemporal precision levels creates promising opportunities not previously available in the therapeutic realm. Nanoparticle carriers that combine receptor-specific ligands conjugated onto their surface can surpass what has been reported thus far in terms of spatial specificity opening the way for treatments that require targeting of specific receptors within a defined region in the brain [122]. Lu et al. designed an angiopep-2-modified small poly(lactic-*co*-glycolic acid) (PLGA) hybrid nanoparticle (NP) drug delivery system for doxorubicin/perfluorooctyl bromide delivery (ANP-D/P). FUS and UCA-mediated BBB disruption at tumor sites in a murine GBM model exhibited strong antitumor efficacy with the longest survival time among all experimental groups indicating the FUS-responsive ANP-D/P presents a novel therapeutic avenue with the potential of clinical translation [123]. Indeed, as it has become clearer that FUS is a safe and valuable clinical tool, an increasing number of clinical studies are underway or have recently been completed. A list of ongoing or completed clinical trials utilizing the FUS-mediated BBB disruption approach can be found in Table 1.

Table 1
Focused ultrasound-mediated blood-brain barrier disruption studies in ongoing or completed clinical trials as listed in ClinicalTrials.gov

Title of study	ClinicalTrials.gov identifier	Condition or disease
Assessment of Safety and Feasibility of ExAblate Blood-Brain Barrier (BBB) Disruption for Treatment of Glioma	NCT03616860	Glioblastoma
Safety of BBB Opening with the SonoCloud	NCT02253212	Glioblastoma Glioma
ExAblate Blood Brain Barrier Disruption for Planned Surgery in Suspected Infiltrating Glioma	NCT03322813	Glioma
Assessment of Safety and Feasibility of ExAblate Blood-Brain Barrier (BBB) Disruption	NCT03551249	Glioblastoma Glioma
Blood-Brain Barrier Disruption Using Transcranial MRI-Guided Focused Ultrasound	NCT02343991	Brain Tumor
ExAblate Blood-Brain Barrier Disruption for Glioblastoma in Patients Undergoing Standard Chemotherapy	NCT03712293	Glioblastoma
Safety of BBB Disruption Using NaviFUS System in Recurrent Glioblastoma Multiforme Patients	NCT03626896	Glioblastoma Glioma
Blood Brain Barrier Disruption Using MRgFUS in the Treatment of Her2-positive Breast Cancer Brain Metastases	NCT03714243	Breast Cancer Brain Metastases
A Study to Evaluate Temporary Blood Brain Barrier Disruption in Patients with Parkinson's Disease Dementia	NCT03608553	Parkinson's Disease
Blood-Brain Barrier Opening Using MR-Guided Focused Ultrasound in Patients with Amyotrophic Lateral Sclerosis	NCT03321487	Amyotrophic Lateral Sclerosis
ExAblate Blood-Brain Barrier Opening for Treatment of Alzheimer's Disease	NCT03739905	Alzheimer's Disease
Blood Brain Barrier Opening in Alzheimer' Disease	NCT03119961	Alzheimer's Disease
ExAblate Blood-Brain Barrier Disruption for the Treatment of Alzheimer's Disease	NCT03671889	Alzheimer's Disease
Blood-Brain-Barrier Opening Using Focused Ultrasound with IV Contrast Agents in Patients with Early Alzheimer's Disease	NCT02986932	Alzheimer's Disease

The various trials have been color-coded to indicate similar conditions or diseases where applicable

6   Conclusions

FUS is an emerging technology with great potential to revolutionize the biomedical field. It has attracted significant interest from a range of different applications pertinent to image-guided and transcranial technologies, including brain tumors, neuromodulation, neurodegenerative indications, traumatic brain injury, and targeted drug delivery in the CNS. FUS is a relatively noninvasive technique compared to other approaches (e.g., convection-enhanced delivery) and does not require anesthesia or incision, which can be particularly risky procedures in immunocompromised patients undergoing chemoradiation treatments or elderly populations.

As the field of FUS moves toward increasing numbers of clinical trials, the primary concern will be safety and effective utilization. The clinical application of FUS-mediated BBB disruption will necessitate the development of robust indices to assess the quality of successful BBB disruption correlating it with the delivery of neurotherapeutics across the CNS. Passive cavitation dose (PCD) estimation applied to the activity of UCAs for the quantification of acoustic emissions will play a major role in this undertaking. PCD correlated with sub- and ultraharmonic dosing can provide not only real-time assessment of UCA activity pertinent to BBB disruption but most importantly provide a safety monitoring scheme to avoid adverse effects during treatment.

7   Future Perspectives

Neuroinflammatory and chronic alterations at the cytoarchitectural level in the acute stages of FUS have not been studied extensively. It is, therefore, imperative to elucidate and extensively investigate these potential effects in preclinical models ranging from small to large animals and nonhuman primates. Behavioral studies need to be an integral part of any future endeavors as no long-term effects of FUS in this respect have been negated or established. Efficiency is yet another aspect that remains critically important. Is FUS efficient enough to replace the existing standard of care practices in a safe and efficacious way? The results and studies around FUS have thus far been highly encouraging, an observation supported by the increasing number of FUS-related clinical trials and peer-reviewed publications.

The brain is a distinct site from an immunological point of view with complex anatomy, highly compartmentalized structure, and stark variations in terms of molecular composition between the individual compartments and intricate interaction between them. These restrictions have impeded the development of effective treatments for brain tumors. FUS being minimally invasive and incisionless can have a major impact on how we approach therapeutic treatments. There is a growing body of evidence for employing image-guided FUS and the integration of acoustic emission of oscillating UCAs to ensure safety and control efficacy of FUS-mediated BBB disruption for the targeted delivery of neurotherapeutics to the tumor microenvironment.

Acknowledgments

This work was supported, in part, by the National Institutes of Health (T32 Training Grant in Cancer Biology 5T32CA154274, R01NS107813, and R21NS113016) and the Focused Ultrasound Foundation.

References

1. Jones EG (1970) On the mode of entry of blood vessels into the cerebral cortex. J Anat 106(Pt 3):507–520

2. Rennels ML, Nelson E (Oct 1975) Capillary innervation in the mammalian central nervous system: an electron microscopic demonstration. Am J Anat 144(2):233–241. https://doi.org/10.1002/aja.1001440208

3. Cohen Z, Bonvento G, Lacombe P, Hamel E (Nov 1996) Serotonin in the regulation of brain microcirculation. Prog Neurobiol 50 (4):335–362

4. Iadecola C (2004) Neurovascular regulation in the normal brain and in Alzheimer's disease. Nat Rev Neurosci 5(5):347–360. https://doi.org/10.1038/nrn1387

5. Somlyo AP, Wu X, Walker LA, Somlyo AV (1999) Pharmacomechanical coupling: the role of calcium, G-proteins, kinases and phosphatases. Rev Physiol Biochem Pharmacol 134:201–234

6. Lo EH, Dalkara T, Moskowitz MA (2003) Mechanisms, challenges and opportunities in stroke. Nat Rev Neurosci 4(5):399–414. https://doi.org/10.1038/nrn1106

7. del Zoppo GJ, Mabuchi T (2003) Cerebral microvessel responses to focal ischemia. J Cereb Blood Flow Metab 23(8):879–894. https://doi.org/10.1097/01.Wcb.0000078322.96027.78

8. Abbott NJ, Rönnbäck L, Hansson E (2006) Astrocyte–endothelial interactions at the blood–brain barrier. Nat Rev Neurosci 7 (1):41–53. https://doi.org/10.1038/nrn1824

9. Arvanitis CD, Ferraro GB, Jain RK (2019) The blood–brain barrier and blood–tumour barrier in brain tumours and metastases. Nat Rev Cancer 20(1):26–41. https://doi.org/10.1038/s41568-019-0205-x

10. Dubois LG et al (2014) Gliomas and the vascular fragility of the blood brain barrier. Front Cell Neurosci 8:418–418. https://doi.org/10.3389/fncel.2014.00418

11. Wolburg H, Noell S, Fallier-Becker P, Mack AF, Wolburg-Buchholz K (2012) The disturbed blood-brain barrier in human glioblastoma. Mol Aspects Med 33 (5-6):579–589. https://doi.org/10.1016/j.mam.2012.02.003

12. Liebner S et al (2000) Claudin-1 and claudin-5 expression and tight junction morphology are altered in blood vessels of human glioblastoma multiforme. Acta Neuropathol 100 (3):323–331. https://doi.org/10.1007/s004010000180

13. Wolburg H et al (Jun 2003) Localization of claudin-3 in tight junctions of the blood-brain barrier is selectively lost during experimental autoimmune encephalomyelitis and human glioblastoma multiforme. Acta Neuropathol 105(6):586–592. https://doi.org/10.1007/s00401-003-0688-z

14. Medawar PB (Feb 1948) Immunity to homologous grafted skin; the fate of skin homografts transplanted to the brain, to subcutaneous tissue, and to the anterior chamber of the eye. Br J Exp Pathol 29(1):58–69

15. Billingham RE, Brent L, Medawar PB, Sparrow EM (1954) Quantitative studies on tissue transplantation immunity. I. The survival times of skin homografts exchanged between members of different inbred strains of mice. Proc R Soc Lond B Biol Sci 143(910):43–58

16. Billingham RE, Brent L, Medawar PB (1953) Actively acquired tolerance of foreign cells. Nature 172(4379):603–606. https://doi.org/10.1038/172603a0

17. Murphy JB, Sturm E (1923) Conditions determining the transplantability of tissues in the brain. J Exp Med 38(2):183–197. https://doi.org/10.1084/jem.38.2.183

18. Shirai Y (1921) On the transplantation of the rat sarcoma in adult heterogenous animals. Jpn Med World 1(14):14–15

19. Waksman BH, Adams RD (1955) Allergic neuritis: an experimental disease of rabbits induced by the injection of peripheral nervous tissue and adjuvants. J Exp Med 102 (2):213–236. https://doi.org/10.1084/jem.102.2.213

20. Ransohoff RM, Engelhardt B (2012) The anatomical and cellular basis of immune surveillance in the central nervous system. Nat Rev Immunol 12(9):623–635. https://doi.org/10.1038/nri3265

21. Shechter R et al (2013) Recruitment of beneficial M2 macrophages to injured spinal cord is orchestrated by remote brain choroid plexus. Immunity 38(3):555–569. https://doi.org/10.1016/j.immuni.2013.02.012

22. Kipnis J, Gadani S, Derecki NC (2012) Pro-cognitive properties of T cells. Nat Rev Immunol 12(9):663–669. https://doi.org/10.1038/nri3280

23. Reifenberger G, Wirsching HG, Knobbe-Thomsen CB, Weller M (2017) Advances in the molecular genetics of gliomas - implications for classification and therapy. Nat Rev Clin Oncol 14(7):434–452. https://doi.org/10.1038/nrclinonc.2016.204

24. Ostrom QT et al (2015) CBTRUS Statistical Report: Primary Brain and Central Nervous System Tumors Diagnosed in the United States in 2008-2012. Neuro Oncol 17(Suppl 4):iv1–iv62. https://doi.org/10.1093/neuonc/nov189

25. Louis DN et al (2016) The 2016 World Health Organization classification of tumors of the central nervous system: a summary. Acta Neuropathol 131(6):803–820. https://doi.org/10.1007/s00401-016-1545-1

26. Mrdjen D et al (2018) High-dimensional single-cell mapping of central nervous system immune cells reveals distinct myeloid subsets in health, aging, and disease. Immunity 48(2):380–395.e6. https://doi.org/10.1016/j.immuni.2018.01.011

27. Bechmann I et al (2001) Turnover of rat brain perivascular cells. Exp Neurol 168(2):242–249. https://doi.org/10.1006/exnr.2000.7618

28. Greter M, Lelios I, Croxford AL (2015) Microglia versus myeloid cell nomenclature during brain inflammation. Frontiers in Immunology, Short Survey 6:249. https://doi.org/10.3389/fimmu.2015.00249

29. Kivisakk P et al (2009) Localizing central nervous system immune surveillance: meningeal antigen-presenting cells activate T cells during experimental autoimmune encephalomyelitis. Ann Neurol 65(4):457–469. https://doi.org/10.1002/ana.21379

30. Ginhoux F et al (2010) Fate mapping analysis reveals that adult microglia derive from primitive macrophages. Science 330(6005):841–845. https://doi.org/10.1126/science.1194637

31. Schulz C et al (2012) A lineage of myeloid cells independent of myb and hematopoietic stem cells. Science 335(6077):86–90. https://doi.org/10.1126/science.1219179

32. Schreiner B, Heppner FL, Becher B (2009) Modeling multiple sclerosis in laboratory animals. Semin Immunopathol 31(4):479–495. https://doi.org/10.1007/s00281-009-0181-4

33. Louveau A et al (2015) Structural and functional features of central nervous system lymphatic vessels. Nature 523(7560):337–341. https://doi.org/10.1038/nature14432

34. Rankin LC, Artis D (2018) Beyond host defense: emerging functions of the immune system in regulating complex tissue physiology. Cell 173(3):554–567. https://doi.org/10.1016/j.cell.2018.03.013

35. Gadani SP, Smirnov I, Smith AT, Overall CC, Kipnis J (2017) Characterization of meningeal type 2 innate lymphocytes and their response to CNS injury. J Exp Med 214(2):285–296. https://doi.org/10.1084/jem.20161982

36. Aspelund A et al (2015) A dural lymphatic vascular system that drains brain interstitial fluid and macromolecules. J Exp Med 212(7):991–999. https://doi.org/10.1084/jem.20142290

37. Louveau A et al (2018) CNS lymphatic drainage and neuroinflammation are regulated by meningeal lymphatic vasculature. Nat Neurosci 21(10):1380–1391. https://doi.org/10.1038/s41593-018-0227-9

38. Antila S et al (2017) Development and plasticity of meningeal lymphatic vessels. J Exp Med 214(12):3645–3667. https://doi.org/10.1084/jem.20170391

39. Da Mesquita S et al (2018) Functional aspects of meningeal lymphatics in ageing and Alzheimer's disease. Nature 560(7717):185–191. https://doi.org/10.1038/s41586-018-0368-8

40. Neuwelt E et al (2008) Strategies to advance translational research into brain barriers. Lancet Neurol 7(1):84–96

41. Kipnis J (2016) Multifaceted interactions between adaptive immunity and the central nervous system. Science 353(6301):766–771. https://doi.org/10.1126/science.aag2638

42. Lathia JD, Mack SC, Mulkearns-Hubert EE, Valentim CL, Rich JN (2015) Cancer stem cells in glioblastoma. Genes Dev 29(12):1203–1217. https://doi.org/10.1101/gad.261982.115

43. Hambardzumyan D, Amankulor NM, Helmy KY, Becher OJ, Holland EC (2009) Modeling Adult Gliomas Using RCAS/t-va Technology. Transl Oncol 2(2):89

44. Weller M et al (2015) Glioma. Nat Rev Dis Primers 1:15017. https://doi.org/10.1038/nrdp.2015.17

45. Phillips HS et al (2006) Molecular subclasses of high-grade glioma predict prognosis, delineate a pattern of disease progression, and resemble stages in neurogenesis. Cancer Cell 9(3):157–173. https://doi.org/10.1016/j.ccr.2006.02.019

46. Ellor SV, Pagano-Young TA, Avgeropoulos NG (2014) Glioblastoma: background, standard treatment paradigms, and supportive care considerations. J Law Med Ethics 42 (2):171–182. https://doi.org/10.1111/jlme.12133

47. Thakkar JP et al (2014) Epidemiologic and molecular prognostic review of glioblastoma. Cancer Epidemiol Biomarkers Prev 23 (10):1985–1996. https://doi.org/10.1158/1055-9965.Epi-14-0275

48. Ostrom QT et al (2014) The epidemiology of glioma in adults: a state of the science review. Neuro Oncol 16(7):896–913. https://doi.org/10.1093/neuonc/nou087

49. Blissitt PA (2014) Clinical practice guideline series update: care of the adult patient with a brain tumor. J Neurosci Nurs 46(6):367–368. https://doi.org/10.1097/jnn.0000000000000088

50. Westphal M, Lamszus K (Oct 2015) Circulating biomarkers for gliomas. Nat Rev Neurol 11(10):556–566. https://doi.org/10.1038/nrneurol.2015.171

51. Schweitzer T, Vince GH, Herbold C, Roosen K, Tonn JC (2001) Extraneural metastases of primary brain tumors. J Neurooncol 53(2):107–114. https://doi.org/10.1023/a:1012245115209

52. Bettegowda C et al (2014) Detection of circulating tumor DNA in early- and late-stage human malignancies. Sci Transl Med 6 (224):224ra24. https://doi.org/10.1126/scitranslmed.3007094

53. Fleischhacker M, Schmidt B (2007) Circulating nucleic acids (CNAs) and cancer--a survey. Biochim Biophys Acta 1775(1):181–232. https://doi.org/10.1016/j.bbcan.2006.10.001

54. Alix-Panabieres C, Schwarzenbach H, Pantel K (2012) Circulating tumor cells and circulating tumor DNA. Annu Rev Med 63:199–215. https://doi.org/10.1146/annurev-med-062310-094219

55. Müller Bark J, Kulasinghe A, Chua B, Day BW, Punyadeera C (2019) Circulating biomarkers in patients with glioblastoma. British Journal of Cancer 122(3):295–305. https://doi.org/10.1038/s41416-019-0603-6

56. Best MG, Sol N, Zijl S, Reijneveld JC, Wesseling P, Wurdinger T (2015) Liquid biopsies in patients with diffuse glioma. Acta Neuropathol 129(6):849–865. https://doi.org/10.1007/s00401-015-1399-y

57. Rennert RC, Hochberg FH, Carter BS (2016) ExRNA in biofluids as biomarkers for brain tumors. Cell Mol Neurobiol 36 (3):353–360. https://doi.org/10.1007/s10571-015-0284-5

58. Muller C et al (2014) Hematogenous dissemination of glioblastoma multiforme. Sci Transl Med 6(247):247ra101. https://doi.org/10.1126/scitranslmed.3009095

59. T. C. G. A. T. R. Network (2008) Comprehensive genomic characterization defines human glioblastoma genes and core pathways. Nature 455(7216):1061–1068. https://doi.org/10.1038/nature07385

60. Parsons DW et al (2008) An integrated genomic analysis of human glioblastoma multiforme. Science 321(5897):1807–1812. https://doi.org/10.1126/science.1164382

61. Chow LML et al (2011) Cooperativity within and among Pten, p53, and Rb pathways induces high-grade astrocytoma in adult brain. Cancer cell 19(3):305–316. https://doi.org/10.1016/j.ccr.2011.01.039

62. Alifieris C, Trafalis DT (2015) "glioblastoma multiforme: pathogenesis and treatment," (in eng). Pharmacol Ther 152:63–82. https://doi.org/10.1016/j.pharmthera.2015.05.005

63. Wilson TA, Karajannis MA, Harter DH (2014) Glioblastoma multiforme: state of the art and future therapeutics. Surg Neurol Int 5:64. https://doi.org/10.4103/2152-7806.132138

64. Young RM, Jamshidi A, Davis G, Sherman JH (2015) Current trends in the surgical management and treatment of adult glioblastoma. Ann Transl Med 3(9):121. https://doi.org/10.3978/j.issn.2305-5839.2015.05.10

65. Brennan CW et al (2013) The somatic genomic landscape of glioblastoma. Cell 155 (2):462–477. https://doi.org/10.1016/j.cell.2013.09.034

66. Verhaak RG et al (2010) Integrated genomic analysis identifies clinically relevant subtypes of glioblastoma characterized by abnormalities in PDGFRA, IDH1, EGFR, and NF1.

Cancer Cell 17(1):98–110. https://doi.org/10.1016/j.ccr.2009.12.020

67. Wang H et al (2015) The challenges and the promise of molecular targeted therapy in malignant gliomas. Neoplasia 17(3):239–255. https://doi.org/10.1016/j.neo.2015.02.002

68. Dirkse A et al (2019) Stem cell-associated heterogeneity in Glioblastoma results from intrinsic tumor plasticity shaped by the microenvironment. Nat Commun 10(1):1787. https://doi.org/10.1038/s41467-019-09853-z

69. Wick W, Kessler T (2018) New glioblastoma heterogeneity atlas — a shared resource. Nature Reviews Neurology 14(8):453–454. https://doi.org/10.1038/s41582-018-0038-3

70. Garnier D, Renoult O, Alves-Guerra M-C, Paris F, Pecqueur C (2019) Glioblastoma stem-like cells, metabolic strategy to kill a challenging target. Front Oncol 9:118. https://doi.org/10.3389/fonc.2019.00118

71. Stupp R et al (2005) Radiotherapy plus concomitant and adjuvant temozolomide for glioblastoma. N Engl J Med 352(10):987–996. https://doi.org/10.1056/NEJMoa043330

72. Wick W, Wick A, Schulz JB, Dichgans J, Rodemann HP, Weller M (2002) Prevention of irradiation-induced glioma cell invasion by temozolomide involves caspase 3 activity and cleavage of focal adhesion kinase. Cancer Res 62(6):1915

73. Nakada M, Nakada S, Demuth T, Tran NL, Hoelzinger DB, Berens ME (2007) Molecular targets of glioma invasion. Cell Mol Life Sci 64(4):458. https://doi.org/10.1007/s00018-007-6342-5

74. Grossman SA et al (2003) Phase III study comparing three cycles of infusional carmustine and cisplatin followed by radiation therapy with radiation therapy and concurrent carmustine in patients with newly diagnosed supratentorial glioblastoma multiforme: Eastern Cooperative Oncology Group Trial 2394. J Clin Oncol 21(8):1485–1491. https://doi.org/10.1200/jco.2003.10.035

75. Ewing JR et al (2006) Model selection in magnetic resonance imaging measurements of vascular permeability: Gadomer in a 9L model of rat cerebral tumor. J Cereb Blood Flow Metab 26(3):310–320. https://doi.org/10.1038/sj.jcbfm.9600189

76. Neuwelt EA, Barnett PA, Bigner DD, Frenkel EP (Jul 1982) Effects of adrenal cortical steroids and osmotic blood-brain barrier opening on methotrexate delivery to gliomas in the rodent: the factor of the blood-brain barrier. Proc Natl Acad Sci U S A 79(14):4420–4423. https://doi.org/10.1073/pnas.79.14.4420

77. Groothuis DR, Fischer JM, Lapin G, Bigner DD, Vick NA (Mar 1982) Permeability of different experimental brain tumor models to horseradish peroxidase. J Neuropathol Exp Neurol 41(2):164–185. https://doi.org/10.1097/00005072-198203000-00006

78. Neuwelt EA et al (1985) Growth of human lung tumor in the brain of the nude rat as a model to evaluate antitumor agent delivery across the blood-brain barrier. Cancer Res 45(6):2827–2833

79. Burger PC (1987) The anatomy of astrocytomas. Mayo Clin Proc 62(6):527–529. https://doi.org/10.1016/s0025-6196(12)65479-2

80. Halperin EC, Burger PC, Bullard DE (1988) The fallacy of the localized supratentorial malignant glioma. Int J Radiat Oncol Biol Phys 15(2):505–509. https://doi.org/10.1016/s0360-3016(98)90036-0

81. Liu H-L, Fan C-H, Ting C-Y, Yeh C-K (2014) Combining microbubbles and ultrasound for drug delivery to brain tumors: current progress and overview. Theranostics 4(4):432–444. https://doi.org/10.7150/thno.8074

82. Korpanty G, Grayburn PA, Shohet RV, Brekken RA (2005) Targeting vascular endothelium with avidin microbubbles. Ultrasound Med Biol 31(9):1279–1283. https://doi.org/10.1016/j.ultrasmedbio.2005.06.001

83. Singhal S, Moser CC, Wheatley MA (1993) Surfactant-stabilized microbubbles as ultrasound contrast agents: stability study of Span 60 and Tween 80 mixtures using a Langmuir trough. Langmuir 9(9):2426–2429. https://doi.org/10.1021/la00033a027

84. Wang W, Moser CC, Wheatley MA (1996) Langmuir trough study of surfactant mixtures used in the production of a new ultrasound contrast agent consisting of stabilized microbubbles. J Phys Chem 100(32):13815–13821. https://doi.org/10.1021/jp9613549

85. Mehta AI et al (2012) Convection enhanced delivery of macromolecules for brain tumors. Curr Drug Discov Technol 9(4):305–310

86. Bruce JN et al (2011) Regression of recurrent malignant gliomas with convection-enhanced delivery of topotecan. Neurosurgery 69 (6):1272–1279; discussion 1279-80. https://doi.org/10.1227/NEU.0b013e3182233e24

87. Pollina J et al (1998) Intratumoral infusion of topotecan prolongs survival in the nude rat intracranial U87 human glioma model. J Neurooncol 39(3):217–225. https://doi.org/10.1023/a:1005954121521

88. Lonser RR et al (2007) Image-guided, direct convective delivery of glucocerebrosidase for neuronopathic Gaucher disease. Neurology 68(4):254–261. https://doi.org/10.1212/01.wnl.0000247744.10990.e6

89. Shahar T, Ram Z, Kanner AA (2012) Convection-enhanced delivery catheter placements for high-grade gliomas: complications and pitfalls. J Neurooncol 107(2):373–378. https://doi.org/10.1007/s11060-011-0751-x

90. Dressaire E, Bee R, Bell DC, Lips A, Stone HA (2008) Interfacial polygonal nanopatterning of stable microbubbles. Science 320 (5880):1198–1201. https://doi.org/10.1126/science.1154601

91. Myrset AH, Nicolaysen H, Toft K, Christiansen C, Skotland T (1996) Structure and organization of albumin molecules forming the shell of air-filled microspheres: evidence for a monolayer of albumin molecules of multiple orientations stabilizing the enclosed air. Biotechnol Appl Biochem 24 (2):145–153

92. Pattle RE (1955) Properties, function and origin of the alveolar lining layer. Nature 175 (4469):1125–1126. https://doi.org/10.1038/1751125b0

93. Lonser RR, Gogate N, Morrison PF, Wood JD, Oldfield EH (1998) Direct convective delivery of macromolecules to the spinal cord. J Neurosurg 89(4):616–622. https://doi.org/10.3171/jns.1998.89.4.0616

94. Sprowls SA et al (2019) Improving CNS delivery to brain metastases by blood-tumor barrier disruption. Trends Cancer 5 (8):495–505. https://doi.org/10.1016/j.trecan.2019.06.003

95. Hynynen K, McDannold N, Vykhodtseva N, Jolesz FA (2001) Noninvasive MR imaging-guided focal opening of the blood-brain barrier in rabbits. Radiology 220(3):640–646

96. Wei KC et al (2013) Focused ultrasound-induced blood-brain barrier opening to enhance temozolomide delivery for glioblastoma treatment: a preclinical study. PLos One 8(3):e58995. https://doi.org/10.1371/journal.pone.0058995

97. White PJ, Clement GT, Hynynen K (2006) Longitudinal and shear mode ultrasound propagation in human skull bone. Ultrasound Med Biol 32(7):1085–1096. https://doi.org/10.1016/j.ultrasmedbio.2006.03.015

98. Jones RM, O'Reilly MA, Hynynen K (2014) Passive mapping of acoustic sources within the human skull cavity with a hemispherical sparse array using computed tomography-based aberration corrections. J Acoust Soc Am 135(4):2208–2209. https://doi.org/10.1121/1.4877211

99. Tanter M, Thomas JL, Fink M (1998) Focusing and steering through absorbing and aberrating layers: application to ultrasonic propagation through the skull. J Acoust Soc Am 103(5 Pt 1):2403–2410. https://doi.org/10.1121/1.422759

100. Hynynen K, Jolesz FA (1998) Demonstration of potential noninvasive ultrasound brain therapy through an intact skull. Ultrasound Med Biol 24(2):275–283. https://doi.org/10.1016/s0301-5629(97)00269-x

101. McDannold N, Arvanitis CD, Vykhodtseva N, Livingstone MS (2012) Temporary disruption of the blood-brain barrier by use of ultrasound and microbubbles: safety and efficacy evaluation in rhesus macaques. Cancer Res 72(14):3652–3663. https://doi.org/10.1158/0008-5472.Can-12-0128

102. Clement GT, Hynynen K (2002) A non-invasive method for focusing ultrasound through the human skull. Phys Med Biol 47 (8):1219–1236. https://doi.org/10.1088/0031-9155/47/8/301

103. Hynynen K et al (2006) Pre-clinical testing of a phased array ultrasound system for MRI-guided noninvasive surgery of the brain--a primate study. Eur J Radiol 59 (2):149–156. https://doi.org/10.1016/j.ejrad.2006.04.007

104. Marquet F et al (2009) Non-invasive transcranial ultrasound therapy based on a 3D CT scan: protocol validation and in vitro results. Phys Med Biol 54(9):2597–2613. https://doi.org/10.1088/0031-9155/54/9/001

105. Chauvet D et al (2013) Targeting accuracy of transcranial magnetic resonance-guided high-intensity focused ultrasound brain therapy: a fresh cadaver model. J Neurosurg 118

(5):1046–1052. https://doi.org/10.3171/2013.1.Jns12559

106. Aubry JF, Tanter M, Pernot M, Thomas JL, Fink M (2003) Experimental demonstration of noninvasive transskull adaptive focusing based on prior computed tomography scans. J Acoust Soc Am 113(1):84–93. https://doi.org/10.1121/1.1529663

107. Miller GW, Eames M, Snell J, Aubry J-F (2015) Ultrashort echo-time MRI versus CT for skull aberration correction in MR-guided transcranial focused ultrasound: In vitro comparison on human calvaria. Med Phys 42 (5):2223. https://doi.org/10.1118/1.4916656

108. Sassaroli E, Hynynen K (2005) Resonance frequency of microbubbles in small blood vessels: a numerical study. Phys Med Biol 50 (22):5293–5305. https://doi.org/10.1088/0031-9155/50/22/006

109. Hosseinkhah N, Hynynen K (2012) A three-dimensional model of an ultrasound contrast agent gas bubble and its mechanical effects on microvessels. Phys Med Biol 57(3):785–808. https://doi.org/10.1088/0031-9155/57/3/785

110. McDannold N, Vykhodtseva N, Hynynen K (2006) Targeted disruption of the blood-brain barrier with focused ultrasound: association with cavitation activity. Phys Med Biol 51(4):793–807. https://doi.org/10.1088/0031-9155/51/4/003

111. Tung YS, Vlachos F, Choi JJ, Deffieux T, Selert K, Konofagou EE (2010) In vivo transcranial cavitation threshold detection during ultrasound-induced blood-brain barrier opening in mice. Phys Med Biol 55 (20):6141–6155. https://doi.org/10.1088/0031-9155/55/20/007

112. Arvanitis CD, Livingstone MS, Vykhodtseva N, McDannold N (2012) Controlled ultrasound-induced blood-brain barrier disruption using passive acoustic emissions monitoring. PLoS One 7(9): e45783. https://doi.org/10.1371/journal.pone.0045783

113. Tung YS, Marquet F, Teichert T, Ferrera V, Konofagou EE (2011) Feasibility of noninvasive cavitation-guided blood-brain barrier opening using focused ultrasound and microbubbles in nonhuman primates. Appl Phys Lett 98(16):163704. https://doi.org/10.1063/1.3580763

114. O'Reilly MA, Hynynen K (2012) Blood-brain barrier: real-time feedback-controlled focused ultrasound disruption by using an acoustic emissions-based controller. Radiology 263 (1):96–106. https://doi.org/10.1148/radiol.11111417

115. Aryal M, Arvanitis CD, Alexander PM, McDannold N (2014) Ultrasound-mediated blood-brain barrier disruption for targeted drug delivery in the central nervous system. Adv Drug Deliv Rev 72:94–109. https://doi.org/10.1016/j.addr.2014.01.008

116. Treat LH, McDannold N, Zhang Y, Vykhodtseva N, Hynynen K (2012) Improved anti-tumor effect of liposomal doxorubicin after targeted blood-brain barrier disruption by MRI-guided focused ultrasound in rat glioma. Ultrasound Med Biol 38 (10):1716–1725. https://doi.org/10.1016/j.ultrasmedbio.2012.04.015

117. Yang FY et al (2012) Pharmacokinetic analysis of 111 in-labeled liposomal Doxorubicin in murine glioblastoma after blood-brain barrier disruption by focused ultrasound. PLoS One 7(9):e45468. https://doi.org/10.1371/journal.pone.0045468

118. Lin CY et al (2015) Focused ultrasound-induced blood-brain barrier opening for non-viral, non-invasive, and targeted gene delivery. J Control Release 212:1–9. https://doi.org/10.1016/j.jconrel.2015.06.010

119. Lin CY et al (2016) Non-invasive, neuron-specific gene therapy by focused ultrasound-induced blood-brain barrier opening in Parkinson's disease mouse model. J Control Release 235:72–81. https://doi.org/10.1016/j.jconrel.2016.05.052

120. Zhao G et al (2018) Targeted shRNA-loaded liposome complex combined with focused ultrasound for blood brain barrier disruption and suppressing glioma growth. Cancer Lett 418:147–158. https://doi.org/10.1016/j.canlet.2018.01.035

121. Mead BP, Mastorakos P, Suk JS, Klibanov AL, Hanes J, Price RJ (2016) Targeted gene transfer to the brain via the delivery of brain-penetrating DNA nanoparticles with focused ultrasound. J Control Release 223:109–117. https://doi.org/10.1016/j.jconrel.2015.12.034

122. Fisher DG, Price RJ (2019) Recent advances in the use of focused ultrasound for magnetic resonance image-guided therapeutic nanoparticle delivery to the central nervous system. Front Pharmacol 10:1348. https://doi.org/10.3389/fphar.2019.01348

123. Luo Z et al (2017) On-demand drug release from dual-targeting small nanoparticles triggered by high-intensity focused ultrasound enhanced glioblastoma-targeting therapy. ACS Appl Mater Interfaces 9 (37):31612–31625. https://doi.org/10.1021/acsami.7b10866

124. Brightman MW (1977) Morphology of blood-brain interfaces. Exp Eye Res 25:1–25. https://doi.org/10.1016/S0014-4835(77)80008-0

125. Pardridge WM (2002) Drug and gene targeting to the brain with molecular trojan horses. Nat Rev Drug Discov 1(2):131–139. https://doi.org/10.1038/nrd725

Chapter 9

Neurosurgical Implant-Based Strategy for Brain Cancer Therapy

Joshua Casaos, Noah Gorelick, and Betty Tyler

Abstract

Glioblastoma has long proven to be a challenging disease in regard to chemotherapeutic delivery as it grows within a privileged site protected by the selective blood-brain barrier. For many years biodegradable polymers were investigated as potential drug delivery systems. However, polymer technology in the drug delivery setting only truly advanced after the polyanhydrides were tested and demonstrated to be safe and effective platforms for controlled delivery of hydrolytically unstable chemotherapeutics. An in-depth series of biochemical and preclinical studies laid the foundation for the translation of these findings into clinical use for the treatment of glioblastoma with carmustine. This chapter provides the historical perspective of the carmustine implants through preclinical safety, intracranial distribution, and efficacy studies to clinical trials in recurrent and newly diagnosed glioblastoma and subsequent FDA approval of this chemotherapeutic delivery concept. Moreover, these studies provide a successful framework on which to expand in this new era of targeted and controlled drug delivery.

Key words BCNU wafers, Blood-brain barrier, Chemotherapy, Local drug delivery, Polyanhydrides

1 Introduction

Glioblastoma and anaplastic astrocytoma, together comprising malignant gliomas, are the most common primary brain tumors in adults [1]. Median survival for this disease is less than 2 years, and only a paucity of FDA-approved treatment advances have been made in the previous decades [2]. There are several challenging factors contributing to this, including: (1) the cellular and genetic heterogeneity of the tumor, which contributes to drug resistance, immunologic tolerance, angiogenesis, self-renewal, and recurrence [3–5]; (2) the propensity of tumoral cells to migrate and invade normal brain parenchyma, limiting the effectiveness of radiation and surgical resection [6]; and (3) the site of privilege that is established by the blood-brain barrier (BBB), making it difficult if not impossible for many systemic therapies to reach the tumor at safe and effective concentrations. The development of Gliadel® is

Vivek Agrahari et al. (eds.), *Nanotherapy for Brain Tumor Drug Delivery*, Neuromethods, vol. 163,
https://doi.org/10.1007/978-1-0716-1052-7_9, © Springer Science+Business Media, LLC, part of Springer Nature 2021

one driven by the need for innovative treatments and resulting in the pioneering of a new era of local therapeutic delivery mechanisms. This chapter highlights the development of the Gliadel® wafer and describes the engineering of the biodegradable and biocompatible polyanhydride wafer, with a focus on the preclinical methods and experiments that provided the basis for clinical translation to its use in the operating room.

2   The Blood-Brain Barrier

The brain is a privileged site due to the BBB. The BBB is a distinct selective barrier which surrounds the majority of the central nervous system (CNS) and separates it from the rest of the body. The BBB is made up of a continuous layer of endothelial cells that form tight junctions. Additionally, this initial layer is supported by interactions with the basement membrane and surrounding cells including astrocytes, neurons, and pericytes. These junctions, combined with the low endocytic activity of the endothelial cells, function to strictly limit the exchange of substances into the CNS [7, 8]. Characteristics such as small size and high degree of hydrophobicity favor free diffusion across the BBB [9]. This barrier is largely impermeable to molecules larger than 400 kD and molecules with a low degree of lipid solubility, which unfortunately encompasses the vast majority of chemotherapeutics [10–12]. Analytical studies have demonstrated that the majority of small and virtually all large molecules are unable to diffuse across the BBB, leaving only very few of the available chemotherapeutics able to penetrate into the CNS [10]. In certain states of disease, such as tumors, infection, or inflammation, the integrity of the BBB can be disrupted allowing for the infiltration of a wider array of substances [13]. This forms the basis for the use of intravenous contrast in the diagnosis of certain CNS pathologies [14]. However, breakdown of the BBB is only marginally compromised and remains largely intact in peripheral tumor areas [15]. Additionally, tumor cells have the ability to migrate to brain regions where the BBB is fully intact [7, 8]. Other limitations include the short half-lives of many systemically administered agents, systemic toxicities of these agents, and the small therapeutic window due to the limited concentrations tolerated systemically [16]. These collective factors limit the effectiveness of systemically delivered chemotherapeutics and sparked the initial drive for the development of local drug delivery systems in the brain.

3   Development of Polyanhydride Wafers

For years various biodegradable polymers were investigated for potential use in drug delivery systems, with the goal of achieving controlled drug release. Initially, hydrophilic polymers studied, such as polylactic acid (PLA) and polyglycolic acid (PGA), were found to be quickly infiltrated and hydrolyzed by water, eroding simultaneously throughout the surface and interior of the polymer matrix via bulk degradation. This resulted in poorly controlled polymer degradation and subsequent drug release that failed to obtain zero order kinetics [17, 18].

Polyanhydride polymers were originally synthesized and used in the textile industry [19]. However, they were ultimately abandoned in that arena due to their hydrolytic instability. It was this instability that led the Langer group to consider these polymers as an ideal candidate for controlled drug release systems. Their propensity for hydrolytic interactions allows for a more heterogeneous erosion system, eroding from the surface inward, and provided a more controlled pattern of drug release [20]. Polyanhydrides presented a hydrophobic backbone capable of protecting drugs from hydrolysis, linked together by water-liable anhydride bonds, which provided a biodegradable drug delivery system capable of eroding in a more controlled manner [21].

Initial studies examined the polyanhydride poly[bis(p-carboxyphenoxy)methane anhydride] synthesized via melt polycondensation [22]. When loaded with the model drug, cholic acid, near zero-order degradation, and release kinetics were observed at both 37 °C and 60 °C [20]. Subsequent studies utilized 1,3-bis(carboxyphenoxy)propane (pCPP) anhydride alone and copolymerized with a more hydrophilic monomer, sebacic acid (SA) [23]. Release kinetics demonstrated that more hydrophobic monomers degraded over several months, while degradation was enhanced by increasing the ratio of copolymerized SA. Extrapolation of degradation rates demonstrated pCPP alone to degrade over a period of 3 years. Meanwhile, increasing the SA content resulted in increased degradation rates by as much as 800 times, as observed when the SA content reached near 80% (pCPP-SA 21:79). Additionally, degradation rates were found to be pH dependent, increasing significantly with increasing pH and decreasing under acidic conditions.

In terms of drug release characteristics, formulation of the drug-loaded polymers was observed to play a significant role, as injection molded polymers demonstrated the best results [23]. Drug-polymer combinations were formulated either via injection molding at high temperatures or by compression molding, pressing the drug-polymer mixtures into circular discs. The model drug p-nitroaniline was used to assess drug release. The

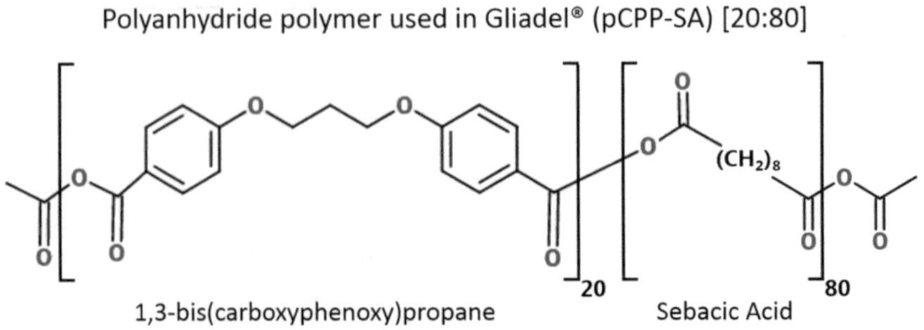

Polyanhydride polymer used in Gliadel® (pCPP-SA) [20:80]

1,3-bis(carboxyphenoxy)propane   Sebacic Acid

Fig. 1 Chemical structure of pCPP-SA. Hydrophobicity of the polymer is attributed to the phenylene groups as well as sebacic acid. Increasing the ratio of sebacic acid results in increased degradation rates

pattern of drug release from the injection molded polymers shadowed the pattern of polymer degradation over a period of 8 months for the pCPP-only polymer and a period of 2 weeks for the more hydrophilic pCPP-SA 21:79 (Fig. 1).

The compression-molded polymer had similar release profiles, though was more susceptible to diffusional escape of the drug from the drug–polymer matrix [24]. Additional studies demonstrated that drug–polymer mixtures could be produced in a variety of forms including sheets, rods, and microspheres [25].

4   Preclinical Experiments

A series of preclinical safety and biocompatibility experiments laid the initial groundwork that investigated the potential use of an implantable polyanhydride-based system to deliver larger molecular weight compounds for the local treatment of brain tumors. The first of these studies evaluated the biocompatibility of blank poly (2-hydroxyethyl methacrylate) (PolyHEMA) and alcohol-washed ethylene-vinyl acetate copolymers (EVAc) in the rabbit cornea [26]. The cornea is a unique environment for biocompatibility studies; it is profoundly sensitive to inflammatory stimuli and, furthermore, allows for regular, noninvasive, in vivo stereomicroscopic evaluation. Neither slit lamp evaluation nor postmortem histological analysis detected edema, neovascularization, or inflammatory cell infiltration in response to either of the implanted polymers; however the lack of biodegradation limited their usefulness [26].

Eventually biologically dissolvable polymers were studied for their biocompatibility and degradation profiles in several experimental animal models (Fig. 2).

Tamargo et al. assessed the inflammatory potential of pCPP-SA (20:80) within the healthy rat brain when they compared the histologic reactions of the polyanhydride with those found in response to Surgicel® (oxidized regenerated cellulose, Ethicon)

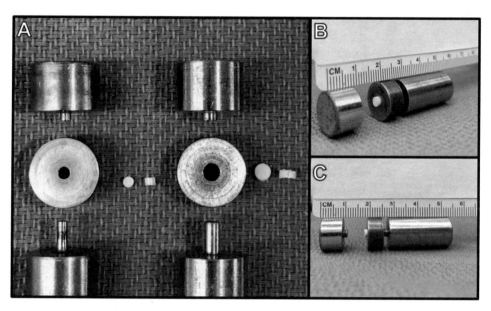

Fig. 2 Preparation of compression-molded pCPP-SA polymer wafers for use in animal models. (**a**) Compression molds for polymer wafer formation: 5 mg (left, for mouse) and 10 mg (right, for rat) wafers. (**b**, **c**) Photographs detailing rat wafer within the compression mold

and Gelfoam® (absorbable gelatin sponge, Pfizer), two products routinely used in brain surgery [27]. They reported that the localized acute and subacute inflammatory reactions caused by the polyanhydride were comparable to those seen in response to Surgicel® and only slightly greater than those elicited by the Gelfoam®. They also found that the polyanhydride degraded completely within 36 days of implantation and caused no behavioral changes, neurological deficits, or signs of toxicity [27]. Next, Brem and colleagues evaluated a more hydrolytically resistant 50:50 formulation of pCPP-SA in both the rabbit brain and cornea. In the brain they compared this polymer to Gelfoam® via contralateral frontal lobe implantations in which each animal served as its own control. Gross and microscopic examination showed comparable local reactions between the brain tissues that had been in contact with the polymer and the Gelfoam®. Complete polymer degradation occurred within 4–5 months, substantially longer than the 36 days seen with the pCPP-SA (20:80) formulation. As seen previously with polyHEMA and EVAc, no inflammatory or neovascular response was elicited by the 50:50 polyanhydride in the rabbit corneal assay [28].

While different polymer-based drug delivery systems were undergoing early testing, experiments were also underway to identify reliable ^{31}P NMR spectroscopy signatures with which to characterize both normal growth patterns of the 9L rodent model of gliosarcoma as well as the tumor's response to the antineoplastic agent 1,3-bis(2-chloroethyl)-1-nitrosourea (BCNU; also known as carmustine) [29]. This method of examining tumor bioenergetics revealed a clear dose-response relationship between increased levels of BCNU and growth delay in the tumor, which were reliably represented by significant changes to ^{31}P NMR spectral parameters [30]. Given this evidence of BCNU's ability to inhibit growth in a gliosarcoma animal model, the next step was to discover which biocompatible polymer system would be best suited to deliver the drug for local treatment.

The first of these studies used a spectrophotoscopic assay to measure the release kinetics of BCNU from an EVAc delivery system in vitro (dissolved in PBS) and in vivo (in blood and brain tissue) following either intraperitoneal (i.p) or intracranial (i.c.) implantation of the polymeric wafers in an F344 rodent model [31]. These studies provided strong evidence that BCNU could indeed be locally released from inert polymers in a safe and predictable manner. Polymers implanted into the rat brain delivered 40 times more BCNU to the ipsilateral hemisphere compared to concentrations found in the contralateral hemisphere and systemic circulation, resulting in a reduction in systemic toxicity. However, despite such impressive evidence of controlled BCNU release, the fact that the EVAc polymer did not degrade limited its potential for clinical translation.

Over the next few years several studies examined various combinations of biodegradable polymers and chemotherapeutic agents to inhibit tumor growth across various experimental models. A fatty acid dimer (FAD) was copolymerized with SA to test the delivery of water-soluble (carboplatin, CB) and hydrolytically unstable (4-hydroperoxycyclophosphamide, 4-HC) chemotherapeutics [32]. When these FAD:SA polymers delivering CB and 4-HC were introduced into rodent brains on the fifth day following i.c. implantation of F98 glioma cells, median survival was extended from 16 to 52 days and from 22 to 35 days, respectively, compared to untreated controls ($p < 0.001$ for both) [32]. This provided encouraging preclinical evidence that water-soluble and hydrolytically unstable drugs could successfully be delivered via polymers directly into the brain to achieve therapeutic results.

Autoradiographic analysis of ^{3}[H]-BCNU in a healthy rabbit model quantified and compared the areas of tissue exposure and local concentrations of the radiolabeled drug delivered via polymeric implantation using the 20:80 formulation of PCPP-SA versus direct injection [33]. Seventy-two hours after polymer implantation as much as 40% of the ipsilateral rabbit brain tissue remained exposed to the radiolabeled BCNU and 15% of the tissue remained exposed after 180–350 h. In stark contrast, direct injection of the same dose of ^{3}[H]-BCNU was initially widely distributed throughout the brain parenchyma; however, exposure quickly dropped to 15% after only 24 h [33]. Temporary local injury characterized by edema and necrosis was noted in a dose-related response to BCNU regardless whether the delivery method was by injection or polymer implantation. Since local tissue injury was not present with polymeric delivery of radiolabeled inulin it was attributed to the high local concentration of the chemotherapeutic agent, underscoring the considerable need for its controlled release.

Despite evidence of BCNU's local toxicity, rodent experiments in the 9L gliosarcoma model highlighted the drug's ability to significantly delay tumor growth as well as a 5.4-fold and 7.3-fold survival extension when delivered locally by either EVAc or pCPP-SA (20:80) polymer systems, respectively. It was observed that the local toxicity of interstitially delivered BCNU was comparable to that found in systemic administration [34]. These results provided the basis for a phase I–II clinical trial in patients with recurrent malignant gliomas to undergo polymeric delivery of BCNU [35].

As translation of polymer-based chemotherapeutics progressed from benchtop to bedside it became increasingly important to examine these drug delivery systems in models that were as similar as possible to the clinical situation. Evaluation of BCNU-impregnated pCPP-SA in the healthy monkey brain conducted in conjunction with radiation therapy sought to assess safety of polymeric delivery within the context of a more realistic clinical setting. Only minor, localized pathological changes were detected and there was no evidence of negative neurological side effects [36]. This supported the safe use of irradiation following polymer implantation in patients. Further assessment of the polyanhydride wafer compared the polymer's degradation in vitro to its breakdown within a number of in vivo settings meant to better replicate various clinical settings. The study concluded that in vivo polymer breakdown was slightly slower than in vitro; however variables including drug loading, presence of tumor, and the increased acidity caused by co-implantation with Surgicel® for hemostasis had no effect on slowing the polymer's degradation [37]. Further pharmacokinetic studies analyzed these differences between in vitro and in vivo release of drug from pCPP-SA. While routine in vitro assays were only able to detect BCNU over several days of degradation, when implanted into the rat and monkey brains, the drug was present for at least 30 days at concentrations higher than those observed with other delivery methods [38, 39]. Scanning electron microscopy (SEM) was also used to visualize morphological characteristics of the polyanhydride during spray dry manufacturing and to analyze the structural changes that took place over the course of in vitro and in vivo erosion. SEM indicated that spray drying resulted in homogeneous distribution of the polymer and drug with no phase separation. Importantly, erosion was confined mainly to the surface of the polymer and took place in a layer-wise fashion [40].

Together, all of these preclinical experiments provided a strong foundation of evidence that showed laminar degradation of a biocompatible polymer could contribute to the highly controlled, local release of a large, hydrolytically unstable and otherwise prohibitively toxic chemotherapeutic agent over a sustained period of time with negligible toxicity. Preclinical experimentation continued to address questions regarding the optimal proportions of pCPP to SA, the maximally safe and effective dose of BCNU that could be incorporated into the polymer, and whether the technology had utility in metastatic models of brain tumors [41–43].

5   Clinical Trials

Given the promising results in preclinical studies, in 1990 an initial phase I–II clinical trial sought to assess the safety of BCNU formulated with the pCPP-SA 20:80 polymer at three doses of BCNU, 1.9%, 3.9%, and 6.4% [35, 44]. These doses corresponded to total BCNU doses of 30, 60, and 100 mg, respectively. For formulation of the polymers, pCPP-SA and BCNU were co-dissolved in methylene chloride, subsequently spray dried into microspheres, and then compression-molded into wafers 1.4 cm in diameter and 1.0 mm thick. Up to eight polymers were implanted in the surgical cavity intraoperatively (Fig. 3).

Twenty-one patients with recurrent glioma were included in the study. To assess for both systemic and local toxicity, patients were followed by neurological examination, Karnofsky Performance Scale (KPS), laboratory chemistries, and urinalysis. The polymers were found to be well tolerated at all three doses with no adverse reactions to the BCNU wafer treatment itself and no signs of systemic toxicity to BCNU. The median survival after

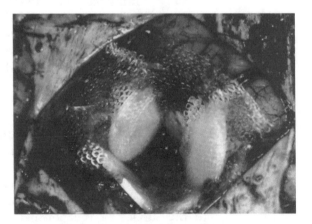

Fig. 3 BCNU impregnated Gliadel® wafers implanted in the human brain. Photograph of Gliadel® wafers implanted in the surgical cavity at the time of resection. Up to eight polymers may be placed, lining the surgical cavity [45]

reoperation was 65, 64, and 48 weeks for the three dosing groups, respectively. Overall, the authors found that the polymers were clinically safe for implantation in the resection cavity at the time of surgical resection, and a BCNU dose of 3.85% was selected for further studies.

Based on these results, a subsequent randomized placebo-controlled trial was carried out to assess the efficacy of BCNU wafers in the treatment of recurrent malignant gliomas [46]. This study involved 222 patients and 27 medical centers. Patients requiring reoperation for malignant gliomas received pCPP-SA 20:80 polymers with or without 3.85% BCNU. Formulation was as described above. Median survival of patients receiving BCNU polymer was 31 weeks compared to 23 weeks for the placebo group, with a p-value of 0.006 after adjusting for prognostic factors. Additionally, the 6-month survival of patients with glioblastoma that received BCNU wafers was 44%, compared to 64% in those that received placebo. No local or systemic adverse reactions related to the BCNU wafers were reported. These findings culminated in the FDA approval of the implantable BCNU-loaded biodegradable polymer (Gliadel®) for the treatment of recurrent high-grade malignant gliomas in 1996 [47].

Given these promising results found in recurrent glioma, studies were undertaken to assess the safety and efficacy of BCNU wafers in primary, newly diagnosed high-grade malignant glioma. In 1995, an initial phase I study demonstrated the use of BCNU wafers followed by radiation therapy postoperatively, in newly diagnosed gliomas. The use of BCNU wafers was deemed safe as there was no increase in adverse events [48]. A subsequent phase III randomized controlled trial was set to recruit 100 patients with newly diagnosed malignant glioma. However, the study was cut short due to a shortage of drug and only 32 patients were included. Nonetheless, the BCNU treatment group had a significantly increased median survival compared to the placebo group (58.1 vs. 39.9 weeks) [49]. A larger separate phase III randomized controlled trial later showed that BCNU wafers provided a survival benefit in newly diagnosed malignant glioma. This study was a multicenter, multinational study of 240 patients that evaluated the efficacy of BCNU wafers vs. placebo, with subsequent external beam radiation, in newly diagnosed glioma [50]. The intent to treat group demonstrated an increased median survival (13.9 vs. 11.6 months), with a 29% death risk reduction rate. These significant findings paved the way for FDA approval of Gliadel® as an option for all patients with newly diagnosed high-grade glioma in 2003 [50]. Subsequently, several multicenter retrospective studies involving various countries have validated these survival benefits in both primary and recurrent malignant glioma [51–53].

In 2005, a landmark study established the currently used treatment regimen for patients with malignant glioma known commonly as the Stupp protocol [2]. It is comprised of radiation therapy plus concomitant oral chemotherapy, with the alkylating agent temozolomide (TMZ). Median survival of TMZ plus radiation therapy was significantly increased compared to radiation alone (14.6 vs. 12.1 months). This resulted in TMZ being added to the currently available nonexperimental options for patients which included surgery, radiation, TMZ, and Gliadel®. However, the use of BCNU wafers in combination with the alkylating agent TMZ quickly became a subject of question. In 2009, a retrospective study demonstrated the use of BCNU wafers in addition to standard therapy (comprised of TMZ and radiation) to be safe and effective [54, 55]. In this study patients that received BCNU wafers in addition to radiation and TMZ had an increased survival benefit compared to those who received radiation and TMZ alone (21.3 vs. 12.4 months). The use of BCNU wafers with TMZ and radiation has been validated with several retrospective studies as well as phase I and II trials demonstrating no association with increased complications in the setting of administration with TMZ and radiation therapy for both newly diagnosed primary and recurrent high-grade malignant gliomas [56–61]. Additionally, a recent systematic review reported little evidence of enhanced toxicity and an increased median overall survival of 3–4 months [62].

Furthermore, the use of BCNU wafers has also been evaluated in other settings, such as in the treatment of elderly patients with glioblastoma. With regard to elderly patients, defined as patients >65 years old, diagnosed with glioblastoma 133 patients who underwent surgery for primary newly diagnosed glioblastoma and received either BCNU wafers or no BCNU wafers were matched and analyzed retrospectively for median survival. BCNU wafer use in the elderly was associated with a significantly increased median survival (8.7 vs. 5.5 months) and was not associated with increased morbidity or mortality [63]. These findings demonstrate that BCNU wafers may be of benefit in patients >65 years old, which comprises a significant portion of the glioblastoma population, though further designed studies are needed.

The role of BCNU wafers has also been studied in the treatment of metastatic brain malignancies, which comprise the majority of CNS tumors [64]. Both preclinical and clinical reports have shed light on the potential for BCNU wafers and local delivery. Preclinical studies have demonstrated that polymeric delivery of BCNU, as well as delivery of carboplatin or camptothecin, individually or in combination with radiation therapy was safe and efficacious in murine intracranial models of tumor that commonly metastasize to the brain. These models included melanoma, lung carcinoma, renal carcinoma, and colon carcinoma [43]. Specifically, BCNU polymer in combination with radiation therapy was found to be effective in prolonging survival in all four cell line models, while carboplatin polymer alone was effective against colon carcinoma and melanoma and when combined with radiation was effective against colon and renal cell carcinomas. Camptothecin was demonstrated to be effective when used in the melanoma model in combination with radiation; however it showed little survival benefit in the other models. Additionally, BCNU polymers were tested and found to be efficacious in extending survival in an animal model of breast carcinoma metastases to the brain [65]. These preclinical findings provided the basis for an initial phase I clinical trial which included 25 patients with solitary brain metastases involving lung, melanoma, breast, and renal carcinoma. Patients underwent surgical resection and polymer placement (with a maximum of eight wafers) and subsequently received whole brain radiation. Results from the study exhibited two adverse events, postulated to be due to wafer placement and both responding to medical therapies. No local recurrence was observed during the study, while 16% of patients relapsed at another location in the brain. Median survival was 33 weeks with 25% overall survival at 2 years [66]. Authors concluded that the combination of BCNU wafers with surgical resection and external beam radiation was safe and well tolerated. Further retrospective studies have validated these findings and demonstrated that surgery in addition to BCNU wafers resulted in less decline in neurocognitive function compared to radiation-associated therapies [67–70]. These findings in the laboratory and in the clinic provide the basis for further clinical investigation into the use of BCNU wafers for the treatment of brain metastasis.

Another factor studied has been the dose of BCNU delivered from the polymers. Preclinical efforts evaluated the potential effects of increasing the dose of BCNU loaded into pCPP-SA 20:80 polymers utilizing a dose escalation study in a rat 9L glioma model [41]. BCNU doses of up to 32% were assessed. A dose of 20% BCNU by weight was found to be optimal in regard to toxicity and antitumor efficacy as it resulted in long-term survival rate of 75% in an intracranial model of gliosarcoma. Clinically, the effect of higher doses of BCNU-loaded wafers was assessed in a dose-escalation clinical study. It was demonstrated that the highest dose of BCNU tolerated was 20% by weight, with no observed increased adverse events [71]. Moreover, at a dose of 28%, side effects of seizures and severe brain edema were observed.

6   Meta-Analysis of Safety and Efficacy of Gliadel® Use

Since the FDA approval of BCNU wafers for newly diagnosed and recurrent high-grade malignant gliomas, two recent meta-analyses and several retrospective studies and systematic reviews have analyzed the safety and efficacy of BCNU wafers [58–61, 72–74]. The majority of these studies have demonstrated an increased survival benefit with no increase in adverse events [75, 76]. Furthermore, two meta-analyses have validated the survival benefit of patients treated with BCNU wafers [72, 73]. In one of these studies, authors analyzed six randomized controlled trials and four cohort studies involving newly diagnosed malignant glioma [72]. The largest of the meta-analyses included 62 publications and 4898 patients. Authors reported an increased median survival, increased 1-year overall survival, and increased 2-year survival in the BCNU wafer group compared to the control group [73]. Furthermore, as mentioned previously, a recent systematic review has reported limited toxicity and increased median survival when BCNU was used in addition to TMZ and radiation therapy [62]. Ultimately, larger and specifically designed trials are necessary to thoroughly understand the effect of BCNU wafer use with TMZ and radiation therapy, as well as to provide better insight into the inclusion of treated patients (who have undergone BCNU wafer implantation and have had tumor recurrence) into other clinical trials [77].

7   Further Studies

The history of Gliadel® development and translation from the laboratory to the operating room has provided not only a new weapon in the armamentarium against malignant gliomas, but perhaps more importantly, it has provided a framework for the development of innovative local drug delivery systems and therapeutics. Polymers provide several advantages as mentioned previously; however, there are also disadvantages to be mentioned. First, wafers only provide drug release for a limited amount of time until the polymer is degraded, about 21 days in animal models [39]. Second, there is not always adequate space in the surgical resection cavity for the rigid dimensions of the carmustine wafers [46, 71]. Furthermore, polymeric technology has been developed and studied extensively for use with only one chemotherapeutic, BCNU, though several studies show promising results utilizing other chemotherapies. Additionally, other barriers exist in the development of locally delivered therapeutics, including: a better understanding of the diffusion of chemotherapeutics in brain tissue, the lack of in vivo imaging technology to study the distribution and efficacy of chemotherapeutics; and the ability to deliver multiple chemotherapeutics to specifically target the heterogeneity of tumor cell populations while avoiding damage to nontumor resident brain tissue [16, 78].

Several additional chemotherapeutics have been explored for polymeric local delivery, including carboplatin, paclitaxel, and TMZ, among others, showing efficacy in prolonging survival in preclinical models [39, 79–82]. Other promising therapeutics are actively being explored for local delivery using polymeric technology, alone and in combination with other therapies that show promise in vitro. Perhaps the most innovative of these developments has bridged the world of immunotherapy and local delivery of therapeutics. Preclinical studies have shown that mice treated with local chemotherapy (BCNU wafers) in combination with systemic immunotherapy, in the form of the immune checkpoint antibody anti-programmed cell death protein 1 (anti-PD-1), exhibited increased survival, increased tumor infiltrating lymphocytes, and immune memory, compared to mice that received systemic BCNU and anti-PD-1 therapy. These studies have ultimately demonstrated the importance of both mode of therapy (systemic vs. local) and timing of systemic chemotherapy in association with immune response to immune modulating agents [83].

Furthermore, the success with Gliadel® has spurred exploration and development of other modes of chemotherapeutic delivery to the brain (Fig. 4) ranging from BBB disruption to drug-loaded microchips and convection enhanced delivery (CED).

Transient BBB disruption, both in chemical and mechanical form, has been investigated to increase the therapeutic delivery of antineoplastic agents to the CNS. Animal models of hyperosmolar mannitol infusion, ultrasound sonication combined with intravascular microbubbles, delivery of the bradykinin analog, lobradimil, and the adenosine receptor agonist, regadenoson, have all shown promising results in BBB disruption using dynamic contrast-enhanced MRI evaluation [84–86]. However, clinical studies investigating the use of lobradimil and regadenoson did not result in increased chemotherapeutic agent or increased effect [85, 87]. A phase I study showed no deleterious effects from intra-arterial cerebral infusion of the antineoplastic agent cetuximab after BBB disruption with mannitol [88]. Pulsed ultrasound with systemically injected microbubbles was shown to be safe and well tolerated in a phase 1/2a trial in patients with GBM receiving carboplatin for treatment [89]. These innovative delivery modalities aim to provide transient and optimized BBB disruption with the goal of increasing the therapeutic payload of intracranial drug delivery for these patients. CED is a technique involving direct infusion of therapeutics into the brain parenchyma. CED has shown promise for the delivery of chemotherapeutics, immunotoxins, and monoclonal antibodies [90, 91].

Targeted Therapy Across the Blood-Brain Barrier

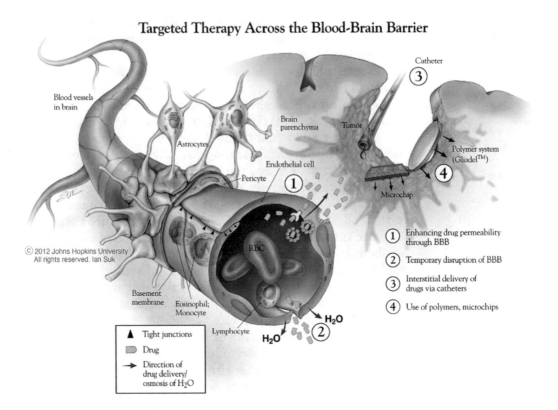

Fig. 4 Evolving approaches for brain tumor drug delivery. The success of Gliadel® wafers has laid the framework for the design and development of local drug delivery systems including BBB permeability and disruption, catheter delivery systems, and microchips. (Printed with permission. ©2018 JHU Neurosurgery–Ian Suk)

Drug-loaded microchips (MicroChips®) can address some of the limitations of polymeric technology, such as dependency on polymer degradation for drug release, and the lack of pulse dosing [92]. Amount and timing of drug release is dependent on the breakdown of nitride membranes which cover implanted drug reservoirs and can be controlled via a computer-based device. Moreover, they can be wirelessly controlled and release single or multiple agents. Preclinical models of malignant gliomas show promising results with extended survival when microchips are loaded with various agents, including TMZ or BCNU [93, 94]. This technology has already been applied in humans for the pulsatile release of parathyroid hormone to treat osteoporosis, though it has yet to be used in any clinical trials for malignant gliomas [95].

8   Future Perspectives

Intracranial therapeutic delivery has been increasingly under investigation for CNS disease using microspheres, nanospheres, liposomes, and gels in animal models. These delivery options include payloads of chemotherapeutic agents, peptide, antibodies, siRNA, as well as tumor-targeting molecules. Poly(lactic-*co*-glycolide) (PLGA) microspheres have been utilized to deliver various chemotherapeutic agents, including temozolomide, paclitaxel, and carmustine, as well as drug "cocktails" of multiple agents with varying mechanisms of action [96–99]. PLGA/nano-hydroxyapatite microspheres delivering temozolomide showed a decrease in proliferation and invasion and an increase in apoptosis in U87 human glioma cells [100]. PLGA when combined with a thermoreversible gelation polymer delivered camptothecin in a C6 rodent glioma model resulted in a statistically significant increase in median survival [101]. Poly(ß-amino ester) nanoparticles' DNA delivery of herpes simplex virus type I thymidine kinase (HSVtk) combined with the prodrug ganciclovir (GCV) showed extended survival in a rodent brain tumor model [102]. While there are many preclinical studies using PLGA to deliver therapeutic compounds intracranially to treat GBM, few studies have evaluated PLGA for GBM treatment in the clinical setting. A phase II study in 2005 investigated the efficacy of PLGA microspheres incorporated with radiosensitizing 5-fluorouracil after tumor resection plus radiotherapy compared to radiotherapy alone in newly diagnosed GBM. No significant difference was found in regard to overall survival, though the trend favored the PLGA group and the study was low powered with only 77 patients [103]. As this biodegradable and biocompatible delivery platform has been shown to be safe and effective with the incorporation of various chemotherapeutics in preclinical studies, there may be clinical opportunities on the horizon.

Liposomal drug delivery has been both investigated preclinically as well as clinically. Preclinical testing of pegylated and non-pegylated liposomes has evaluated delivery of peptides, genes, and chemotherapeutic agents [104, 105]. A phase I study of nonpegylated liposomal doxorubicin in children with recurrent or refractory high-grade glioma determined the maximum recommended dose to be further evaluated for efficacy in a phase II trial [106]. A phase I trial also determined the dosing regimen for intravenous liposomal irinotecan in patients with recurrent high-grade glioma [107–109]. A liposomal formulation was used to intracranially deliver the HSV-1-*tk* gene in patients with recurrent GBM with simultaneous systemically delivered ganciclovir [110]. This targeted gene therapy was well tolerated with two of the eight patients showing more than 50% tumor volume reduction and six of the eight patients displaying focal treatment effects. While this study demonstrated feasibility and safety it also highlighted the inherent heterogeneity of the GBM tumor and the need for imaging in standardizing gene therapy studies.

9   Conclusion

The ideation and subsequent design of BCNU-impregnated polymeric polyanhydride wafers have demonstrated the feasibility of using controlled drug delivery systems to overcome intrinsically and anatomically challenging features of glioblastoma. Biochemical studies were vital in identifying a polymer that was both biocompatible and pharmacologically capable of controlled drug delivery. Preclinical studies utilized animal models to demonstrate safety and efficacy in glioma and brain tumor models. This series of preclinical experiments paved the way for translation to the clinical operating room. Altogether, the development of Gliadel® provided a novel therapeutic option for patients with glioblastoma and contributes an experimental framework for further work to be done in this area.

References

1. Ostrom QT et al (2014) The epidemiology of glioma in adults: a "state of the science" review. Neuro-Oncology 16(7):896–913. https://doi.org/10.1093/neuonc/nou087

2. Stupp R et al (2005) Radiotherapy plus concomitant and adjuvant temozolomide for glioblastoma. N Engl J Med 352 (10):987–996. https://doi.org/10.1056/NEJMoa043330

3. Furnari FB et al (2007) Malignant astrocytic glioma: genetics, biology, and paths to treatment. Genes Dev 21(21):2683–2710. https://doi.org/10.1101/gad.1596707

4. Weller M et al (2013) Molecular neuro-oncology in clinical practice: a new horizon. Lancet Oncol 14(9):e370–e379. https://doi.org/10.1016/S1470-2045(13)70168-2

5. Liu A et al (2016) Genetics and epigenetics of glioblastoma: applications and overall incidence of IDH1 mutation. Front Oncol 6:16. https://doi.org/10.3389/fonc.2016.00016

6. Gruber ML, Hochberg FH (1990) Systematic evaluation of primary brain tumors. J Nucl Med 31(6):969–971

7. Groothuis DR et al (1982) Permeability of different experimental brain tumor models to horseradish peroxidase. J Neuropathol Exp Neurol 41(2):164–185

8. Neuwelt EA et al (1982) Effects of adrenal cortical steroids and osmotic blood-brain barrier opening on methotrexate delivery to gliomas in the rodent: the factor of the blood-brain barrier. Proc Natl Acad Sci U S A 79 (14):4420–4423

9. Banks WA (2009) Characteristics of compounds that cross the blood-brain barrier. BMC Neurol 9(Suppl 1):S3. https://doi.org/10.1186/1471-2377-9-S1-S3

10. Ghose AK, Viswanadhan VN, Wendoloski JJ (1999) A knowledge-based approach in designing combinatorial or medicinal chemistry libraries for drug discovery. 1. A qualitative and quantitative characterization of known drug databases. J Comb Chem 1(1):55–68

11. Pardridge WM (2005) The blood-brain barrier: bottleneck in brain drug development. NeuroRx 2(1):3–14. https://doi.org/10.1602/neurorx.2.1.3

12. Smith QR et al (1988) Kinetics and distribution volumes for tracers of different sizes in the brain plasma space. Brain Res 462(1):1–9

13. Pollay M, Roberts PA (1980) Blood-brain barrier: a definition of normal and altered function. Neurosurgery 6(6):675–685

14. Avsenik J, Bisdas S, Popovic KS (2015) Blood-brain barrier permeability imaging using perfusion computed tomography. Radiol Oncol 49(2):107–114. https://doi.org/10.2478/raon-2014-0029

15. Wen PY, Kesari S (2008) Malignant gliomas in adults. N Engl J Med 359(5):492–507. https://doi.org/10.1056/NEJMra0708126

16. Chaichana KL, Pinheiro L, Brem H (2015) Delivery of local therapeutics to the brain: working toward advancing treatment for malignant gliomas. Ther Deliv 6(3):353–369. https://doi.org/10.4155/tde.14.114

17. Heller J, Baker R (1980) Theory and practice of controlled drug delivery from bioerodible polymers. In: Controlled release of bioactive materials. Elsevier, Amsterdam, pp 1–17

18. Pitt G et al (1981) Aliphatic polyesters II. The degradation of poly (DL-lactide), poly (-ε-caprolactone), and their copolymers in vivo. Biomaterials 2(4):215–220

19. Conix A (1958) Aromatic polyanhydrides, a new class of high melting fiber-forming polymers. J Polym Sci A Polym Chem 29 (120):343–353

20. Rosen HB et al (1983) Bioerodible polyanhydrides for controlled drug delivery. Biomaterials 4(2):131–133

21. Chasin M et al (1990) Polyanhydrides as drug delivery systems. Biodegrad Poly Drug Deliv Syst 45:43–70

22. Domb A, Langer R (1987) Polyanhydrides. I. Preparation of high molecular weight polyanhydrides. J Polym Sci A Polym Chem 25(12):3373–3386

23. Leong K, Brott B, Langer R (1985) Bioerodible polyanhydrides as drug-carrier matrices. I: characterization, degradation, and release characteristics. J Biomed Mater Res A 19 (8):941–955

24. Leong K et al (1986) Bioerodible polyanhydrides as drug-carrier matrices. II. Biocompatibility and chemical reactivity. J Biomed Mater Res A 20(1):51–64

25. Mathiowitz E, Langer R (1987) Polyanhydride microspheres as drug carriers I. hot-melt microencapsulation. J Control Release 5 (1):13–22

26. Langer R, Brem H, Tapper D (1981) Biocompatibility of polymeric delivery systems for macromolecules. J Biomed Mater Res A 15(2):267–277

27. Tamargo RJ et al (1989) Brain biocompatibility of a biodegradable, controlled-release polymer in rats. J Biomed Mater Res A 23 (2):253–266

28. Brem H et al (1989) Biocompatibility of a biodegradable, controlled-release polymer in the rabbit brain. Sel Cancer Ther 5(2):55–65

29. Steen RG et al (1988) In vivo 31P nuclear magnetic resonance spectroscopy of subcutaneous 9L gliosarcoma: effects of tumor growth and treatment with 1, 3-bis (2-chloroethyl)-1-nitrosourea on tumor bioenergetics and histology. Cancer Res 48(3):676–681

30. Grant Steen R et al (1989) In vivo31P nuclear magnetic resonance spectroscopy of rat 9l gliosarcoma treated with BCNU: dose response of spectral changes. Magn Reson Med 11(2):258–266

31. Yang MB, Tamargo RJ, Brem H (1989) Controlled delivery of 1, 3-bis (2-chloroethyl)-1-nitrosourea from ethylene-vinyl acetate copolymer. Cancer Res 49(18):5103–5107

32. Domb A et al (1991) Controlled delivery of water soluble and hydrolytically unstable anticancer drugs for polymeric implants. Polym Prepr 32(2):219–220

33. Grossman SA et al (1992) The intracerebral distribution of BCNU delivered by surgically implanted biodegradable polymers. J Neurosurg 76(4):640–647

34. Tamargo RJ et al (1993) Interstitial chemotherapy of the 9L gliosarcoma: controlled release polymers for drug delivery in the brain. Cancer Res 53(2):329–333

35. Brem H et al (1991) Interstitial chemotherapy with drug polymer implants for the treatment of recurrent gliomas. J Neurosurg 74 (3):441–446. https://doi.org/10.3171/jns.1991.74.3.0441

36. Brem H et al (1994) Biodegradable polymers for controlled delivery of chemotherapy with and without radiation therapy in the monkey brain. J Neurosurg 80(2):283–290

37. Wu MP et al (1994) In vivo versus in vitro degradation of controlled release polymers for intracranial surgical therapy. J Biomed Mater Res A 28(3):387–395

38. Fung LK et al (1996) Chemotherapeutic drugs released from polymers: distribution of 1, 3-bis (2-chloroethyl)-l-nitrosourea in the rat brain. Pharm Res 13(5):671–682

39. Fung LK et al (1998) Pharmacokinetics of interstitial delivery of carmustine, 4-hydroperoxycyclophosphamide, and paclitaxel from a biodegradable polymer implant in the monkey brain. Cancer Res 58 (4):672–684

40. Dang W, Daviau T, Brem H (1996) Morphological characterization of polyanhydride biodegradable implant Gliadel® during in vitro and in vivo erosion using scanning electron microscopy. Pharm Res 13(5):683–691

41. Sipos EP et al (1997) Optimizing interstitial delivery of BCNU from controlled release polymers for the treatment of brain tumors. Cancer Chemother Pharmacol 39 (5):383–389

42. Ewend M et al. (2019) Local delivery of BCNU from biodegradable polymer is superior to radiation therapy in treating

intracranial melanoma metastases. In: SUR-GICAL FORUM-CHICAGO-. pp. 564–565

43. Ewend MG et al (1996) Local delivery of chemotherapy and concurrent external beam radiotherapy prolongs survival in metastatic brain tumor models. Cancer Res 56 (22):5217–5223

44. Brem H (1990) Polymers to treat brain tumours. Biomaterials 11(9):699–701

45. Brem H, Langer R (1996) Polymer-based drug delivery to the brain. Sci Med 3:52–61

46. Brem H et al (1995) Placebo-controlled trial of safety and efficacy of intraoperative con-trolled delivery by biodegradable polymers of chemotherapy for recurrent gliomas. The polymer-brain tumor treatment group. Lan-cet 345(8956):1008–1012

47. Drapeau A, Fortin D (2015) Chemotherapy delivery strategies to the central nervous sys-tem: neither optional nor superfluous. Curr Cancer Drug Targets 15(9):752–768

48. Brem H et al (1995) The safety of interstitial chemotherapy with BCNU-loaded polymer followed by radiation therapy in the treatment of newly diagnosed malignant gliomas: phase I trial. J Neuro-Oncol 26(2):111–123

49. Valtonen S et al (1997) Interstitial chemo-therapy with carmustine-loaded polymers for high-grade gliomas: a randomized double-blind study. Neurosurgery 41(1):44–49

50. Westphal M et al (2003) A phase 3 trial of local chemotherapy with biodegradable car-mustine (BCNU) wafers (Gliadel wafers) in patients with primary malignant glioma. Neuro-Oncology 5(2):79–88. https://doi.org/10.1093/neuonc/5.2.79

51. Menei P et al (2010) Biodegradable carmustine wafers (Gliadel) alone or in combination with chemoradiotherapy: the French experience. Ann Surg Oncol 17(7):1740–1746. https://doi.org/10.1245/s10434-010-1081-5

52. Hart MG et al (2011) Chemotherapy wafers for high grade glioma. Cochrane Database Syst Rev 3:CD007294. https://doi.org/10.1002/14651858.CD007294.pub2

53. Aoki T et al (2014) A multicenter phase I/II study of the BCNU implant (Gliadel((R)) wafer) for Japanese patients with malignant gliomas. Neurol Med Chir (Tokyo) 54 (4):290–301

54. McGirt MJ et al (2009) Gliadel (BCNU) wafer plus concomitant temozolomide ther-apy after primary resection of glioblastoma multiforme. J Neurosurg 110(3):583–588. https://doi.org/10.3171/2008.5.17557

55. McGirt MJ, Brem H (2010) Carmustine wafers (Gliadel) plus concomitant temozolomide therapy after resection of malignant astrocytoma: growing evidence for safety and efficacy. Ann Surg Oncol 17 (7):1729–1731. https://doi.org/10.1245/s10434-010-1092-2

56. Gururangan S et al (2001) Phase I study of Gliadel wafers plus temozolomide in adults with recurrent supratentorial high-grade glio-mas. Neuro-Oncology 3(4):246–250. https://doi.org/10.1093/neuonc/3.4.246

57. Salmaggi A et al (2013) Prospective study of carmustine wafers in combination with 6-month metronomic temozolomide and radiation therapy in newly diagnosed glioblas-toma: preliminary results. J Neurosurg 118 (4):821–829. https://doi.org/10.3171/2012.12.JNS111893

58. Pan E, Mitchell SB, Tsai JS (2008) A retro-spective study of the safety of BCNU wafers with concurrent temozolomide and radio-therapy and adjuvant temozolomide for newly diagnosed glioblastoma patients. J Neuro-Oncol 88(3):353–357. https://doi.org/10.1007/s11060-008-9576-7

59. Bock HC et al (2010) First-line treatment of malignant glioma with carmustine implants followed by concomitant radiochemotherapy: a multicenter experience. Neurosurg Rev 33 (4):441–449. https://doi.org/10.1007/s10143-010-0280-7

60. Salmaggi A et al (2011) Loco-regional treat-ments in first-diagnosis glioblastoma: litera-ture review on association between Stupp protocol and Gliadel. Neurol Sci 32(Suppl 2):S241–S245. https://doi.org/10.1007/s10072-011-0797-8

61. Miglierini P et al (2012) Impact of the per-operatory application of GLIADEL wafers (BCNU, carmustine) in combination with temozolomide and radiotherapy in patients with glioblastoma multiforme: effi-cacy and toxicity. Clin Neurol Neurosurg 114(9):1222–1225. https://doi.org/10.1016/j.clineuro.2012.02.056

62. Ashby LS, Smith KA, Stea B (2016) Gliadel wafer implantation combined with standard radiotherapy and concurrent followed by adjuvant temozolomide for treatment of newly diagnosed high-grade glioma: a system-atic literature review. World J Surg Oncol 14 (1):225. https://doi.org/10.1186/s12957-016-0975-5

63. Chaichana KL et al (2011) The efficacy of carmustine wafers for older patients with glio-blastoma multiforme: prolonging survival. Neurol Res 33(7):759–764. https://doi.org/10.1179/1743132811Y.0000000006

64. Milojkovic Kerklaan B et al (2016) Strategies to target drugs to gliomas and CNS metastases of solid tumors. J Neurol 263 (3):428–440. https://doi.org/10.1007/s00415-015-7919-9

65. Ewend MG et al (1998) Local delivery of chemotherapy prolongs survival in experimental brain metastases from breast carcinoma. Neurosurgery 43(5):1185–1193

66. Ewend MG et al (2007) Treatment of single brain metastasis with resection, intracavity carmustine polymer wafers, and radiation therapy is safe and provides excellent local control. Clin Cancer Res 13 (12):3637–3641. https://doi.org/10.1158/1078-0432.CCR-06-2095

67. Mu F et al (2015) Tumor resection with carmustine wafer placement as salvage therapy after local failure of radiosurgery for brain metastasis. J Clin Neurosci 22(3):561–565. https://doi.org/10.1016/j.jocn.2014.08.020

68. Ene CI et al (2016) Safety and efficacy of carmustine (BCNU) wafers for metastatic brain tumors. Surg Neurol Int 7(Suppl 11):S295–S299. https://doi.org/10.4103/2152-7806.181987

69. Brem S et al (2013) Preservation of neurocognitive function and local control of 1 to 3 brain metastases treated with surgery and carmustine wafers. Cancer 119 (21):3830–3838. https://doi.org/10.1002/cncr.28307

70. Abel TJ et al (2013) Gliadel for brain metastasis. Surg Neurol Int 4(Suppl 4):S289–S293. https://doi.org/10.4103/2152-7806.111305

71. Olivi A et al (2003) Dose escalation of carmustine in surgically implanted polymers in patients with recurrent malignant glioma: a new approaches to brain tumor therapy CNS consortium trial. J Clin Oncol 21 (9):1845–1849. https://doi.org/10.1200/JCO.2003.09.041

72. Xing WK et al (2015) The role of Gliadel wafers in the treatment of newly diagnosed GBM: a meta-analysis. Drug Des Devel Ther 9:3341–3348. https://doi.org/10.2147/DDDT.S85943

73. Chowdhary SA, Ryken T, Newton HB (2015) Survival outcomes and safety of carmustine wafers in the treatment of high-grade gliomas: a meta-analysis. J Neuro-Oncol 122 (2):367–382. https://doi.org/10.1007/s11060-015-1724-2

74. Attenello FJ et al (2008) Use of Gliadel (BCNU) wafer in the surgical treatment of malignant glioma: a 10-year institutional experience. Ann Surg Oncol 15 (10):2887–2893. https://doi.org/10.1245/s10434-008-0048-2

75. Perry J et al (2007) Gliadel wafers in the treatment of malignant glioma: a systematic review. Curr Oncol 14(5):189–194

76. Zhang YD et al (2014) Efficacy and safety of carmustine wafers in the treatment of glioblastoma multiforme: a systematic review. Turk Neurosurg 24(5):639–645. https://doi.org/10.5137/1019-5149.JTN.8878-13.1

77. Kleinberg L (2012) Polifeprosan 20, 3.85% carmustine slow-release wafer in malignant glioma: evidence for role in era of standard adjuvant temozolomide. Core Evid 7:115–130. https://doi.org/10.2147/CE.S23244

78. Woodworth GF et al (2014) Emerging insights into barriers to effective brain tumor therapeutics. Front Oncol 4:126. https://doi.org/10.3389/fonc.2014.00126

79. Olivi A et al (1996) Interstitial delivery of carboplatin via biodegradable polymers is effective against experimental glioma in the rat. Cancer Chemother Pharmacol 39 (1-2):90–96

80. Brem S et al (2007) Local delivery of temozolomide by biodegradable polymers is superior to oral administration in a rodent glioma model. Cancer Chemother Pharmacol 60 (5):643–650. https://doi.org/10.1007/s00280-006-0407-2

81. Mangraviti A et al (2017) HIF-1alpha-targeting Acriflavine provides long term survival and radiological tumor response in brain cancer therapy. Sci Rep 7(1):14978. https://doi.org/10.1038/s41598-017-14990-w

82. Lin SH, Kleinberg LR (2008) Carmustine wafers: localized delivery of chemotherapeutic agents in CNS malignancies. Expert Rev Anticancer Ther 8(3):343–359. https://doi.org/10.1586/14737140.8.3.343

83. Mathios D et al (2016) Anti-PD-1 antitumor immunity is enhanced by local and abrogated by systemic chemotherapy in GBM. Sci Transl Med 8(370):370ra180. https://doi.org/10.1126/scitranslmed.aag2942

84. Blanchette M, Fortin D (2011) Blood-brain barrier disruption in the treatment of brain tumors. In: The Blood-Brain and Other Neural Barriers. Springer, New York, pp 447–463

85. Jackson S et al (2016) The effect of regadenoson-induced transient disruption of the blood–brain barrier on temozolomide delivery to normal rat brain. J Neuro-Oncol 126(3):433–439

86. Marty B et al (2012) Dynamic study of blood–brain barrier closure after its disruption using ultrasound: a quantitative analysis. J Cereb Blood Flow Metab 32(10):1948–1958

87. Jackson S et al (2018) The effect of an adenosine A 2A agonist on intra-tumoral concentrations of temozolomide in patients with recurrent glioblastoma. Fluids Barriers CNS 15(1):2

88. Chakraborty S et al (2016) Superselective intraarterial cerebral infusion of cetuximab after osmotic blood/brain barrier disruption for recurrent malignant glioma: phase I study. J Neuro-Oncol 128(3):405–415

89. Carpentier A et al (2016) Clinical trial of blood-brain barrier disruption by pulsed ultrasound. Sci Transl Med 8(343):343re342

90. Raghavan R et al (2006) Convection-enhanced delivery of therapeutics for brain disease, and its optimization. Neurosurg Focus 20(4):E12. https://doi.org/10.3171/foc.2006.20.4.7

91. Mangraviti A, Tyler B, Brem H (2015) Interstitial chemotherapy for malignant glioma: future prospects in the era of multimodal therapy. Surg Neurol Int 6(Suppl 1):S78–S84. https://doi.org/10.4103/2152-7806.151345

92. Richards Grayson AC et al (2003) Multi-pulse drug delivery from a resorbable polymeric microchip device. Nat Mater 2(11):767–772

93. Kim GY et al (2007) Resorbable polymer microchips releasing BCNU inhibit tumor growth in the rat 9L flank model. J Control Release 123(2):172–178. https://doi.org/10.1016/j.jconrel.2007.08.003

94. Masi BC et al (2012) Intracranial MEMS based temozolomide delivery in a 9L rat gliosarcoma model. Biomaterials 33(23):5768–5775. https://doi.org/10.1016/j.biomaterials.2012.04.048

95. Farra R et al (2012) First-in-human testing of a wirelessly controlled drug delivery microchip. Sci Transl Med 4(122):122ra121. https://doi.org/10.1126/scitranslmed.3003276

96. Zhang Z et al (2018) Development of a novel morphological paclitaxel-loaded PLGA microspheres for effective cancer therapy: in vitro and in vivo evaluations. Drug Deliv 25(1):166–177

97. Zhang Y-H et al (2011) Temozolomide/PLGA microparticles: a new protocol for treatment of glioma in rats. Med Oncol 28(3):901–906

98. Zhu T et al (2014) BCNU/PLGA microspheres: a promising strategy for the treatment of gliomas in mice. Chin J Cancer Res 26(1):81

99. Allhenn D et al (2013) A "drug cocktail" delivered by microspheres for the local treatment of rat glioblastoma. J Microencapsul 30(7):667–673

100. Zhang D et al (2012) The effect of temozolomide/poly (lactide-co-glycolide)(PLGA)/nano-hydroxyapatite microspheres on glioma U87 cells behavior. Int J Mol Sci 13(1):1109–1125

101. Ozeki T et al (2012) Combination therapy of surgical tumor resection with implantation of a hydrogel containing camptothecin-loaded poly (lactic-co-glycolic acid) microspheres in a C6 rat glioma model. Biol Pharm Bull 35(4):545–550

102. Mangraviti A et al (2015) Polymeric nanoparticles for nonviral gene therapy extend brain tumor survival in vivo. ACS Nano 9(2):1236–1249

103. Menei P et al (2005) Local and sustained delivery of 5-fluorouracil from biodegradable microspheres for the radiosensitization of malignant glioma: a randomized phase II trial. Neurosurgery 56(2):242–248

104. Wei X et al (2015) Liposome-based glioma targeted drug delivery enabled by stable peptide ligands. J Control Release 218:13–21

105. P-j Y et al (2014) OX26/CTX-conjugated PEGylated liposome as a dual-targeting gene delivery system for brain glioma. Mol Cancer 13(1):191

106. Chastagner P et al (2015) Phase I study of non-pegylated liposomal doxorubicin in children with recurrent/refractory high-grade glioma. Cancer Chemother Pharmacol 76(2):425–432

107. Clarke JL et al (2017) A phase 1 trial of intravenous liposomal irinotecan in patients with recurrent high-grade glioma. Cancer Chemother Pharmacol 79(3):603–610

108. Zhao M et al (2018) Targeted therapy of intracranial glioma model mice with curcumin nanoliposomes. Int J Nanomedicine 13:1601

109. Chen X et al (2016) Synergistic combination of doxorubicin and paclitaxel delivered by blood brain barrier and glioma cells dual targeting liposomes for chemotherapy of brain glioma. Curr Pharm Biotechnol 17(7):636–650

110. Voges J et al (2003) Imaging-guided convection-enhanced delivery and gene therapy of glioblastoma. Ann Neurol 54(4):479–487

Chapter 10

Liposome-Templated Hydrogel Nanoparticles for Targeted Delivery of CRISPR/Cas9 to Brain Tumors

Zeming Chen, Ann T. Chen, and Jiangbing Zhou

Abstract

Gene therapy has emerged as an exciting and innovative field for next-generation cancer treatments. In particular, genome editing using clustered regularly interspaced short palindromic repeats (CRISPR)/ CRISPR-associated protein 9 (Cas9) has the potential to significantly advance cancer gene therapy. However, clinical translation of CRISPR/Cas9 for brain cancer treatment requires the development of approaches able to simultaneously overcome cellular barriers for delivery of genetic materials into cells and the blood-brain barrier for drug delivery to the brain. In this chapter, we discuss how to synthesize liposome-templated hydrogel nanoparticles (LHNPs) for targeted delivery of CRISPR/Cas9 to brain tumors. We demonstrate that LHNPs can knock out GFP expression in U87MG-GFP brain tumor-bearing mice. Thus, LHNPs are capable of not only highly efficient gene editing but also penetrating the blood-brain barrier for gene therapy in the brain. Overall, LHNPs have the potential to serve as a powerful CRISPR/Cas9 delivery vehicle for clinical applications.

Key words Gene therapy, Brain cancer, CRISPR/Cas9, Liposomes, Nanogels, Drug delivery

1 Introduction

Due to recent advances in molecular biology and biochemistry, gene therapy has become an increasingly feasible therapeutic option and an attractive alternative to using traditional pharmacological procedures for cancer treatment [1–3]. Gene therapy can be utilized for a wide variety of purposes, including correcting mutated genes, altering the tumor microenvironment, and increasing the sensitivity of tumor cells to drugs. However, delivery of gene therapy, particularly clustered regularly interspaced short palindromic repeats (CRISPR)/CRISPR-associated protein 9 (Cas9)-based gene therapy, to brain tumors is challenging. It requires overcoming not only cellular barriers to transfer genetic materials to cells but also the blood-brain barrier (BBB) that prevents the penetration of most agents to the brain.

Vivek Agrahari et al. (eds.), *Nanotherapy for Brain Tumor Drug Delivery*, Neuromethods, vol. 163,
https://doi.org/10.1007/978-1-0716-1052-7_10, © Springer Science+Business Media, LLC, part of Springer Nature 2021

1.1 Direct Genetic Modification for Cancer Gene Therapy

Direct genetic modification of cancer cells has been widely used for gene therapy and is often accomplished by expressing or suppressing specific genes in target cells. Many prior efforts have been focused on expressing tumor suppressor genes, which function in normal cells but are mutated in tumor cells. For example, the p53 tumor suppressor gene is a common target for direct cancer gene therapy, as mutations in this gene occur in almost half of human tumors [4]. Other targets, such as PTEN [5] and miRNAs [6, 7], have significantly increased survival in experimental animal models. However, their success in clinical settings has been largely limited. Another approach is to inhibit oncogenes that are normally inactivated or are activated at low levels in normal cells. Oncogene activation may result from gene amplification, mutation, or chromosomal rearrangements and often leads to increased proliferation and survival of cancer cells [8, 9]. Thus, oncogene inhibition is a promising choice for cancer gene therapy. However, oncogene-targeting gene therapies have achieved modest success in experimental and clinical studies [9–13]. One potential explanation is that oncogene inhibition has been traditionally accomplished by using RNA interference (RNAi), a posttranslational gene-specific silencing mechanism, which has several notable limitations. First, RNAi targets mRNAs, which results in transient gene knockdown, leaves the original copy of the oncogene intact, and, thus, increases the chance of cancer recurrence [14]. Second, small interfering RNAs (siRNAs) are unstable and highly susceptible to extracellular enzymatic degradation [15]. Third, siRNA treatment may lead to off-target gene silencing and undesireable side effects due to high accumulation of cytoplasmic siRNA [16]. Short hairpin RNAs (shRNAs) are a viable alternative to using siRNAs for RNAi. Unlike siRNAs, shRNAs are first processed in the nucleus and then converted into siRNAs in the cytoplasm of target cells. One major advantage of shRNAs is that they are able to perform RNAi over an extended period of time compared to that of siRNAs. Prolonged expression of shRNAs can potentially be achieved through integration of the expression cassette into chromosomes using certain virus- or transposon-based delivery approaches. However, these options are unsafe and typically associated with a high risk of insertional mutagenesis [17]. A major hurdle for translating direct genetic modification-based treatments for cancer gene therapy has been the lack of safe and efficient delivery vehicles, as viral vectors, which are often used in current clinical settings, often suffer from safety concerns [18].

1.2 CRISPR/Cas9 as a Promising Gene Editing System

Recently, the CRISPR/Cas9 system has evolved as a highly promising and versatile tool for gene editing [19]. Mechanistically, an engineered single guide RNA (sgRNA) directs the Cas9 nuclease to a specific genomic region complementary to the sgRNA, where the nuclease induces site-specific DNA cleavages. These cleavages induce error-prone DNA repair, and as a result, insertions and deletions can be introduced at the target loci, leading to disruption in gene expression. Unlike RNAi, which uses posttranscriptional modification to transiently silence genes, the CRISPR/Cas9 system is able to alter genes directly at the transcriptional level. Therefore, CRISPR/Cas9-based gene editing is heritable. Genetic changes resulting from CRISPR/Cas9-based gene editing can be passed on to its progeny for sustained physiological effects [20].

Despite its strong clinical potential and advantages over RNAi, translation of the CRISPR/Cas9 system requires the development of safe and efficient delivery platforms. Current formulations of CRISPR/Cas9-based gene therapy often use viral vectors, such as adenoviruses [21], adeno-associated viruses [22], and retroviruses [23]. For these approaches, risk of carcinogenesis and limitations in encapsulation efficiency as drawbacks must be considered [24]. More importantly, immunogenicity is a common problem associated with viral delivery [25]. To overcome these limitations, nonviral vectors have emerged as a promising alternative. Recently, several nonviral delivery approaches have been explored for their potential to deliver CRISPR/Cas9 in vivo. These include hydrodynamic injection [26], cell penetrating peptides [27], and synthetic nanoparticles [28]. Although they demonstrated significantly reduced immunogenicity, these approaches had relatively lower delivery efficiency compared to their viral counterparts. Furthermore, the delivery efficiency was even more reduced when using these methods for delivery across the BBB. Table 1 summarizes the approaches that have been explored for delivery of CRISPR/Cas9-based gene therapy.

Table 1
Common CRISPR/Cas9 delivery vehicles and their advantages and disadvantages

Delivery method	Advantages	Disadvantages
Adenoviruses	High delivery efficiency	High immunogenicity
Adeno-associated viruses	High specificity	Limited delivery capacity
Retroviruses	Integration into target genome; stable expression	High risk of insertional mutagenesis
Hydrodynamic injection	Nonviral; variety of cargo can be delivered	High injection volumes needed; nonspecific interactions
Cell penetrating peptides	Nonviral; ease of synthesis	Nonspecific cell targeting; variable delivery efficiency
Synthetic nanoparticles	Nonviral; easy modification	Variable delivery efficiency; variable specificity

Viral-based delivery approaches traditionally have higher delivery efficiency, but also have higher risk of immunogenicity and toxicity. Nonviral delivery approaches are a safer alternative, but have relatively lower delivery efficiency compared to their viral counterparts.

**1.3 The BBB
as a Major Barrier
for Delivery
of Therapeutics
to the Brain**

The BBB is a highly selective barrier that prevents influx of most compounds from the blood to the brain. It maintains the highly regulated central nervous system (CNS) milieu and prevents infections from blood-borne pathogens. The BBB is composed of endothelial cells, astrocytic end-feet, and pericytes. Tight junctions formed between the endothelial cells limit the paracellular flux of hydrophilic molecules and large molecules, such as antibodies and antibody-drug conjugates, but permit free diffusion for small molecules which are lipophilic and smaller than 400–500 kDa [29, 30]. Even molecules that can transverse the endothelial cell membrane are often sent back into the capillary lumen by efflux transporters. The BBB is often partially disrupted in brain tumors; however, the degree of disruption is often insufficient to allow for delivery of pharmacologically significant quantities of therapeutics to the brain for tumor treatment. Thus, effective treatment of brain tumors requires the development of novel approaches to overcoming the BBB [18, 31].

**1.4 Overview
of LHNPs**

Recently, we developed liposome-templated hydrogen nanoparticles (LHNPs) for targeted delivery of Cas9 proteins and sgRNAs for efficient gene editing to brain tumors [14]. LHNPs are composed of chemically modified polyetherimide (PEI) that form a hydrogel core for encapsulation of Cas9 protein and a liposome shell for nucleic acid loading. LHNPs are capable of highly efficient protein encapsulation and able to co-deliver proteins and nucleic acids to tumor tissues for tumor growth inhibition. The surface of LHNPs was conjugated with mHph3 [32], a cell penetration peptide, and iRGD, a peptide with high affinity for $\alpha_v\beta_3/\alpha_v\beta_5$ integrins that are highly expressed in most tumors [33, 34], for enhanced transfection efficiency and tumor targeting, respectively. Furthermore, LHNPs were synthesized to encapsulate Lexiscan (LEX), a small molecule known to enhance BBB permeability [35]. A schematic diagram of a LHNP is shown in Fig. 2c. A major advantage of LHNPs is that they are able to simultaneously co-deliver Cas9 protein and sgRNA, which maximizes their efficiency and reduces their off-target effects [36, 37]. After intravenous administration, LHNPs are able to reach brain tumors through blood vessel leakage and engagement with integrins that are expressed in both blood vessels and tumor cells. After being internalized by tumor cells, LHNPs escape endosomes and release their payload of Cas9 protein and sgRNA. Because sgRNA is delivered in the form of DNA, specifically minicircle DNA as described in this chapter, it can form complexes with the Cas9 protein. The resulting Cas9/sgRNA complex can diffuse into the nucleus due to the nuclear-localization-signal peptides on the Cas9 protein. Once in the nucleus, the Cas9/sgRNA complex can then search for the target DNA locus and induce specified deletions or modifications for gene editing (Fig. 1).

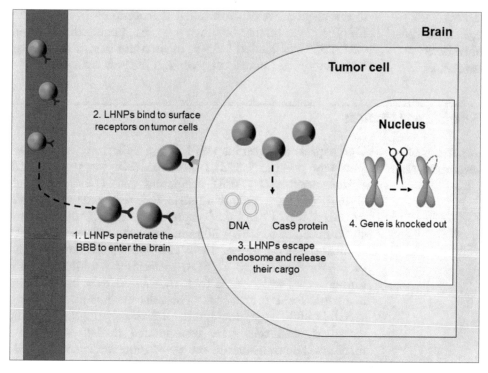

Fig. 1 Schematic diagram of LHNP-mediated CRISPR/Cas9 delivery to brain tumor cells

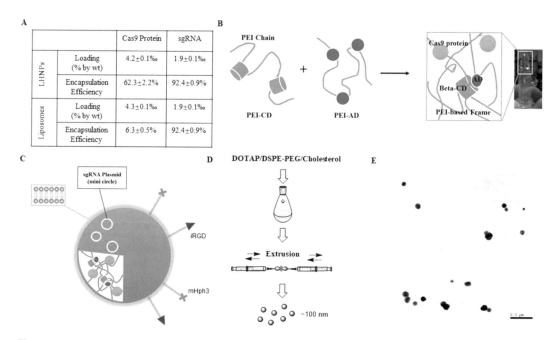

Fig. 2 Fabrication and characterization of LHNPs. (**a**) Loading and encapsulation efficiencies of Cas9 protein and sgRNA minicircle in LHNPs ($n = 4$). (**b**) CD-AD-based host–guest interaction mediates the formation of PEI hydrogel. (**c**) Schematic diagram of an LHNP and its synthesis. (**d**) A representative TEM image of LHNPs. Scale bar: 500 nm

1.5 *LHNPs*	In this chapter, we describe detailed procedures for the synthesis of
for Targeted Delivery	LHNPs. We further characterize the gene editing effect of
of CRISPR/Cas9	CRISPR/Cas9-loaded LHNPs in vitro and in vivo using a sgRNA
to Brain Tumors	targeting green fluorescent protein (GFP)-encoding gene.

2   Materials and Methods

2.1   *Generation of Cas9-Expressing Cells*

Cas9 expression plasmid, pMJ806, was a gift from Jennifer Doudna (Addgene plasmid # 39312) [19]. pMJ806 was transformed into *E. coli* Rosetta II (DE3) competent cells (Novagen, Millipore Sigma). Briefly, 50 μL of Rosetta II (DE3) cells was thawed on ice, mixed with 1 μL of Cas9 expression plasmid (100–150 ng/μL), and transferred to ice for 30 min. Next, the cells were transformed via heat shock for 45 s at 42 °C and then placed immediately back on ice for 3 min. 500 μL SOC medium was added to the transformed cells, which were then transferred to a horizontal shake incubator for 1 h at 37 °C. The cells were spun down at a low speed ($<2000 \times g$) for 1 min. Afterward, the supernatant was discarded, and cells were resuspended in 100 μL fresh SOC medium. The transformed cell suspension was then pipetted onto LB agar plates containing selection antibiotics, and the plates were incubated overnight at 37 °C.

2.2   *Cas9 Protein Expression and Purification*

Single colonies of Cas9-expressing cells were picked and transferred to 14 mL polypropylene round-bottom tubes containing 5 mL $2\times$ YT medium with proper antibiotic and incubated at 37 °C for 16 h. When $OD_{600nm} \geq 0.6$, the cell suspension was transferred to an Erlenmeyer flask containing 500 mL $2\times$ YT medium and proper antibiotic. After incubation at 37 °C for 4–6 h, cells were collected by centrifugation at $4000 \times g$ for 15 min. Cas9 protein was extracted from bacteria by ultrasonication and purified using His60 Ni Superflow Resin columns (Clontech Laboratories, Inc.). Briefly, the unpurified protein solution was run through a gravity column containing His60 Ni resin. The protein became bound to the His60 Ni resin and was then subsequently washed by equilibration buffer (50 mM sodium phosphate, 300 mM sodium chloride, 20 mM imidazole; pH 7.4), wash buffer (50 mM sodium phosphate, 300 mM sodium chloride, 40 mM imidazole; pH 7.4), and elution buffer (50 mM sodium phosphate, 300 mM sodium chloride, 300 mM imidazole; pH 7.4), respectively. Lastly, elution buffer was collected and dialyzed in phosphate buffered saline (PBS, pH 7.4) to remove imidazole.

2.3   *Mini Circle sgRNA Preparation*

Mini circle DNA was produced according to previously reported procedures [38]. The expression cassette for GFP-targeting sgRNA (sgGFP) was amplified by polymerase chain reaction (PCR) from lentiCRISPR—EGFP sgRNA 2, a gift from Feng Zhang (Addgene plasmid # 51761) [39]. The forward and reverse primers used for PCR amplification were GTCGGTAGATCTGAGGGCC-TATTTCCCATGAT and GTCATTGTCGACAAAAAAGCACC-GACTCGGTG, respectively. The cassette was then cloned into the BglII and Sal1 restriction sites of vector pMC.BESPX [38]. To produce mini circle sgGFP, the sgGFP pMC.BESPX construct was transformed into *E. coli* strain ZYCY10P3S2T (System Biosciences). A single clone of transformed *E. coli* was selected and cultured at 37 °C in 400 mL Terrific Broth supplemented with kanamycin (Thermo Scientific). After 24 h, minicircle induction media comprising 400 mL lysogeny broth, 16 mL 1 N NaOH, and 400 μL 20% L-arabinose (Sigma-Aldrich) was added to the culture. After 6 h incubation at 32 °C, the *E. coli* culture was collected through centrifugation. Minicircle-sgGFP, which is simplified as sgGFP in the following description, was purified using EndoFree Plasmid Kits (Qiagen).

2.4 LHNP Preparation

LHNPs have a core-shell nanostructure, with cyclodextrin (CD)/adamantane (AD) grafted onto PEI composing the core and cationic 1,2-dioleoyl-3-trimethylammonium-propane chloride salt (DOTAP) composing the shell.

To synthesize PEI-CD, 78 mg branched polyethylenimine (M.W. 25,000, Sigma-Aldrich) was dissolved in ddH$_2$O, and 77 mg randomized mono-Tos-cyclodextrin (M.W. 1230, Trappsol) was added into the solution. The mixture was heated to 60 °C and stirred for 8 h. Afterward, the reactant was dialyzed with ddH$_2$O to remove unreacted cyclodextrin, and the resulting PEI-CD was collected by lyophilization. To synthesize PEI-AD, 70 mg branched polyethylenimine was dissolved in chloroform, and 5 mg 1-adamantyl isocyanate (M.W. 177.24, Sigma-Aldrich) was added to the solution. The mixture was heated to 70 °C and stirred for 6 h. After the reaction returned to room temperature, chloroform was removed by rotor evaporation. The remaining product was redissolved in ddH$_2$O. Unreacted 1-admantyl isocyanate was removed by centrifugation at 4000 rpm (3550 × g) for 10 min. The resulting PEI-AD was collected by lyophilization.

To synthesize LHNPs, DOTAP (Avanti Polar Lipids), cholesterol (Avanti Polar Lipids), and 1,2-distearoyl-sn-glycero-3-phosphoethanolamine (DSPE)-polyethylene glycol (PEG)2000-maleimide (MAL) (Avanti Polar Lipids) were dissolved in 2:1 chloroform/methanol solution at a concentration of 14.3, 26.0, and 6.8 μmol/mL, respectively. Typically, 700 μL DOTAP, 400 μL cholesterol, and 40 μL DSPE-PEG2000-MAL were mixed in 7 mL glass vial by vortex, and then solvents were removed by airflow. Next, the resulting product was resuspended in 1 mL aqueous solution containing 490 μL PEI-AD (10 mg/mL in ddH$_2$O), 490 μL PEI-CD (10 mg/mL in ddH$_2$O), and 20 μL Cas9 protein solution (2 mg/mL in PBS). Iterations of 30 s vortexing followed by 5 s idle sitting at room temperature were repeated five times. The resulting multilamellar liposomes were extruded five times through a 200 nm polycarbonate membrane (Whatman) and seven times through a 100 nm membrane using a Avanti Mini Extruder (Avanti Polar Lipids). Immediately after extrusion, the mixture was diluted into 35 mL ddH$_2$O and subjected to centrifugation at 15,000 rpm (13,300 × g) for 30 min. Liposomes were then collected, washed with ddH$_2$O twice, and resuspended in ddH$_2$O at 6 mg/mL. To load DNA, 1 mL Lipo-gel solution was incubated with 1.2 μg DNA and incubated on ice for 1 h.

To enhance the delivery efficiency of Cas9/sgRNA, mHph3, a cell penetration peptide, was conjugated to the surface of LHNPs [32]. To achieve this, 30 μg peptide was added into the LHNP suspension and reacted for 60 min. To improve the delivery of LHNPs specifically to tumors, iRGD, a modified form of RGD peptide with high affinity for $\alpha_v\beta_3/\alpha_v\beta_5$ integrins and neuropilin-1 [33, 34], was conjugated to the surface of LHNPs. For conjugation of both mHph3 and iRGD, 30 μg mHph3 peptide was added into the LHNP suspension. After reacting for 30 min, 30 μg iRGD was added and reacted for additional 30 min. Unreacted peptides were removed by centrifugation at 15,000 rpm (13,300 × g) for 30 min. To enhance the delivery of LHNPs to brain tumors, LHNPs were modified to excapsulate LEX, a small molecule known to have the ability to transiently enhance BBB permeability [35, 40]. For this purpose, the same synthesis procedure was followed except that 0.1 mg LEX was added to the lipid mixture prior to solvent evaporation.

2.5 Characterization of Encapsulation Efficiency and Drug Release

For characterization of encapsulation efficiency, 1 mg LHNPs were lysed in 10 μL dimethyl sulfoxide (DMSO) and diluted in 90 μL PBS. The amounts of Cas9 protein and DNA were determined based on the standard bicinchoninic acid (BCA) assay and Pico-Green assay, respectively, using commercial kits (Thermo Fisher Scientific). LHNPs without encapsulants were used as background. For characterization of drug release, LHNPs (3–5 mg) were suspended in 1 mL PBS (pH 7.4) and incubated at 37 °C with gentle shaking. At each sampling time, the LHNP suspension was centrifuged for 15 min at 15,000 rpm (13,300 × g). The supernatant was removed for quantification of Cas9 protein or DNA and replaced with an equivalent volume of PBS for continued monitoring of release. Detection of Cas9 protein or DNA was conducted using the same BCA and PicoGreen assays described above.

2.6 Preparation of LHNPs for Transmission Electron Microscopy (TEM)

After synthesis, LHNPs were applied to holey carbon-coated copper grids (Electron Microscopy Sciences) and dried by air flow. Images were captured using a TEM microscope (FEI Tecnai TF20 TEM).

3 Results and Discussion

3.1 Characterization of LHNPs

Due to their high gene delivery efficiency and favorable biological safety profile, DOTAP liposomes have been previously utilized in clinical trials [41, 42]. However, we found that DOTAP liposomes were unable to efficiently encapsulate Cas9 protein, with a loading efficiency limited to 6% (Fig. 2a). To improve the Cas9 encapsulation efficiency, we formed hydrogel within the liposome core, in which the Cas9 protein is embedded (Fig. 2b, c). The hydrogel was assembled by mixing PEI-CD (25 kDa) and PEI-AD (25 kDa), which react through the CD-AD-mediated host–guest interaction. The resulting LHNPs contained Cas9 within the hydrogel core and DOTAP lipid on the shell. sgGFP was then loaded into the lipid layer through charge–charge interaction (Fig. 2b, c). LHNPs were synthesized through the standard extrusion procedures with minor modifications and were further conjugated with mHph3 and iRGD for enhanced transfection efficiency and tumor targeting (Fig. 2d). Using transmission electron microscopy (TEM), we found that LHNPs are spherical in shape with a diameter of ~95 nm (Fig. 2e). Compared to DOTAP liposomes, formulation of hydrogel in the core significantly increased the Cas9 loading efficiency to 62%. The resulting LNHPs encapsulated Cas9 protein with 4.2% by weight.

3.2 Gene Knockout in U87MG-GFP Cells

We engineered human U87MG glioblastoma cells to express GFP (U87MG-GFP) through lentiviral transduction (LV-CMV-GFP, Kerafast). To determine the gene knockout efficiency, U87MG-GFP cells were plated at a density of 10,000 cells in 3 mL/well in 6-well plates with a 18 mm × 18 mm cover slip (Fisher Scientific) inside each well. After overnight incubation, cells were treated with 1 mg/mL LHNPs. After 5 days, the GFP signal in cells was determined by a Leica SP8 confocal laser microscope. As shown in Fig. 3, treatment with Cas9/sgGFP-loaded LHNPs significantly reduced the GFP fluorescence intensity in cells, whereas empty LHNPs did not.

3.3 Gene Knockout in U87MG-GFP Brain Tumor-Bearing Mice

To establish intracranial U87MG-GFP mouse xenografts, 5–6-week-old female C57BL6 mice were anesthetized via intraperitoneal injection of ketamine and xylazine. Twenty thousand U87MG-GFP cells in 2 μL of PBS were injected into the right striatum 2 mm lateral and 0.5 mm posterior to the bregma and 3 mm below the dura using a stereotactic apparatus using an UltraMicroPump (UMP3) (World Precision Instruments, FL, USA) [35]. All procedures were approved by the Institutional Animal Care and Utilization Committee (IACUC) of Yale University. Three weeks after tumor inoculation, Cas9/sgGFP-loaded LHNPs were administered to mice through tail-vein injection.

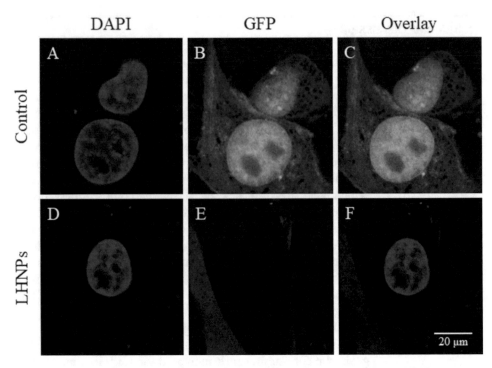

Fig. 3 Confocal microscopy analyses of U87MG-GFP cells that were treated with blank LHNPs (**a–c**) or LHNPs with encapsulation of Cas9 and sgGFP. Blue: DAPI. Green: GFP. Scare bar, 20 μm

Each mouse received 1 mg LHNPs in 200 μL PBS daily for 4 continuous days. Control mice received the same amount of empty LHNPs. Two days after the last injection, mice were euthanized and perfused with PBS and 4% paraformaldehyde (PFA). The brains were isolated and imaged using an in vivo imaging system (IVIS). Next, the brains were incubated in 4% PFA overnight and then for 2 additional days in PBS containing 30% sucrose. 20-μm-thick sections were obtained using a vibratome (Leica). Tumor-containing brain sections were mounted and imaged by a Leica SP8 confocal laser microscope. Ex vivo imaging of the brains (Fig. 4a) and microscopic analysis (Fig. 4b, c) of brain tissues revealed that systemic treatment of Cas9/sgGFP-loaded LHNPs significantly reduced the expression of GFP in tumors.

4   Conclusion

In this study, we synthesized LHNPs and characterized their use for targeted delivery of CRISPR/Cas9 for gene editing in brain tumors. The gene editing efficiency was quantified by measuring the GFP intensity. We demonstrated that LEX-loaded LHNPs mediate efficient gene editing in vitro in cells and in vivo in intracranial tumors. This finding is consistent with our recent report

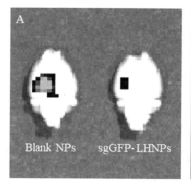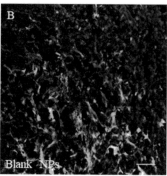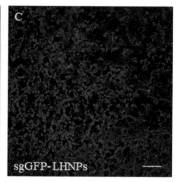

Fig. 4 In vivo characterization of LHNPs for systemic delivery of Cas9/sgGFP for gene editing in brain tumors. Ex vivo imaging of the brains (**a**) and microscopic analysis of brain tumors isolated from mice received the indicated treatment (**b**, **c**). Scare bar, 60 μm

[14], in which we showed that delivery of PLK1-targeting Cas9/sgRNA efficiently decreases the expression of PLK1 and prolongs the survival of tumor-bearing mice. Despite its high efficiency, the LHNP-based delivery system has two limitations. First, LHNPs are composed of multiple components and, thus, are complicated to synthesize. Second, LHNPs are unstable in solution and need to be freshly prepared for each application. To maximize the therapeutic capabilities of LHNPs for clinical translation, further studies are needed to reduce their complexity and enhance their stability. Overall, LHNPs are a powerful tool for targeted delivery of CRISPR/Cas9 to brain tumors and have the potential to greatly improve the clinical management of brain cancer patients.

Acknowledgments

This work was supported by NIH Grants R01NS095817 and UG3NS115597.

References

1. Cheng CJ, Bahal R, Babar IA et al (2015) MicroRNA silencing for cancer therapy targeted to the tumour microenvironment. Nature 518:107–110

2. Esensten JH, Bluestone JA, Lim WA (2017) Engineering therapeutic T cells: from synthetic biology to clinical trials. Annu Rev Pathol 12:305–330

3. Rabbani PS, Zhou A, Borab ZM et al (2017) Novel lipoproteoplex delivers Keap1 siRNA based gene therapy to accelerate diabetic wound healing. Biomaterials 132:1–15

4. Oren M (1999) Regulation of the p53 tumor suppressor protein. J Biol Chem 274:36031–36034

5. Zhang W, Fine H (2006) Mechanisms of Gliomagenesis. In: Janigro O (ed) The cell cycle in the central nervous system, vol VI. Humana Press, Totowa, NJ, pp 449–462

6. Kefas B, Godlewski J, Comeau L et al (2008) microRNA-7 inhibits the epidermal growth factor receptor and the Akt pathway and is down-regulated in glioblastoma. Cancer Res 68:3566–3572

7. Corsten MF, Miranda R, Kasmieh R et al (2007) MicroRNA-21 knockdown disrupts glioma growth in vivo and displays synergistic cytotoxicity with neural precursor cell delivered S-TRAIL in human gliomas. Cancer Res 67:8994–9000

8. Croce CM (2008) Oncogenes and cancer. N Engl J Med 358:502–511

9. Bishop JM (1991) Molecular themes in oncogenesis. Cell 64:235–248

10. Muralidharan R, Babu A, Amreddy N et al (2017) Tumor-targeted nanoparticle delivery of HuR siRNA inhibits lung tumor growth in vitro and in vivo by disrupting the oncogenic activity of the RNA-binding protein HuR. Mol Cancer Ther 16:1470–1486

11. Davis ME, Zuckerman JE, Choi CH et al (2010) Evidence of RNAi in humans from systemically administered siRNA via targeted nanoparticles. Nature 464:1067–1070

12. Zuckerman JE, Gritli I, Tolcher A et al (2014) Correlating animal and human phase Ia/Ib clinical data with CALAA-01, a targeted, polymer-based nanoparticle containing siRNA. Proc Natl Acad Sci U S A 111:11449–11454

13. Zuckerman JE, Davis ME (2015) Clinical experiences with systemically administered siRNA-based therapeutics in cancer. Nat Rev Drug Discov 14:843–856

14. Chen ZM, Liu FY, Chen YK et al (2017) Targeted delivery of CRISPR/Cas9-mediated cancer gene therapy via liposome-templated hydrogel nanoparticles. Adv Funct Mater 27:1703036

15. Gavrilov K, Saltzman WM (2012) Therapeutic siRNA: principles, challenges, and strategies. Yale J Biol Med 85:187–200

16. Aagaard L, Rossi JJ (2007) RNAi therapeutics: principles, prospects and challenges. Adv Drug Deliv Rev 59:75–86

17. Wang SL, Yao HH, Qin ZH (2009) Strategies for short hairpin RNA delivery in cancer gene therapy. Expert Opin Biol Ther 9:1357–1368

18. Zhou J, Atsina KB, Himes BT et al (2012) Novel delivery strategies for glioblastoma. Cancer J 18:89–99

19. Jinek M, Chylinski K, Fonfara I et al (2012) A programmable dual-RNA-guided DNA endonuclease in adaptive bacterial immunity. Science 337:816–821

20. Unniyampurath U, Pilankatta R, Krishnan MN (2016) RNA interference in the age of CRISPR: will CRISPR interfere with RNAi? Int J Mol Sci 17:291

21. Wang D, Mou H, Li S et al (2015) Adenovirus-mediated somatic genome editing of Pten by CRISPR/Cas9 in mouse liver in spite of Cas9-specific immune responses. Hum Gene Ther 26:432–442

22. Senis E, Fatouros C, Grosse S et al (2014) CRISPR/Cas9-mediated genome engineering: an adeno-associated viral (AAV) vector toolbox. Biotechnol J 9:1402–1412

23. Niu D, Wei HJ, Lin L et al (2017) Inactivation of porcine endogenous retrovirus in pigs using CRISPR-Cas9. Science 357:1303–1307

24. Wang LY, Li FF, Dang L et al (2016) In vivo delivery Systems for Therapeutic Genome Editing. Int J Mol Sci 17:626

25. Bessis N, Garciacozar FJ, Boissier MC (2004) Immune responses to gene therapy vectors: influence on vector function and effector mechanisms. Gene Ther 11:S10–S17

26. Xue W, Chen SD, Yin H et al (2014) CRISPR-mediated direct mutation of cancer genes in the mouse liver. Nature 514:380

27. Ramakrishna S, Dad AK, Beloor J et al (2014) Gene disruption by cell-penetrating peptide-mediated delivery of Cas9 protein and guide RNA. Genome Res 24:1020–1027

28. Zuris JA, Thompson DB, Shu Y et al (2015) Cationic lipid-mediated delivery of proteins enables efficient protein-based genome editing in vitro and in vivo. Nat Biotechnol 33:73–80

29. Chacko AM, Li CS, Pryma DA et al (2013) Targeted delivery of antibody-based therapeutic and imaging agents to CNS tumors: crossing the blood-brain barrier divide. Expert Opin Drug Deliv 10:907–926

30. Patel T, Zhou J, Piepmeier JM et al (2012) Polymeric nanoparticles for drug delivery to the central nervous system. Adv Drug Deliv Rev 64:701–705

31. Deeken JF, Loscher W (2007) The blood-brain barrier and cancer: transporters, treatment, and Trojan horses. Clin Cancer Res 13:1663–1674

32. Chen Z, Liu F, Chen Y et al (2017) Targeted delivery of CRISPR/Cas9-mediated cancer gene therapy via liposome-templated hydrogel nanoparticles. Adv Funct Mater 27:1703036

33. Sugahara KN, Teesalu T, Karmali PP et al (2009) Tissue-penetrating delivery of compounds and nanoparticles into tumors. Cancer Cell 16:510–520

34. Zhou J, Patel TR, Fu M et al (2012) Octafunctional PLGA nanoparticles for targeted and efficient siRNA delivery to tumors. Biomaterials 33:583–591

35. Han L, Kong DK, Zheng MQ et al (2016) Increased nanoparticle delivery to brain tumors by autocatalytic priming for improved treatment and imaging. ACS Nano 10:4209–4218

36. Kouranova E, Forbes K, Zhao G et al (2016) CRISPRs for optimal targeting: delivery of CRISPR components as DNA, RNA, and protein into cultured cells and single-cell embryos. Hum Gene Ther 27:464–475

37. Liang X, Potter J, Kumar S et al (2015) Rapid and highly efficient mammalian cell engineering via Cas9 protein transfection. J Biotechnol 208:44–53

38. Kay MA, He CY, Chen ZY (2010) A robust system for production of minicircle DNA vectors. Nat Biotechnol 28:1287–1289

39. Shalem O, Sanjana NE, Hartenian E et al (2014) Genome-scale CRISPR-Cas9 knockout screening in human cells. Science 343:84–87

40. Carman AJ, Mills JH, Krenz A et al (2011) Adenosine receptor signaling modulates permeability of the blood-brain barrier. J Neurosci 31:13272–13280

41. Lu C, Stewart DJ, Lee JJ et al (2012) Phase I clinical trial of systemically administered TUSC2(FUS1)-nanoparticles mediating functional gene transfer in humans. PLoS One 7: e34833

42. Porteous DJ, Dorin JR, Mclachlan G et al (1997) Evidence for safety and efficacy of DOTAP cationic liposome mediated CFTR gene transfer to the nasal epithelium of patients with cystic fibrosis. Gene Ther 4:210–218

Chapter 11

Methods to Separate, Characterize, and Encapsulate Drug Molecules into Exosomes for Targeted Delivery and Treatment of Glioblastoma

Yen Nguyen, Sandeep Kaur, Hae Shim, Vaibhav Mundra, and Venkatareddy Nadithe

Abstract

There have been many drug delivery strategies for the effective targeted delivery of drugs across the blood-brain barrier (BBB) cells to overcome challenges posed by BBB. However, drug delivery to the brain is still a hurdle that has yet to be solved. Due to the tight junctions and high selectivity of the BBB, most active and passive strategies deliver an insufficient or insignificant amount of drug across the protective BBB shield. Recently, exosomes, "biological nanoparticles" with the inherent homing capability to brain cells, have been shown to deliver drugs efficiently by preserving their therapeutic activity. Many different drug molecules are loaded into exosomes, belonging to the category of small synthetic drug molecules (doxorubicin, rhodamine) or large protein-based molecules (catalase) or nucleic acid-based drugs such as small interfering RNA (siRNA). In this chapter, we will focus on describing exosome isolation, characterization, and drug loading methods that are suitable for studying and treating glioblastoma.

Key words Exosome, Isolation, Characterization, Loading, Blood-brain barrier, Drug delivery

1 Introduction

Exosomes are extracellular vesicles that are approximately 40–100 nm in diameter and are composed of a cellular membrane with numerous adhesive proteins on their surface [1]. They also bear marker proteins such as CD9, CD63, and CD81, which are distinctive from other particles shed from the plasma membrane [2]. Their formation starts with inward budding of late endosomes', or multivesicular bodies (MVB), membrane inside viable cells forming vesicles in the lumen which are called intraluminal vesicles (ILV). The protein complex is a part of these vesicles' formation, such as the endosomal sorting complex required for transport (ESCRT), and monoubiquitination is a signal to direct protein complexes into MVB. These complexes help to concentrate

Vivek Agrahari et al. (eds.), *Nanotherapy for Brain Tumor Drug Delivery*, Neuromethods, vol. 163,
https://doi.org/10.1007/978-1-0716-1052-7_11, © Springer Science+Business Media, LLC, part of Springer Nature 2021

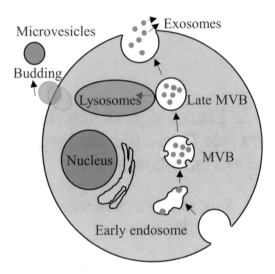

Fig. 1 Formation of exosome and microvesicle. Exosome is derived from endosome formed from plasma membrane. As early endosome becomes late endosomes, inward budding occurs and forms multivesicular bodies (MVB) containing numerous intraluminal vesicles (ILV). MVB can either get degraded by lysosomes or fuse with the membrane to release ILV called exosomes. Microvesicles, on the other hand, originate from the budding of the plasma membrane [3]. (Reproduced from (Acta Pharmaceutica Sinica B, 2016) with permission from Elsevier))

proteins into ILV. These vesicles are then secreted out of the cell as exosomes (Fig. 1) [3]. Secretions can be constitutive or triggered by the changes in intracellular calcium level in mast cells and erythroleukemia cell line [2].

Exosomes are shown as highly attractive drug delivery vehicles due to their involvement in the cell to cell communication. They can transfer biological information by traveling short or long distances within the body [4]. Additionally, exosomes have the advantage of loading drug cargo and have the potential for targeted delivery [5]. Exosomes can be modified to express a targeting moiety on their surface by modifying the parental cells. Various adhesive proteins on the surface of exosomes facilitate interaction with the cell membrane and delivery of molecules to target cells. The presence of exosomes in brain tumor cells from glioma mice further indicates that exosomes are capable of crossing the blood-brain barrier [4]. Their composition of cellular membranes with a wide range of adhesion proteins such as integrins and biomarkers gives them the specificity and precision of drug delivery that other artificial delivery systems fail to achieve. Additionally, exosomes bear surface proteins specific to their origin and will be more selective to interact with those cell lines. Thus, utilizing exosomes, a nanocarrier is a natural way to deliver drugs to a specific target [6]. Furthermore, identifying mechanisms for the successful

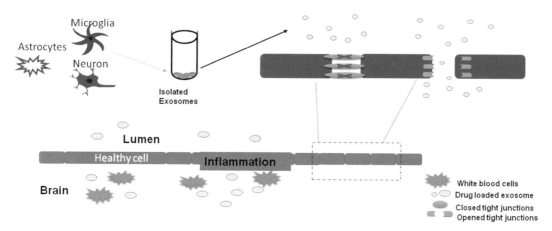

Fig. 2 Drug-loaded exosomes with the ability to cross BBB and to treat brain-related diseases

crossing of exosomes through the BBB is the focus of many research groups, with hope for the successful delivery of drugs to treat neurological disorders.

The blood-brain barrier is a tightly sealed membrane with a function to protect a delicate organ from outside invaders. Many pharmaceutical products are developed yet are restricted from entry into the brain. This matter poses the biggest challenge for the development of exosomes as effective drug delivery molecules (Fig. 2). Passive diffusion of molecules through the blood-brain barrier (BBB) requires high lipophilicity, neutral charge, and sizes of less than 400–500 Da [7]. However, for pharmaceutical molecules that do not satisfy those requirements, various methods have been tested in vitro and showed promising results. One proposed mechanism is a two-step delivery method. The first round of treatment is meant to damage the integrity of the BBB to allow therapeutic drug molecules from the second treatment round to enter into the brain [8]. Another potential delivery mechanism relies on the increase of BBB's permeability during inflammation. Liposaccharides are used to induce inflammation during the experiment, resulting in an increase in the expression of adhesion proteins and uptake of inflammatory molecules through diapedesis [9]. An alternative administration method, such as intranasal delivery, has shown some success in delivering small molecules into the brain through the olfactory route. Small molecules travel in the olfactory tract to get into the brain without the hassle of crossing the BBB [4]. However, this strategy is only applicable for the very potent drug since only a small volume can be delivered at a time [10]. Once the drug delivery systems have been administered, various methods of uptake are proposed to be responsible for the delivery across BBB, such as adsorptive-mediated transcytosis with a cell-penetrating peptide or receptor-mediated transport with the invagination of molecules [7].

In this chapter, the materials and methods of exosome isolation, characterization, and drug loading will be discussed by selecting exosome cell sources relevant to neurological disease such as glioblastomas. Exosomes derived from cell lines involved in those neurological conditions will selectively target its origin [6]. Pharmaceutical drugs described in this chapter for drug loading into these exosomes are selected from group of small molecules, protein, peptides, or siRNA. Most of the methods described in this chapter are broad in nature and can be applied to many other cells for exosome source or drug loading.

2 Methods of Isolation

Exosomes for drug delivery can be isolated from different cells and body fluids such as blood. There are many methods of isolation that are reported in literature (Table 1). Among them, we described the methods that are routinely used in the field of drug delivery with applications specific to neurological diseases.

2.1 Differential Centrifugation

Among the different exosome isolation methods described in this chapter, differential centrifugation involves simple steps to isolate exosomes. This method is also one of the cost-effective and time-efficient methods to isolate exosomes. However, the major drawback observed with this method of isolation is potential cellular impurities within isolated exosomes. To prevent experimental interference in preclinical studies, it is essential to purify further and improve the quality of exosomes isolated using differential centrifugation techniques. In the described method below, exosomes are isolated by centrifuging the cultured media at two different centrifugal forces [11].

Table 1
Exosome isolation, characterization, drug loading, and labeling methods

Isolation of exosomes	Characterization of exosomes	Drug loading methods	Labeling exosome membranes
Differential centrifugation	Nanoparticle tracking analysis	–	–
Sequential centrifugation	Dynamic light scattering method	Incubation	Fluorescent labels
Precipitation	Western blotting method	Sonication	PKH lipophilic membrane dyes
Density gradient method	Flow cytometry	Electroporation	–
Immunoisolation	Transmission electron	–	–

2.1.1 *Materials*

1. Murine anaplastic astrocytoma cells (SMA560) (Sigma-Aldrich, SCC179).

2. Medium: MEM (Richter's modification) (Thermo Fisher Cat. No. 10373-017) supplemented with 10% FBS (Cat. No. ES--009--B).

3. CO_2 incubator, tissue culture plates.

4. Ultracentrifuge and ultracentrifugation tubes.

5. Phosphate buffered saline (PBS) solution.

2.1.2 *Method of Isolation*

1. Astrocytoma cells are cultured in the "exosome-free" FBS (fetal bovine serum) media for 48 h and undergone overnight centrifugation at $100,000 \times g$.

2. After 48 h, cultured media is collected and centrifuged at $800 \times g$ for 10 min at $4\,°C$.

3. Collected supernatant is centrifuged at $10,000 \times g$ for 10 min.

4. The process is repeated by using supernatant and centrifuging at $100,000 \times g$ for 3 h.

5. Collected pellet (exosome) is suspended in PBS and centrifuged at $100,000 \times g$ for 60 min.

6. Collected pellet is resuspended in small volume of PBS (approximately 0.2 to 0.5 mL).

2.2 Sequential Centrifugation

This method is one of the most common methods used to isolate exosomes. The supernatant should be centrifuged at least three centrifugal forces to separate exosomes. Sequential centrifugation may help isolate exosomes with excellent purity than differential centrifugation because of using more than three centrifugation steps at different centrifugal forces. It is also one of the cost-effective methods of isolation. Most of the cellular vesicles that are not exosome-like are removed during these steps. However, the technique exposes exosomes to multiple centrifugation steps and thus related stress during isolation steps, which may result in low yield. This method will also not wholly remove cellular impurities and may still have minor impurities present in the exosome isolates. The following steps describe the isolation of exosomes from a glioblastoma cell line expressing characteristics of human brain cancer [12].

2.2.1 *Materials*

1. U-373 MG (Uppsala) cell line (Sigma-Aldrich, 08061901).

2. U373 (human glioma cell line) in EMEM (EBSS) + 2 mM glutamine + 1% nonessential amino acids (NEAA) + 1 mM sodium pyruvate + 10% fetal bovine serum (FBS), and 100 units/mL of penicillin/streptomycin (Invitrogen CA).

3. Cell seeding at $2-4 \times 10,000$ cells/cm^2.

4. CO_2 incubator (5% CO_2 incubator at 37 °C).

5. Use 0.25% trypsin or trypsin/EDTA for detaching cells in media.

6. Phosphate buffered saline (PBS) (Invitrogen CA).

7. Ultracentrifuge and ultracentrifugation tubes.

2.2.2 Method of Isolation

1. Culture U 373 glioma cells in EMEM media containing 10% exosome-free FBS and 100 units/mL penicillin/streptomycin at 37 °C in a 5% CO_2 incubator to 70–80% confluence.

2. Collect culture media and carry out sequential centrifugations as described below at 4 °C.

3. Centrifuge media at 600 × g for 10 min and collect supernatant media by discarding pelleted cell debris.

4. Perform a second centrifugation at 3000 × g for 10 min and collect and use the supernatant in the next step.

5. Collected supernatant is centrifuged a third time at 10,000 × g for 30 min to remove other large vesicles and cell debris.

6. After the third centrifugation, only the supernatant is collected and centrifuged one last time at 150,000 × g for 3–4 h.

7. Collect the pellet and resuspend in PBS solution.

8. Centrifuge this mixture at 150,000 × g one more time for 1 h for collecting pure exosomes.

9. Suspend isolated exosomes in PBS.

2.3 Precipitation Method

For biological samples such as plasma samples, isolation technique precipitation is suitable. In this method of isolation, the sample is subjected to low centrifugation force. Once the exosomes are precipitated using this technique, it requires exosome suspending medium such as PBS buffer that is equal to the initial plasma sample. Additional cellular impurities may be present in precipitated exosome products. In this method, exosomes are isolated using the technique of precipitation. The process below describes the isolation of exosomes from a plasma sample [13].

2.3.1 Materials

1. Blood samples from glioblastoma patients.

2. ExoQuick Exosome Precipitation Solution (EXOQ5TM-1, System Biosciences).

3. ExoQuick solution 5 U/mL thrombin (System Biosciences).

4. Incubator and tissue culture plates.

5. Centrifuge and centrifuge tubes.

6. PBS solution.

2.3.2 Method of Isolation

1. Collect blood samples in EDTA tube from glioblastoma patients.
2. To separate plasma, centrifuge the blood sample at $2000 \times g$ for 15 min at 4 °C.
3. Plasma is transferred to clean tube and centrifuged at $14,000 \times g$ for 15 min at 4 °C.
4. The supernatant is incubated with 5 U/mL thrombin for 5 min.
5. Mixture is centrifuged at $10,000 \times g$ for 5 min to collect plasma.
6. Add 63 μL of ExoQuick solution for every 250 μL of plasma.
7. Incubate at 4 °C for 30 min.
8. Then, centrifuge at $1500 \times g$ for 30 min at 4 °C.
9. Collect pelleted exosomes and resuspended in PBS.

2.4 Density Gradient Method

Although the three of the methods described above are cost-effective and convenient methods to isolate exosomes, the major problem is the low quality of exosome. Additionally, these isolation methods need centrifugation forces above $100,000 \times g$ and multiple steps that put external stress on exosomes. To circumvent these issues, the density gradient method can be used to isolate exosomes with high quality and purity [14]. In this method, uniform-sized exosomes are separated based on their density that is primarily associated with their sizes. Even though this method of isolation is not a cost-effective and straightforward isolation method, it helps to remove impurities and achieve high-quality exosomes. Exosomes are isolated based on the differences in the density of the isolating medium. The process below describes the isolation of exosomes from glioblastoma patients' blood samples, the most common form of malignant brain tumor [14].

2.4.1 Materials

1. Whole blood sample from glioblastoma patient.
2. Conical tube 15 mL (Falcon No. 352097).
3. Sterile cryogenic vial (Corning Incorporated No. 430488).
4. OptiPrep Density Gradient Medium (Sigma-Aldrich No. D1556).
5. Ultra-Clear centrifuge tube (Beckman Coulter No. 344060).
6. EDTA (ethylenediaminetetraacetic acid), 2.5 M sucrose, PBS, water.
7. Tricine at pH 7.8.
8. Eppendorf centrifuge 5810 No. 0012529-rotor A-4-81).
9. Beckman Coulter Optima LE-80K ultracentrifuge--rotor SW 40 Ti No. 99U 10480.

2.4.2 Method of Isolation

1. Whole blood sample is collected from glioblastoma patients in EDTA tubes centrifuged briefly for 10 min at $1811 \times g$.

2. Supernatant plasma sample is transferred to a 15 mL conical tube and centrifuged for 15 min to remove cellular debris and any remaining erythrocytes by spinning the tube at $1811 \times g$.

3. Plasma sample is transferred to sterile cryogenic vials and stored at $-20\ ^{\circ}C$ before isolation.

4. To prepare OptiPrep diluent, mix 5 mL of 2.5 M sucrose, 0.1 g (6 mM) of EDTA, 10.08 g (120 mM) of tricine (pH = 7.8), and 45 mL of water.

5. Prepare 50% OptiPrep solution by adding 45 mL of OptiPrep density gradient medium with 9 mL of OptiPrep diluent.

6. Prepare buffer A by mixing 100 mL of 2.5 M sucrose, 0.34 g EDTA, 3.58 g (20 mM) of tricine (pH = 7.8), and 900 mL of water.

7. Prepare 10% OptiPrep solution by mixing 17.6 mL of buffer A with 4.4 mL of 50% OptiPrep solution.

8. For isolation of exosomes, mix 1 mL of centrifuged plasma with 1 mL of 50% OptiPrep solution in an Ultra-Clear centrifuge tube (this will become 25% OptiPrep solution).

9. Cautiously layer 11 mL of 10% OptiPrep solution on the above mixed solution inside the Ultra-Clear centrifuge tube.

10. Centrifuge the tube using Beckman Coulter Ultracentrifuge-rotor at $102,445 \times g$ for 90 min.

11. Collect the top 10 mL of supernatant and transfer it to a new tube and centrifuge it for an additional 16 h at $102,445 \times g$.

12. Discard the supernatant and collect pelleted exosomes from the bottom of the centrifuge tube.

13. Suspend exosomes in PBS.

2.5 Immunoisolation

To achieve higher quality or purity of the exosomes immunoisolation method should be used. Exosomes obtained directly by using this approach or isolated by one of the above techniques and purified by this method will help obtain exosomes of the highest quality and purity. However, the efficiency of separation depends on many factors, such as the size of beads and sample incubation time, and can be used only with cell-free samples. In this method of isolation, exosomes are captured on a surface by using antibodies for the exosome surface markers [15, 16].

2.5.1 Materials

1. Exosomes from human melanoma brain metastasis (H3) cell line that are CD9+.

2. Human CD9 exosome Flow Detection Kit (Dynal, Thermo Fisher Scientific).

3. Dynabeads (2.7 μm).

4. Antihuman CD9-RPE clone ML-13 (BD Biosciences, Norway).

5. Orbital shaker.

6. Test tube rotator mixer.

7. 0.1% bovine serum albumin (BSA).

8. PBS and elution buffer.

2.5.2 Method of Isolation

1. Mix 100 μL of exosome sample with 20 μL of anti-CD9 coated Dynabeads in test tube.

2. Incubate the mixture overnight at 4 °C using rotating mixer.

3. Wash bead-bound exosomes three times with 0.1% BSA in 0.1 μL PBS.

4. Set an orbital shaker at 1000 rpm.

5. In the orbital shaker, incubate bead-bound exosomes with antihuman CD9-RPE clone ML-13 for 45 min at room temperature away from light.

6. Wash bead-bound exosomes two times in 0.1% BSA in PBS.

7. Add elution buffer to separate exosomes from beads.

8. Centrifuge the elution at $100,000 \times g$ and collect pelleted purified exosomes.

9. Suspend exosomes in PBS.

2.6 Isolation of RNA from Exosomes

Exosomes separated from neural cell sources carry RNA. If we are interested in separating only the RNA from exosome reagents and steps described in this method should be used. In this method, cellular debris and exosome constituents are removed, and pure RNA within exosomes or RNA associated with exosomes will be isolated that, in some circumstances, will be useful to understand the neural disease progression or prevention. RNA content can be isolated from exosomes directly from cultured cell media or using exosomes obtained from other methods. Exosome RNA content can be separated from exosomes directly from cultured cell media using below commercially available kit [17].

2.6.1 Materials

1. Exosomes isolated by using any one of the above methods.

2. TRIzol LS reagent (Thermo Fisher Scientific, Cat No: 10296-028).

3. Chloroform, RLT buffer, absolute ethanol, sodium acetate (Ambion).

4. RNeasy Mini Kit (Qiagen, Cat No: 74106).

5. Centrifuge.

<table>
<tr><td>

2.6.2 Method of Isolation

</td><td>

1. Add TRIzol LS reagent (500 µL) and chloroform (200 µL) to exosome sample.

2. Shake sample thoroughly for 30 s.

3. Incubate at room temperature for 10 min.

4. The top aqueous layer is collected and mixed with RLT buffer (3.5 × volume), absolute ethanol (2.5 × volume), and sodium acetate [3 M, pH 5.5] (0.1 × volume).

5. The above mixture is poured into spin column supplied by manufacturer with RNeasy Kit.

6. Then wash the sample three times with wash buffer.

7. Use eluting buffer to elute total RNA (including micro RNA, mRNA).

</td></tr>
</table>

3 Characterization of Exosomes

Exosome characterization is necessary to confirm the identity and morphology before or after drug loading. The most important techniques used to characterize the exosomes involve size, morphology, and the exosome-specific surface biomarkers. The characterization techniques listed (Table 1) must be carried out to confirm that the exosome pellets are indeed distinct exosome vesicles and not cellular debris or exosome fragments.

3.1 Nanoparticle Tracking Analysis (NTA)

This characterization method is used to determine the size and concentration (number) of exosomes by measuring the Brownian motion of exosome particles. The particles in the suspension are viewed, and scattered light of the particles is used to calculate the diameter. This technique is currently used as the gold standard method in exosome-related work to measure their size [13, 18, 19].

3.1.1 Materials

1. Exosomes from bone marrow-derived DCs (BMDCs).

2. Standard size silica nanoparticles (100 nm).

3. NanoSight (LM10, LM20, NS200, NS500, Malvern Instruments Ltd., UK).

4. PBS.

3.1.2 Method

1. Calibrate the instrument with standard 100 nm silica microspheres.

2. Measure 1 µL of exosomes.

3. Dilute exosomes to 1:100 with PBS to get a uniform distribution of exosomes in the solution to achieve a particle count range in between 2×10^8 and 1×10^9 per mL.

4. Load the exosome sample into NanoSight sample chamber.

5. Adjust the instrument settings until exosomes appear as sharp dots of light.

6. Visually check and repeat the measurements until the five size measurement profiles are in agreement.

7. Record at least 900 particles for each measurement [20].

8. Analyze the results.

3.2 Dynamic Light Scattering Method

This is another method to determine the mean size of exosomes that would help determine the average particle size of exosomes. Determination of exosomes sizes in conjugation with confirmation with surface markers will help characterize exosomes [21].

3.2.1 Materials

1. Exosomes from the medulloblastoma cell line (D283MED).

2. Cuvettes.

3. Nicomp 370 submicron particle sizer (Agilent Technologies, Inc., USA).

3.2.2 Methods

1. Isolate exosomes from the medulloblastoma cell line.

2. Exosomes are aliquoted into 400 μL samples.

3. Transfer to a cuvette for size analysis of using dynamic light scattering analysis.

4. Nicomp Zpw software is used to apply fluctuation rate of scattered light and the channel width for each measured sample is adjusted.

5. To calculate the final size distribution for each sample, the number-weight Gaussian setting is selected.

3.3 Western Blotting Method

This method helps to qualitatively and quantitatively determine the integral protein molecules of exosomes and confirm the specificity of exosomes. Characterization methods that only determine size is not a definitive technique to tell that they are exosomes. However, incubating exosome protein with specific antibodies helps to determine the surface markers that are specific to exosomes [11, 14, 18].

3.3.1 Materials

1. XCell SureLock Mini-Cell, XCell II Blot Module (Invitrogen).

2. Buffers for western blot; sample buffer, running buffer, transfer buffer, blocking buffer, protein marker, dye.

3. BCA protein assay (Thermo Scientific, USA).

4. RIPA buffer and BCA protein assay kit (Thermo Scientific, USA).

5. Polyacrylamide gels (12.5%, Bio-Rad No. 3450015).

6. PVDF membranes (Bio-Rad No. 162--0175).

7. CD81, CD63; (System Biosciences, USA).

8. Secondary antibodies conjugated to horseradish peroxidase.

3.3.2 Method	1. Assemble and fill the gel running unit (XCell SureLock Mini-Cell) with running buffer.

2. Suspend exosomes in PBS and lyse by adding an equal volume of RIPA buffer for protein quantification by using BCA protein assay kit.

3. Exosomes are normalized by using protein content determined.

4. Load protein content (20 µg per sample) into polyacrylamide gels.

5. Turn on the unit to 120 V and allow the gel to run for 60–90 min until the dye reaches front of the gel bottom.

6. Transfer onto PVDF membrane and incubate for (1 h) with blocking buffer followed by overnight incubation with primary antibodies for CD9, CD63, and CD81.

7. After subsequent 1-h incubation with secondary antibodies conjugated to horseradish peroxidase visualize the gels by chemiluminescence.

3.4 Flow Cytometry

Apart from using tedious western blot analysis, a fluorescence-based flow cytometric method described below will also help to confirm and quantify the isolated exosomes. Because exosomes are too small for analysis using flow cytometry method, they are first attached to beads and analyzed further by using FACS (fluorescence-activated cell sorting) [18, 22].

3.4.1 Materials

1. Aldehyde/sulfate latex beads (Invitrogen).

2. PBS and 100 mM of glycine (Sigma-Aldrich).

3. Laboratory centrifuge.

4. Fluorescein isothiocyanate (FITC)-labeled antibodies for exosome surface markers (e.g., CD9 (BioLegend), FITC-MHC II (BioLegend).

5. FACS buffer/tubes.

6. FACSCalibur and CellQuest software (BD Biosciences, CA).

3.4.2 Method

1. Follow manufacturer's instructions to prepare 4 µM aldehyde/sulfate latex beads.

2. Add 30 µg of exosomes to 100 µL of beads and incubate at room temperature for 10 min and for another 2 h after adding 900 µL of PBS.

3. To terminate the reaction, add 100 mM of glycine and centrifuge at $3000 \times g$ for 10 min to collect exosomes as pellets.

4. Resuspend the beads in 5 mL PBS and centrifuge at $3000 \times g$ for 10 min and repeat this step three times to wash beads.

5. Incubate exosome-coated beads in the dark with FITC-labeled antibodies at 10 μg/mL at 4 °C for 60 min by slow shaking.

6. Resuspend the beads in 100 μL FACS buffer and analyze by flow cytometry for determination of exosome surface markers.

4   Transmission Electron

This is a microscopic technique where exosomes are characterized and observed visually using a microscope. Two different sample preparation and characterization procedures are described below [17, 20, 22]. This technique in conjugation with the information on expression of surface markers on these vesicles can confirm the exosome identity and nature.

4.1   Copper Grid Method

4.1.1   Materials

1. Exosomes from human plasma.
2. Copper grid.
3. 0.125% Formvar in chloroform.
4. 1% v/v uranyl acetate in ddH_2O.
5. JEOL 101 I transmission electron microscope.

4.1.2   Method

1. Coat a copper grid with 0.125% Formvar in chloroform.
2. Place freshly isolated exosomes onto the copper grid and set aside for 10 min.
3. Stain the copper grid using 1% v/v uranyl acetate in ddH_2O.
4. Immediately examine the exosomes under JEOL 101 I transmission electron microscope.

4.2   Glutaraldehyde Method

4.2.1   Materials

1. Exosomes from human plasma.
2. Cold 2.5% v/v glutaraldehyde in 0.1 M PBS.
3. Centrifuge.
4. Ethanol and Epon.
5. Uranyl acetate.
6. Reynold's lead citrate.

4.2.2   Method

1. In this method exosomes are fixed.
2. Centrifuge isolated exosomes in PBS at 100,000 × g for 45 min.
3. Discard the supernatant and add cold 2.5% v/v glutaraldehyde in 0.1 M PBS to fix the pelleted exosomes.
4. After an hour, rinse fixed exosomes using PBS.
5. The pellet is dehydrated through a graded series of ethanol and embedded in Epon.

6. Use uranyl acetate and Reynold's lead citrate to stain them as ultrathin sections (65 nm).

7. Use a JEOL 101 I transmission electron microscope to examine the exosomes.

---

5   Drug Loading into Exosomes

Many types of drug molecules, such as small molecules, proteins, and nuclear-based drugs, are loaded (Table 1 and Fig. 3) into the exosomes purified and characterized by one of the above methods. These different methods of drug loading are described by using exosomes derived from brain cells. The three primary methods of drug loading are simple incubation, temporarily disrupting, and allowing entry of molecules into exosomes by sonication, and electroporation that is widely reported in the literature is described below (Fig. 4).

5.1   Incubation

Incubation is a straightforward drug loading method where a suitable drug of choice concentration is mixed with exosomes and incubated for a few hours (3–24 h). This method can be easily applied and optimized for many different types of small molecule drugs. Here we described the incubation method of drug loading by using a low molecular weight dye [23].

5.1.1   Materials

1. Rhodamine 123 (Sigma-Aldrich, MO, USA).

2. 1.5 mL Eppendorf tubes.

3. Water bath (for incubation at 37 °C).

4. Exosomes isolated from human neuronal glioblastoma-astrocytoma U-87 MG cells.

5. Ultracentrifuge and tubes.

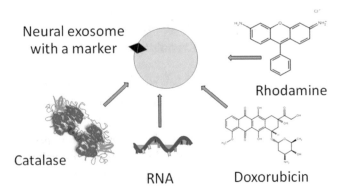

Fig. 3 Exosome and various types of drug molecules described in drug-loading methods

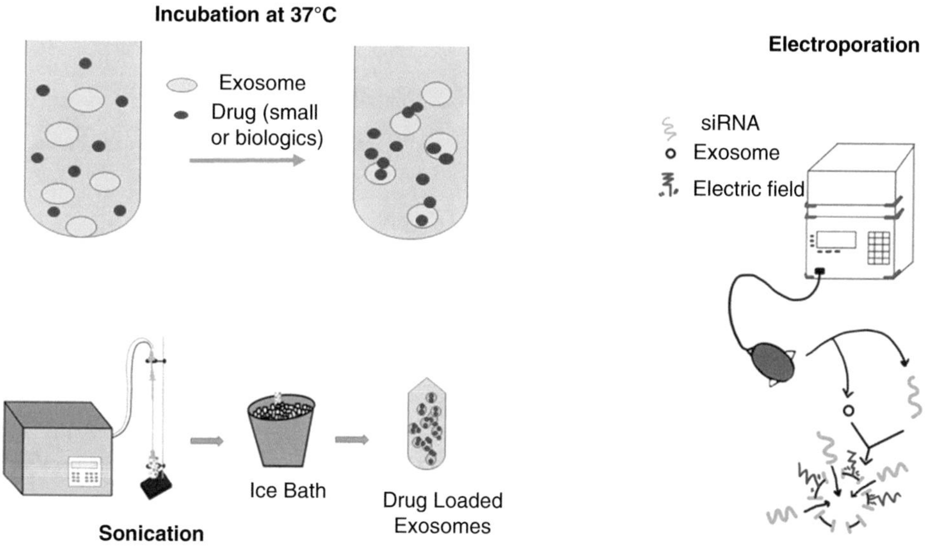

Fig. 4 Schematic representation of exosome drug-loading method; (**a**) incubation, (**b**) sonication, and (**c**) electroporation

5.1.2 Method of Loading

1. A solution of rhodamine 123 (dye with a molecular weight of 380 daltons) at a concentration of 2 mg/mL is added to exosomes at a concentration of 200 μg/mL (exosome protein content) in a 1.5 mL Eppendorf tube.

2. The mixture is incubated at 37 °C for 2 h.

3. Free dye is separated by centrifuging the sample for 1 h at $100,000 \times g$.

4. Rhodamine-loaded exosomes are collected at the bottom of the tube.

5. A density gradient method of separation such as OptiPrep density gradient separation described in exosome isolation can also be used to separate free drug from drug-loaded exosomes.

5.2 Sonication

Sonication is a different method to load drug molecules into exosomes. Sonication helps to agitate exosome particles and incorporate drug molecules by applying ultrasonic energy. This method is generally used to load large-sized drug molecules and can be applied to small drug molecules as well. This method is not very suitable if the drug of choice is sensitive and prone to degradation in sonication conditions. The following process describes the loading of a large molecule, such as an enzyme. Catalase (240 kDa polypeptide, four peptide chains, with each chain containing more than 500 amino acids) is an enzyme that is loaded to exosomes by using below described sonication method [4].

<table>
<tr><td>

5.2.1 Materials

</td><td>

1. Qsonica Sonicator Q700 (Fisher Scientific, USA).

2. Ice.

3. Bovine Catalase (Calbiochem, San Diego).

4. Exosomes from Raw 264.7 macrophages (ATCC, Cat # TIB-71).

5. Sepharose 6 BCL (Sigma-Aldrich).

</td></tr>
<tr><td>

5.2.2 Method of Loading

</td><td>

1. Place 250 µL (0.5 mg/mL of total protein) of exosomes and 0.5 mg/mL of catalase into a tube.

2. The mixture of catalase with exosomes is sonicated by running probe at 2 kHz frequency, 500 V, 20% power, six cycles by 4 s pulse/2 s pause.

3. Subsequently, place the mixture on ice to cool down for 2 min.

4. The mixture of catalase and exosome is sonicated a second time by using the same parameters.

5. Loaded exosomes and catalase are separated by using gel filtration chromatography with Sepharose 6 BCL (Sigma-Aldrich).

</td></tr>
</table>

6   Electroporation

This method allows drug loading rapidly that lets the permeabilization of drug molecules into exosomes by applying an electrical field. This method cannot be used if the drug molecule is sensitive to electric current. This method is routinely used to load biological molecules such as siRNA (short interfering RNA). Here we described the loading of two different categories of drug molecules: a small molecule and a nucleic acid-based drug into exosomes. The molecular weight of the small molecule drug doxorubicin is 580 Daltons. The average MW of a siRNA (21 bp duplex), a nucleic acid-based drug, is approximately 13,300 Daltons.

<table>
<tr><td>

6.1   Loading of Doxorubicin into Exosomes Using Electroporation [24]

6.1.1 Materials

</td><td>

1. Doxorubicin (Sigma-Aldrich).

2. Exosomes from mesenchymal cells.

3. Electroporation buffer and 0.4 cm cuvettes (Bio-Rad, USA).

4. Multiporator (Eppendorf, Hauppauge, USA).

5. 100-kDa Amicon filter.

6. PBS.

</td></tr>
<tr><td>

6.1.2 Method of Loading

</td><td>

1. 100 µg of doxorubicin (DOX) is mixed with 100 µg of exosomes and 200 µL of electroporation buffer at 4 °C in the cuvette.

2. Using a Multiporator, perform electroporation at 600 V in 0.4 cm cuvettes.

</td></tr>
</table>

3. Incubate the cuvettes at 37 °C for 30 min.

4. Free DOX is separated by running the electroporated mixture through a 100-kDa Amicon filter.

5. Collect the drug-loaded exosomes from the filter and suspend in PBS.

6.2 Loading of siRNA into Exosomes Using Electroporation [18]

6.2.1 Materials

1. siRNA.

2. Exosomes from bone marrow dendritic cells.

3. Electroporation buffer: mixture of 1.15 mM potassium phosphate (pH 7.2), 25 mM potassium chloride, and 21% (v/v) OptiPrep.

4. Gene Pulse XCell electroporation system (Bio-Rad).

5. Electroporation cuvettes, gap width 0.4 cm (Sigma-Aldrich).

6.2.2 Method of Loading

1. Mix the exosomes and siRNA at weight ratio of 1:1 (w/w) in electroporation buffer without exceeding the exosome concentration above 0.5 μg/μL.

2. Place the above mixture in 0.4 mm electroporation cuvettes.

3. Electroporate at 400 mV and 125 μF capacitance by pulsing for 10–15 ms. For higher volumes of samples, only electroporate 400 μL of mixtures each time.

4. Suspend exosomes in 20 mL of PBS with 1% w/v of BSA.

5. Separate siRNA-loaded exosomes by centrifuging the sample at 120,000 × g for 70 min at 40 °C.

6. Discard the supernatant and suspend the exosomes in PBS.

7 Methods of Labeling Exosomes

For tracking exosome both in in vitro and in vivo studies, a fluorescent method and a lipophilic dye that integrates into the membrane of exosome are described in this chapter. These dyes are commercially available and widely reported in the literature for studying the intracellular uptake and localization and the in vivo fate of exosomes.

7.1 Fluorescent Labeling

This method is used to label exosomes with fluorescent labels and tracks their movement/location both in vitro and in vivo. Depending on the drug molecule fluoresce, exosomes can be labeled either with a dye that can be detected in red or green fluorescent channels [25].

7.1.1 Materials (Exo-Red or Exo-Green Kit)

1. Exo-Red and Exo-Green (System Biosciences, SBI).

2. ExoQuick-TC.

3. System Biosciences Exo-Glow manual.

4. Eppendorf tube.

5. Centrifuge.

7.1.2 Method
of Fluorescent Labeling

1. Add 50 μL of 10x Exo-Red or Exo-Green to 500 μL of exosomes suspended in PBS in a 1.5 mL Eppendorf tube.

2. Mix the content by inverting or flicking the tube (Do not vortex).

3. Incubate the exosomes and labeling mixture at 37 °C for 10 min.

4. Add 100 μL of the ExoQuick-TC to the labeled exosome sample.

5. Invert six times to mix and place labeled exosome on ice (or at 4 °C) for 30 min.

6. Using a microcentrifuge spin the sample at 14,000 rpm for 3 min.

7. Discard the supernatant and collet the pelleted labeled exosomes from the bottom.

8. Suspend the labeled exosome pellet in 500 μL 1× PBS.

9. These fluorescent labeled exosomes can be sued for in vitro and in vivo tracking.

7.2 PKH Lipophilic Membrane Dyes

This method uses lipophilic dyes that are able to insert their aliphatic portion into the lipid of exosomes. The dye-embedded exosomes are isolated using a sucrose solution to separate from the remaining free-floating dye nanoparticles. This dye can also be used to labeled exosomes for in vitro and in vivo tracking [26].

7.2.1 Materials

1. PKH26 Red Fluorescent Cell Linker Kits for General Cell Membrane Labeling (Sigma-Aldrich).

2. Diluent C.

3. Particle-free Dulbecco's phosphate-buffered saline (DPBS; Sigma-Aldrich).

4. Centrifuge (TLA-55; Beckman Coulter) (MLS-50; Beckman Coulter) (MLA-50; Beckman Coulter).

7.2.2 Method

1. Dilute PKH26 dye with 100 μL diluent C to get a concentration of 8 μM.

2. Incubate PKH26 in diluent C in an ultrasonic water bath at 37° C for 15 min.

3. Add 10 μg of exosomes, 20 μg DPBS, and 80 μL diluent C together.

4. Add this mixture to the dye solution.

5. Incubate this dye solution for 5 min while gently pipetting to mix.

6. Dilute exosomes to 1 mL with DPBS.

7. Centrifuge the mixture at 100,000 × g for 70 min at 4° C 9TLA-55).

8. Transfer the pellet to a new centrifuge tube and resuspend in 50 μL of DPBS. Dilute the exosome in 400 μL of DPBS.

9. Place exosomes onto a 20–60% discontinuous sucrose gradient.

10. Centrifuge at 100,000 × g for 18 h at 4 °C (MLS-50).

11. Combine fraction 3–6 (1.08–1.15 g/mL) in a centrifuge tube and dilute with DPBS to get 30 mL.

12. Combine fraction 7–10 (1.17–1.23 g/mL) in another centrifuge tube and dilute with DPBS to get 30 mL.

13. Centrifuge each tube at 100,000 × g for 70 min at 4° C (MLA-50).

14. Resuspend pellets in 50 μL DPBS.

8 Conclusion

The targeted delivery of drugs to specific tissues and cells has long been a goal of medical science. Although no definitive methods have been found to achieve this goal to date, the use of exosomes for targeted delivery of drugs may be the solution. The use of exosomes for drug delivery purposes shows promising results in delivering drugs across the blood-brain barrier. The precise understanding of the exosome delivery mechanisms is not yet fully understood. Exosomes may be tagged with specific molecules that target an exact tissue or neural cell type, allowing the delivery of a drug to a particular area in the brain. The first step to achieve this goal is to isolate exosomes with the homing capability to cells in the brain and encapsulate a pharmacological drug inside them. This chapter has described the most common methods used for exosome isolation, drug encapsulation, and exosome labeling. With the unique advantage in crossing the BBB, these cell-derived biological nanosystems may prove to be beneficial in the treatment of neurological diseases and becoming the next breakthrough therapy in medicine and for the future of neuronal disease treatments. Due to increasing evidence on exosome's inherent ability to cross the blood-brain barrier, in brain cancers such as glioblastomas, additional medical research and further advances in the field of exosome drug delivery may be able to use full capabilities and unique qualities of exosomes.

References

1. Zhang J, Li S, Li L, Li M, Guo C, Yao J, Mi S (2015) Exosome and exosomal microRNA: trafficking, sorting, and function. Genomics Proteomics Bioinformatics 13(1):17–24

2. Record M, Subra C, Silvente-Poirot S, Poirot M (2011) Exosomes as intercellular signalosomes and pharmacological effectors. Biochem Pharmacol 81(10):1171–1182

3. Ha D, Yang N, Nadithe V (2016) Exosomes as therapeutic drug carriers and delivery vehicles across biological membranes: current perspectives and future challenges. Acta Pharm Sin B 6 (4):287–296

4. Haney MJ, Klyachko NL, Zhao Y, Gupta R, Plotnikova EG, He Z, Patel T, Piroyan A, Sokolsky M, Kabanov AV, Batrakova EV (2015) Exosomes as drug delivery vehicles for Parkinson's disease therapy. J Control Release 207:18–30

5. Jia G, Han Y, An Y, Ding Y, He C, Wang X, Tang Q (2018) NRP-1 targeted and cargo-loaded exosomes facilitate simultaneous imaging and therapy of glioma in vitro and in vivo. Biomaterials 178:302–316

6. Batrakova EV, Kim MS (2015) Using exosomes, naturally-equipped nanocarriers, for drug delivery. J Control Release 219:396–405

7. Wong KH, Riaz MK, Xie Y, Zhang X, Liu Q, Chen H, Bian Z, Chen X, Lu A, Yang Z (2019) Review of current strategies for delivering Alzheimer's disease drugs across the blood-brain barrier. Int J Mol Sci 20(2):381

8. Jarmalaviciute A, Pivoriunas A (2016) Exosomes as a potential novel therapeutic tools against neurodegenerative diseases. Pharmacol Res 113(Pt B):816–822

9. Yuan D, Zhao Y, Banks WA, Bullock KM, Haney M, Batrakova E, Kabanov AV (2017) Macrophage exosomes as natural nanocarriers for protein delivery to inflamed brain. Biomaterials 142:1–12

10. Dong X (2018) Current strategies for brain drug delivery. Theranostics 8(6):1481–1493

11. Graner MW, Cumming RI, Bigner DD (2007) The heat shock response and chaperones/heat shock proteins in brain tumors: surface expression, release, and possible immune consequences. J Neurosci 27(42):11214–11227

12. Kore RA, Abraham EC (2014) Inflammatory cytokines, interleukin-1 beta and tumor necrosis factor-alpha, upregulated in glioblastoma multiforme, raise the levels of CRYAB in exosomes secreted by U373 glioma cells. Biochem Biophys Res Commun 453(3):326–331

13. Domenis R, Cesselli D, Toffoletto B, Bourkoula E, Caponnetto F, Manini I, Beltrami AP, Ius T, Skrap M, Di Loreto C, Gri G (2017) Systemic T cells immunosuppression of glioma stem cell-derived exosomes is mediated by Monocytic myeloid-derived suppressor cells. PLoS One 12(1):e0169932

14. Cumba Garcia LM, Peterson TE, Cepeda MA, Johnson AJ, Parney IF (2019) Isolation and analysis of plasma-derived exosomes in patients with glioma. Front Oncol 9:651

15. Clayton A, Court J, Navabi H, Adams M, Mason MD, Hobot JA, Newman GR, Jasani B (2001) Analysis of antigen presenting cell derived exosomes, based on immuno-magnetic isolation and flow cytometry. J Immunol Methods 247(1-2):163–174

16. Guerreiro EM, Vestad B, Steffensen LA, Aass HCD, Saeed M, Ovstebo R, Costea DE, Galtung HK, Soland TM (2018) Efficient extracellular vesicle isolation by combining cell media modifications, ultrafiltration, and size-exclusion chromatography. PLoS One 13(9): e0204276

17. Prendergast EN, de Souza Fonseca MA, Dezem FS, Lester J, Karlan BY, Noushmehr H, Lin X, Lawrenson K (2018) Optimizing exosomal RNA isolation for RNA-Seq analyses of archival sera specimens. PLoS One 13(5):e0196913

18. El-Andaloussi S, Lee Y, Lakhal-Littleton S, Li J, Seow Y, Gardiner C, Alvarez-Erviti L, Sargent IL, Wood MJ (2012) Exosome-mediated delivery of siRNA in vitro and in vivo. Nat Protoc 7(12):2112–2126

19. Zhang M, Jin K, Gao L, Zhang Z, Li F, Zhou F, Zhang L (2018) Methods and Technologies for Exosome Isolation and Characterization. Small Methods 2(9):1–10

20. Muller L, Muller-Haegele S, Mitsuhashi M, Gooding W, Okada H, Whiteside TL (2015) Exosomes isolated from plasma of glioma patients enrolled in a vaccination trial reflect antitumor immune activity and might predict survival. Onco Targets Ther 4(6):e1008347

21. Epple LM, Griffiths SG, Dechkovskaia AM, Dusto NL, White J, Ouellette RJ, Anchordoquy TJ, Bemis LT, Graner MW (2012) Medulloblastoma exosome proteomics yield functional roles for extracellular vesicles. PLoS One 7(7):e42064

22. Muller L, Hong CS, Stolz DB, Watkins SC, Whiteside TL (2014) Isolation of biologically-

active exosomes from human plasma. J Immunol Methods 411:55–65

23. Yang T, Martin P, Fogarty B, Brown A, Schurman K, Phipps R, Yin VP, Lockman P, Bai S (2015) Exosome delivered anticancer drugs across the blood-brain barrier for brain cancer therapy in *Danio rerio*. Pharm Res 32 (6):2003–2014

24. Gomari H, Forouzandeh Moghadam M, Soleimani M, Ghavami M, Khodashenas S (2019) Targeted delivery of doxorubicin to HER2 positive tumor models. Int J Nanomedicine 14:5679–5690

25. Mead B, Tomarev S (2017) Bone marrow-derived mesenchymal stem cells-derived exosomes promote survival of retinal ganglion cells through miRNA-dependent mechanisms. Stem Cells Transl Med 6(4):1273–1285

26. Puzar Dominkus P, Stenovec M, Sitar S, Lasic E, Zorec R, Plemenitas A, Zagar E, Kreft M, Lenassi M (2018) PKH26 labeling of extracellular vesicles: characterization and cellular internalization of contaminating PKH26 nanoparticles. Biochim Biophys Acta Biomembr 1860(6):1350–1361

Chapter 12

Establishing Orthotopic Xenograft Glioblastoma Models for Use in Preclinical Development

João Basso, José Sereno, Ana Fortuna, Miguel Castelo-Branco, and Carla Vitorino

Abstract

Glioblastoma multiforme is the most common, aggressive and lethal type of brain tumor, characterized by an aggressive, heterogenic and highly angiogenic behavior. Establishing mice models that mimic the etiology, biology and histopathology of human glioblastomas is of extreme importance, as they are a crucial tool to understand the tumor initiation, formation, angiogenesis, progression and metastasis. Orthotopic xenograft mice models remain in the frontline of neuro-oncology as an experimental system to identify novel therapeutic targets and to determine the efficacy of different therapeutic agents and/or nanosystems. The present chapter describes a protocol for establishing brain tumor xenografts in mice following a single injection of glioblastoma cells.

Key words Glioblastoma, Orthotopic, Xenograft, Stereotaxic surgery, Mice, Preclinical development

1 Introduction

Glioblastoma multiforme encompasses a group of grade IV lesions caused predominantly by astrocytic cell differentiation and is considered the most common, aggressive and lethal type of brain tumor [1]. Moreover, its poor prognosis is marked by an infiltrative growth pattern into neighboring structures, thereby preventing a successful complete resection, as well as by drug resistance mechanisms [2]. Glioblastoma proliferation and invasion is usually limited to the Central Nervous System (CNS), more specifically within the white matter and deeper gray matter of the brain. The distribution of the tumor may be related to its origin. In fact, primary tumors present a variable distribution, affecting the frontal, occipital, parietal and temporal lobes, whereas secondary tumors, which usually derive from anaplastic astrocytomas or oligodendrogliomas, often develop within the frontal lobe of patients [3, 4]. Rapid tumor growth and proliferation of cancer stem cells, as well as drug resistance mechanisms, are on the base of the poor outcomes. In

Vivek Agrahari et al. (eds.), *Nanotherapy for Brain Tumor Drug Delivery*, Neuromethods, vol. 163,
https://doi.org/10.1007/978-1-0716-1052-7_12, © Springer Science+Business Media, LLC, part of Springer Nature 2021

fact, the overall survival of patients diagnosed with glioblastoma and submitted to a triple therapeutic approach is reduced (15 and 31 months for primary and secondary glioblastoma, respectively) [5, 6]. Glioblastoma treatment relies on surgical resection, followed by cycles of radiotherapy and chemotherapy with temozolomide, given orally [7].

Over the last decades, several animal cancer models have been developed as a strategy to understand glioblastoma initiation, formation, angiogenesis, progression and, although rare, metastasis [8]. More practical and translational approaches focus on the use of such models as living platforms to evaluate the efficacy of potential treatments, including both chemotherapeutic and other drugs or compounds. Due to the lack of therapeutic efficacy of the standard approach, which includes the use of temozolomide, at least 37 repurposed drugs have been tested using animal models of glioblastoma [9]. Note that drug repositioning is a strategy that can potentially reduce R&D programs up to 3–5 years, increasing the probability of success in relation to the traditional approach [10].

Furthermore, countless strategies for the delivery of other anticancer or repurposed drugs and novel compounds in technologically developed drug delivery systems, including microspheres, lipid, polymeric and metallic nanoparticles, liposomes, extracellular vesicles, micelles, dendrimers and hydrogels, among others, have been developed and had their preclinical efficacy tested using animal models [11–13]. These in vivo approaches include genetically engineered models, viral vector-mediated transduction models, chemical carcinogen-induced models and allograft or xenograft models [8]. The latter are usually developed using permanent human glioblastoma cell lines and immunocompromised mice, although more recent approaches aim at the administration of tissue spheroids or patient-derived tumors into the flanks or brains of mice (Fig. 1). Importantly, the use of heterotopic models, i.e., with subcutaneous flank induction of tumors, does not take into consideration the presence of the blood-brain barrier that limits the access of therapeutics to the CNS. Orthotopic patient-derived xenografts (PDXs) are becoming the gold standard strategy of glioblastoma mice models, as they have significant advantages over cell line-derived tumors (Table 1) [14, 15]. Nonetheless, they are yet to become widely available to the scientific community and have a time- and manipulation-limited use, when compared to tumor cell lines [16, 17].

The development of orthotopic models of glioblastoma is generally accomplished through stereotaxic surgeries. In fact, the animal brain is placed within a three-dimensional coordinate system, in which nonvisualized anatomic structures are identified and targeted by establishing 3D Cartesian coordinates. Interestingly, this technique was first described in 1908, by Horsley and Clarke [18], as an

A Tumor cell lines   **B** Patient derived xenografts

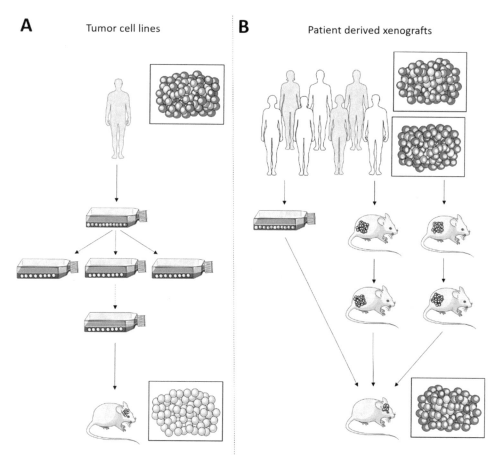

Fig. 1 Development of tumor cell line and patient-derived xenograft mice models of glioblastoma. (**a**) Tumor cell lines often lead to cell isolation and a high number of cell passages and, therefore, to the development of homogeneous tumors. (**b**) Patient-derived xenografts lead to the development of highly heterogeneous tumors, although they can only be used with a limited number of in vitro or in vivo passages, due to cell isolation, medium interference, and mouse stromal cell infiltration. (Adapted from Ref [14] with permission from Impact Journals, LLC)

approach to induce lesions in the dentate nucleus of the cerebellum of a monkey, and has been adapted to rodents and smaller animals. In this minimally invasive procedure, the rodent is carefully immobilized and a small orifice is drilled in the skull of the animal, allowing the further injection and deposition, at a specific location, of a known density of cells, that posteriorly grow and develop into a brain tumor.

Cancerous structures may be identified and characterized by different in vivo noninvasive techniques, including ultrasound (US), photoacoustic tomography (PAT), magnetic resonance imaging (MRI), bioluminescence imaging (BLI), computed tomography (CT), positron emission tomography (PET) and single photon emission computed tomography (SPECT) (Table 2).

Table 1
Comparison between cancer cell lines and patient-derived glioblastoma orthotopic xenograft models [14, 15]

	Advantages	Limitations
Cell line-derived tumors	Continuous self-renewal Easy to culture (low media requirements) Time and cost efficient Good for high-throughput screenings Accumulated experimental data Immunity response, when using mice cancer cell lines Longer-term culture is possible	Lack of tumor heterogeneity No tumor microenvironment No immunity response, when using human cancer cell lines Propagation of in vitro mutations and epigenetic alterations In vitro artificial selection
Patient-derived xenografts	Retains mutations and tumor heterogeneity Takes into consideration a (mouse/human) tumor microenvironment Patient tumor sample biobanks may be possible Metastatic modeling Strong genetic and epigenetic relation with original tumor	Technically challenging, expensive and time-consuming No immunity response (requires immunodeficient mice) Difficult acquisition Tumor microenvironment may be different than those found in human patients Lower culture and implantation success rate Cell preparation with a low number of passages Conventional transfection is problematic (e.g., luciferase or other fluorescent proteins) Lack of definition and standardization (tumor location, growth factor and hormones, host immune system, etc.)

Among the aforementioned imaging techniques, MRI and BLI are the most explored approaches in orthotopic xenograft models of glioblastoma. The advantages of MRI, that also include tumor localization, volumetric measurement and physiologic imaging, prompt its use worldwide. In addition, it is a feasible approach for PDX models. As for BLI, it remains a commonly and good approach to identify tumors early, monitor tumor growth and measure responses to therapeutic interventions. Nonetheless, BLI does not always correlate well with MRI results, particularly in cases with very low BLI signals or very large brain tumor volumes [23].

This chapter is the result of an extensive literature research and experimental approach on stereotaxic surgery and glioblastoma, thus combining the practical advantages of different protocols already established and published. It provides thorough step-by-step and extensive insights on cell manipulation, stereotaxic surgery and in vivo MRI that allow an easy and successful development of an in vivo preclinical model of glioblastoma for unexperienced

Table 2
Noninvasive imaging techniques for the characterization of tumor animal models [19–22]

Imaging technique	Spatial resolution	Advantages	Disadvantages
Ultrasound (US)	50–500 μm	Low cost Ease of procedure No radiation exposure	Low resolution Loss of signal in bone-protected structures
Photoacoustic tomography (PAT)	50–400 μm	High resolution No ionizing radiation Imaging of biochemical processes	Good tissue penetration Loss of signal in bone-protected structures
Magnetic resonance imaging (MRI)	25–100 μm	High resolution (2D and 3D) No ionizing radiation Does not require contrast agents Imaging of physiological and anatomical details (inflammation, connectivity, swelling, white/gray matter, tissue perfusion/permeability, necrosis) Metabolism—MR spectroscopy of different nuclei, such as ^1H, ^{13}C, and ^{31}P	High cost Moderate sensitivity Cannot be conducted in animals with metallic components (e.g., prosthesis)
Bioluminescence imaging (BLI)	2–5 mm	High sensitivity Low cost Fast real-time scanning Continuous repeated detection Functional information No radiation exposure	Low resolution Limited tissue penetration Requires luciferase-expressing cells
Computed tomography (CT)	50–200 μm	High spatial resolution Low radiation exposure Strong penetration depth Distinction between different tissues Low cost	Requires high radiation dose Requires contrast agent Low sensitivity of contrast agents Tissue nonspecificity
Positron emission tomography (PET)	0.5–2 mm	High spatial resolution Imaging of biochemical processes	Requires radiation Low resolution Lack of an anatomical reference frame
Single photon emission computed tomography (SPECT)	0.5–2 mm	High spatial resolution Imaging of biochemical processes	Requires radiation Low resolution Less sensitive than PET Lack of an anatomical reference frame

researchers. Despite being designed for the orthotopic xenograft implantation of U-87 MG cells, as they are the most common and characterized glioblastoma cell line, this protocol is easily transposed to other cancer cell lines, including those of human (such as Hs 683, LN-308, LN-319, LNT-229, SF-295, T98G, and U-251/U-373) and of mouse or rat origin (such as GL-261, 9L, BT4C, C6, F98, RG2T9, and RT-2), as well as for PDXs [8, 24].

2   Materials

2.1 Cell Suspension Preparation

1. U-87 MG cells in culture.
2. DMEM/F12 medium: DMEM/Ham's F-12 medium with L-glutamine (365.3 mg/L) and riboflavin (0.05 mg/L), fortified with 10% (v/v) of FBS and $NaHCO_3$.
3. FBS-free DMEM/F12 medium.
4. Trypsin solution.

2.2 Animal Anesthesia

1. Buprenorphine (0.1 mg/kg).
2. Ketamine (10 mg/kg).
3. Xylazine (100 mg/kg).
4. Disposable 1 mL syringes and 25 G needles.
5. Skin disinfectant (Skin Prep 2/70, 2% (w/v) chlorhexidine gluconate in 70% (v/v) isopropyl alcohol).

2.3 Animal Placement on the Stereotaxic Instrument

1. Stereotaxic apparatus with ear bar fixation.
2. Micromotor drill and bits.
3. Electric heating blanket.
4. Physiological monitoring system (SA Instruments SA, Stony Brook, NY, USA).
5. Eye ointment gel (Lubrithal®).

2.4 Surgical Procedure

1. Skin disinfectant (Skin Prep 2/70, 2% (w/v) chlorhexidine gluconate in 70% (v/v) isopropyl alcohol).
2. Sterile cotton swabs.
3. Surgery kit containing microsurgical dissection forceps, scissors and a disposable scalpel.
4. Timer.
5. Hamilton microsyringe (50 μL) with 24 G blunt tip needle.
6. Self-adhesive resin cement (RelyX™ Unicem 2 with Clicker™ Dispenser).

2.5 Surgery Recovery

1. Suture kit containing microsurgical dissection forceps, scissors and a needle holder.

2. Nonabsorbable suture with needle (Supramid DS19 4/0 met 1.5).

3. Heating pad.

2.6 Tumor Imaging with MRI

1. MRI instrument (BioSpec 94/20 USR, Bruker, with Paravision 6.0.1 software) with a volume coil for excitation (with 86/112 mm of inner/outer diameter, respectively) and quadrature mouse surface coil for signal detection.

2. Isoflurane (B. Braun).

3. Mouse bed and heating cover (BRUKER).

4. Physiological monitoring system (SA Instruments SA, Stony Brook, NY, USA).

---

3 Methods

3.1 Cell Suspension Preparation

U-87 MG glioma cells can be cultured in 75 cm^2 culture flasks with DMEM/F12 medium, supplemented with 10% (v/v) of FBS, 100 U/mL of penicillin G sodium and 100 μg/mL of streptomycin sulfate. U-87 MG cells should be incubated and maintained at 37 °C and 5% CO_2 in the previous days of the surgical procedure. On the day of the procedure, the medium is removed from a culture flask, with the cells being dissociated with 4 mL of trypsin, at 37 °C, and incubated for 5 min at 37 °C with 5% CO_2. Following trypsinization, 6 mL of DMEM/F12 medium are added and the culture suspension homogenized, to guarantee a complete detachment of the adherent cells from the flask. The cell suspension is transferred to a 15 mL falcon tube and centrifuged at $280 \times g$ for 5 min at 25 °C. The supernatant is then carefully removed, and the pellet resuspended in a small volume of FBS-free DMEM/F12 medium, at 37 °C (*see* **Note 1**). The cell density can be adjusted through dilution according to experimental needs (*see* **Note 2**). The obtained cell suspension may be kept on ice until further use and is briefly agitated periodically to prevent cell adhesion to the Eppendorf® walls. Trypan blue assay should be previously conducted to maximize the number of procedures per preparation of the cell suspension while not compromising cell viability.

3.2 Animal Anesthesia

Before beginning the surgical procedure (30 min), nude mice are weighted and injected with buprenorphine (0.1 mg/kg) intraperitoneally to minimize pain and postsurgery discomfort, following a standard technique of mouse immobilization (*see* **Note 3**). In order to maintain anesthesia and sedation, ketamine (100 mg/kg) and xylazine (10 mg/kg) must be administered intraperitoneally. The anesthetic effect should be observed within 5 min and is expected to last from 30 to 40 min (*see* **Note 4**).

3.3 Animal Placement on the Stereotaxic Instrument

The maintenance of body temperature during any procedure that requires sedation is critical, as it may delay the animal recovery to consciousness or induce postoperative complications. The animal is placed in prone position over a heating blanket to maintain the body temperature at 37 ± 1 °C, with the mouse incisor teeth kept secure on the bite bar. Once placed in position, each ear is carefully fixed, promoting a complete immobilization of the mouse head and ensuring a correct cell inoculation in the predefined zone of the brain, usually in the striatum, thus avoiding other brain structures (*see* **Note 5**). Body temperature (rectal probe) and respiratory frequency (thorax sensor) are appropriately monitored in real time. In order to prevent corneal drying during surgery, both eyes of the mouse are lubricated, and direct light exposure should be minimized throughout the entire procedure. The anesthetic effect is, once again, confirmed by toe and tail pinching.

3.4 Surgical Procedure

As a standard approach, all material, including the drill bit, surgery kit, and the external surface of the Hamilton® syringe needle, must be sterilized using a suitable device between surgeries. If not feasible due to time limitations, these materials must be disinfected after each animal with 70% (v/v) of ethanol or isopropyl alcohol or using an antibacterial, antiviral, and antifungal sanitizing solution.

Right before the incision, the skin is disinfected with 2% (w/v) chlorhexidine in 70% (v/v) of isopropyl alcohol in a sterile cotton swab, to minimize the risk of infection. The midline scalp incision is then performed with a ten-blade scalpel, having approximately 1.0 cm of length, and the skin retracted. The incision should be as small as possible, to reduce pain and healing time, while allowing a good visualization of the skull sutures. The connective tissue membranes overlying the skull are also cut, exposing both the coronal and sagittal sutures (Fig. 2a). Following the identification of the bregma coordinates $(x, y, z) = (0, 0, 0)$, i.e., the junction of the coronal and sagittal sutures (*see* **Note 6**), the skull is drilled at $(x, y) = (2.1, 0.5)$, corresponding to 2.1 mm right and 0.5 mm anterior to the bregma (*see* **Notes 7** and **8**). After replacing the drill with the Hamilton® syringe, the needle is adjusted to the predefined (x, y) coordinates. Then, the support is rotated, the cell suspension is gently homogenized, and an excess of volume (e.g., 10 µL) is aspirated into the syringe (*see* **Note 9**). The external portion of the needle is carefully wiped with 70% (v/v) of isopropyl alcohol in a sterile cotton swab, thus preventing the contamination of the incision site with tumor cells and, consequently, extracranial tumor growth. The loaded needle in the support is then rotated back to the injection site, and the z coordinate set to 0 (at the base of the skull). Once the predefined depth $(z = -3)$ is achieved, the suspension is slowly injected into the brain (Fig. 2b) (*see* **Note 9**). This method proved to be fast and reduces the residence time of the cells inside the syringe and needle. The injection of the 5 µL of the

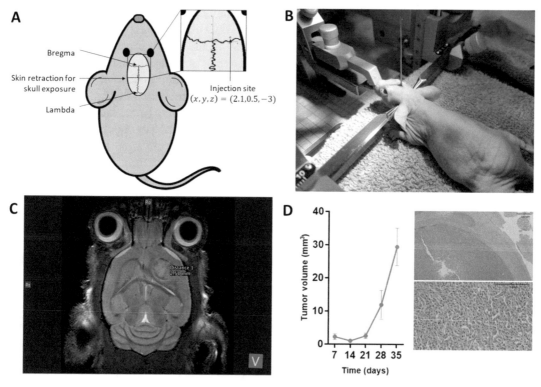

Fig. 2 Overview of the intracranial glioblastoma stereotactic cell injection. (**a**) Identification of cranial sutures (bregma and lambda) and injection site, at $(x, y, z) = (2.1, 0.5, -3)$. (**b**) Intracranial injection of glioblastoma cells following general anesthesia and animal placement on the stereotaxic apparatus, with mouth and ears fixed. A small hole was drilled at the injection (x, y) coordinates. (**c**) Anatomical MR image, at 28 days post implantation, reveals a large cancerous mass within the frontal cortex of the mouse. (**d**) Tumor volume evaluation of glioblastoma-bearing nude mice up to 35 days ($n = 5$), confirmed by histological analysis (H&E staining)

cell suspension is performed slowly, at a rate of 1 μL every 30 s. Once the injection is complete, the operator waits for 2 min before gently withdrawing the needle from the brain (0.5 mm every 30 s), to prevent backflow of the injected cells through the needle tract. The eventual accumulation of fluid should be removed with a sterile cotton swab and the hole closed with self-adhesive resin cement.

3.5 Surgery Recovery

Following the surgical procedure, the animal is placed over a heating pad at 41 °C, the incision closed with two to three simple interrupted stiches and wiped with 2% (w/v) chlorhexidine in 70% (v/v) of isopropyl alcohol in a sterile cotton swab. The mouse is then monitored until it recovers mobility, which may take 30 to 45 min. Buprenorphine (0.1 mg/kg) is readministered 6 h postsurgery and in the following 48 h for pain control. The animal is daily checked to prevent infections, with a complete cicatrization being expected after the stiches fall off around the 3rd–4th day after surgery.

Throughout the experience, the animals must be closely observed on a daily basis, concerning their physical appearance (body weight and coat condition), body function (food intake), presence of loose stools, diarrhea or blood in diarrhea and behavior (handling, aggression, abnormal gait and posture and reluctance to move), as established in [25]. Predetermined human endpoints for animal euthanasia must be also defined (e.g., lack of feeding or diarrhea for more than 48 h and a weight loss superior to 20% of the initial body weight).

3.6 Tumor Imaging with MRI

Tumor growth and dimensions can be assessed by magnetic resonance imaging with a 9.4 T Bruker device (*see* **Note 10**). The mouse is anesthetized with 4% of isoflurane gas, mixed with medical oxygen at a rate of 1 L/min for 3 min. The animal is then transferred to the MRI bed (previously warmed to maintain body temperature) and given maintenance anesthesia during the image acquisition (2% isoflurane gas, at 1 L/min). After placing the body temperature and respiratory sensors, the magnet of the 12 mm diameter transmit coil is positioned over the center of the head, thus enhancing the signal to noise ratio. For structural analysis, T2-weighted images are acquired in sagittal, coronal and axial planes, using a rapid acquisition with relaxation enhancement sequence (T2 Turbo RARE), defined with the following parameters: TR = 3800 ms, TE = 33 ms, 10 averages, pixel size of 0.078 mm × 0.078 mm and slice thickness of 0.5 mm, without spacing between slices (total head volume, 256 pixels × 256 pixels × 34 slices). Sagittal plane with 1 average (2 min), coronal with 3 averages (4 min) and axial plan with 10 averages (20 min). After the image acquisition, the isoflurane is removed and the animal kept with medicinal oxygen in the MRI heated bed until it awakes. Finally, the animal is transferred to a recovery cage. The image acquisition protocol is repeated weekly and the tumor diameter evaluated (*see* **Note 11**).

4 Notes

1. In order to evaluate cell density and viability, the operator should appropriately dilute an aliquot of the suspension with medium and add an equal volume of a 0.4% (w/v) trypan blue solution. Cell counting should be performed in a hemocytometer according to in-house protocols. Cell density should only consider live cells and is then calculated according to (Eq. 1):

$$\text{Cell density (cells/mL)} = \text{Mean cell count per set} \times 10^4 \times \text{Dilution Factor} \tag{1}$$

As an alternative, patient-derived tumor spheres cultured with stem cell medium (100 mL of DMEM/F12 medium, 1 mL of penicillin (5000 U/mL)/streptomycin (5 mg/mL)/neomycin (10 mg/mL) solution (100× stock), 2 mL of B-27 supplement (50× stock), added right before use with 100 μL of EGF (20 ng/mL, 1000× stock) and 100 μL of FGF2 (10–20 ng/mL, 1000× stock) may be used. Following a 5-min centrifugation at 200 × g, the supernatant is removed and the pellet covered with a cell detachment solution of proteolytic and collagenolytic enzymes at room temperature. The cells are then diluted in ice-cold stem cell medium and the cell density adjusted accordingly.

2. A volume of injection higher than 5 μL is not encouraged, due to a considerable increase in intracranial pressure and consequent significant tissue damage. Cell density and injection volume may range from 30,000 to 500,000 cells and 1–5 μL. The aqueous media described in literature is also variable, being FBS-free DMEM or PBS the most commonly described. Before doing the surgical procedure, the authors recommend the conduction of preliminary tests to better understand the inoculation and growth rate of the tumor.

3. The mouse is immobilized with the left thumb and forefinger by grabbing the loose skin on the back of the neck, as close to the ears as possible. With the mouse lying on its back within the palm of the hand, the tail is restrained using the little finger and an intraperitoneal injection is performed on the right side of the abdomen, thus avoiding the cecum. The injection volume for an intraperitoneal administration can be estimated through (Eq. 2):

$$V_{inj} \ (mL) = \frac{Weight \ (kg) \times Dose \ (mg/kg)}{Concentration \ of \ working \ solution \ (mg/mL)} \quad (2)$$

A dose of 0.1 mg/kg of Buprenorphine is recommended for procedures that induce severe pain [26]. Consider the dilution to a working solution of 0.01 mg/mL, in saline, prepared from a commercial buprenorphine medicine (e.g., 0.3 mg/mL).

4. For a long-induced anesthesia, ketamine (100 mg/kg) and xylazine (10 mg/kg) may be used [26]. Consider diluting a commercial ketamine medicine (e.g., 100 mg/mL) to a working solution of 20 mg/mL and xylazine (e.g., 20 mg/mL) to a working solution of 2 mg/mL, before premixing in a syringe, and administer intraperitoneally (maximum volume for intraperitoneal administration should not exceed 2 mL). As an alternative, isoflurane may be used as general anesthetic drug (oxygen flow of 1.0 L/min, induction (4% of isoflurane); maintenance (2% of isoflurane)).

The anesthetic effect should be assessed before positioning the animal on the stereotaxic apparatus. Any reaction to noxious stimuli, such as toe and tail pinch, may indicate a light anesthesia. If the expected time to sedation was exceeded, 25% of the initial dose of ketamine and xylazine may be readministered. In a more extreme situation, if the animal is starting to gain mobility, the authors suggest a 50% of the initial dose of ketamine and xylazine administration. During the procedure and recovery, vital signs (such as heartbeat, respiratory rate and depth, body temperature) and reflexes (such as toe and tail pinch, palpebral moves) should be monitored.

5. Several inbred strains and C57BL/6 mice skulls have a 15° inclination relative to the brain, which may promote pronounced mistargets in deeper regions of the brain. This should be taken into consideration when placing the animal onto the apparatus.

6. The identification of bregma is not often simple, due to non-linear and unclear cranial sutures. If necessary, the authors recommend a visual definition of parallel lines that closely approximate to the sutures, thus reducing bregma definition variability. In addition, the approximate angle of these imaginary lines should be kept sensibly fixed. Due to the low thickness of the skull, avoid pressuring with metal objects, which may lead to deformation and mistargeting.

7. Cells are injected into the striatum of the frontal cortex of the mouse, in a vascularized region where glioblastoma commonly develops in humans. The coordinates, in relation to bregma, are $(x, y, z) = (2.1, 0.5, -3.0)$ (Fig. 3), although different approaches may be possible.

8. During drilling, there is a risk of brain injury, as the thickness of the mouse skull in only about 0.3 mm. The skull should be perforated slowly and alterations on drill noise or in trepidation may indicate a complete perforation of the bone. The hole should be as small as the diameter of the Hamilton syringe needle, in order to minimize cells and fluid extravasation.

9. The Hamilton syringe may contain air bubbles, which can be hard to remove. A higher injection volume can promote a better visualization of the meniscus and assures a correct volume injection. After the injection, the remaining 5 µL should be discarded. After reaching the z coordinate, consider inserting the needle up to 10% of the predefined depth, in order to allow a better deposition of cells and retract back to the defined z coordinate before injection.

10. As an alternative, luciferase-expressing glioblastoma cells may be monitored by bioluminescence, a noninvasive technique based on the emission of visible light after the oxidative

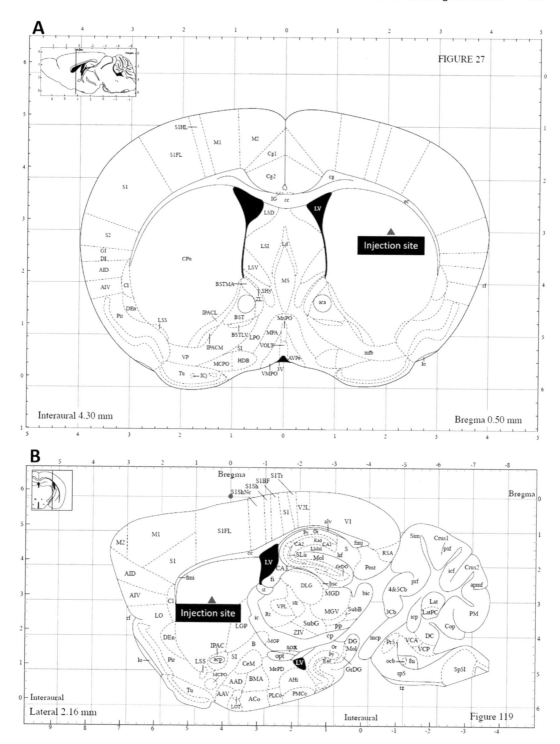

Fig. 3 Identification of the injection site at Cartesian coordinates (x, y, z) = (2.1, 0.5, −3), in relation to bregma, in a coronal (**a**) and sagittal (**b**) view. This location corresponds to the striatum of the frontal cortex of the mouse, according to the Paxinos & Franklin Mouse Brain Atlas. (Adapted from Ref. [27], with permission from Academic Press, Elsevier)

carboxylation of luciferin by luciferase. This technique is cheaper than MRI, may be done without the assistance of a well-trained technician and is not time consuming. However, it requires the use of a luciferase-expressing cell line (thus not being applicable to PDXs), an intraperitoneal administration of luciferin before the analysis, and the risk of loss of luciferase expression (resulting in an underestimation of the number of cancer cells) should not be discarded. A pointed pitfall of this technique relies on the time-dependent emission of light, since consecutive measurements after the administration of luciferin reflect different signals and tumor dimensions. As a consequence, a comparative analysis between mice/time points should be conducted at the same predetermined time (maximal BLI signal) after the luciferin administration. For a detailed protocol of bioluminescence imaging of brain tumors, the reader is referred to [28].

11. The MR images may be saved as Digital Imaging and Communications in Medicine files (DICOM) and manually segmented in the 3D Slicer software for 3D reconstruction.

5 Conclusions

Glioblastoma remains one of the most aggressive and problematic cancers to treat, due to the limitations of the standard treatment approaches. Therefore, there is an increasing need in developing novel therapeutic strategies that aim at circumventing drug delivery and resistance mechanism issues. The development of preclinical in vivo glioblastoma models to assess the efficacy of new therapies is crucial to allow the translation to clinical trials and consequent distribution in clinic. This chapter may be a valuable tool for unexperienced researchers in stereotaxic surgery and glioblastoma mice model development, by detailing an easy step-by-step protocol, with increased reproducibility and minimal animal losses.

Acknowledgments

João Basso acknowledges the PhD research grant SFRH/BD/149138/2019 assigned by Fundação para a Ciência e a Tecnologia (FCT), the Portuguese Agency for Scientific Research. The authors also acknowledge FCT for the financial support through the Research Projects no. 016648 (Ref. POCI-01-0145-FEDER-016648), Pest UID/NEU/04539/2013, COMPETE (Ref. POCI-01-0145-FEDER-007440), and CENTRO-01-0145-FEDER-030752 and the Coimbra Chemistry Centre, supported by FCT, through the Project UID/QUI/00313/2020.

References

1. Louis DN, Perry A, Reifenberger G, von Deimling A, Figarella-Branger D, Cavenee WK, Ohgaki H, Wiestler OD, Kleihues P, Ellison DW (2016) The 2016 World Health Organization classification of tumors of the central nervous system: a summary. Acta Neuropathol 131(6):803–820. https://doi.org/10.1007/s00401-016-1545-1

2. Ohgaki H, Kleihues P (2005) Population-based studies on incidence, survival rates, and genetic alterations in astrocytic and oligodendroglial gliomas. J Neuropathol Exp Neurol 64(6):479–489. https://doi.org/10.1093/jnen/64.6.479

3. Zlatescu MC, TehraniYazdi A, Sasaki H, Megyesi JF, Betensky RA, Louis DN, Cairncross JG (2001) Tumor location and growth pattern correlate with genetic signature in oligodendroglial neoplasms. Cancer Res 61(18):6713–6715

4. Ohgaki H, Kleihues P (2013) The definition of primary and secondary glioblastoma. Clin Cancer Res 19(4):764–772. https://doi.org/10.1158/1078-0432.CCR-12-3002

5. Yan H, Parsons DW, Jin G, McLendon R, Rasheed BA, Yuan W, Kos I, Batinic-Haberle I, Jones S, Riggins GJ (2009) IDH1 and IDH2 mutations in gliomas. N Engl J Med 360(8):765–773. https://doi.org/10.1056/NEJMoa0808710

6. Nobusawa S, Watanabe T, Kleihues P, Ohgaki H (2009) IDH1 mutations as molecular signature and predictive factor of secondary glioblastomas. Clin Cancer Res 15(19):6002–6007. https://doi.org/10.1158/1078-0432.CCR-09-0715

7. Davis ME (2016) Glioblastoma: overview of disease and treatment. Clin J Oncol Nurs 20(5):S2–S8. https://doi.org/10.1188/16.CJON.S1.2-8

8. Miyai M, Tomita H, Soeda A, Yano H, Iwama T, Hara A (2017) Current trends in mouse models of glioblastoma. J Neurooncol 135(3):423–432. https://doi.org/10.1007/s11060-017-2626-2

9. Basso J, Miranda A, Sousa J, Pais A, Vitorino C (2018) Repurposing drugs for glioblastoma: from bench to bedside. Cancer Lett 428:173–183. https://doi.org/10.1016/j.canlet.2018.04.039

10. Ashburn TT, Thor KB (2004) Drug repositioning: identifying and developing new uses for existing drugs. Nat Rev Drug Discov 3(8):673–683

11. Zhou J, Atsina K-B, Himes BT, Strohbehn GW, Saltzman WM (2012) Novel delivery strategies for glioblastoma. Cancer J 18(1):89–99. https://doi.org/10.1097/PPO.0b013e318244d8ae

12. Zottel A, Videtič Paska A, Jovčevska I (2019) Nanotechnology meets oncology: nanomaterials in brain cancer research, diagnosis and therapy. Materials 12(10):1588. https://doi.org/10.3390/ma12101588

13. Basso J, Miranda A, Nunes S, Cova T, Sousa J, Vitorino C, Pais A (2018) Hydrogel-based drug delivery nanosystems for the treatment of brain tumors. Gels 4(3):62. https://doi.org/10.3390/gels4030062

14. Sulaiman A, Wang L (2017) Bridging the divide: preclinical research discrepancies between triple-negative breast cancer cell lines and patient tumors. Oncotarget 8(68):113269–113281. https://doi.org/10.18632/oncotarget.22916

15. Namekawa T, Ikeda K, Horie-Inoue K, Inoue S (2019) Application of prostate cancer models for preclinical study: advantages and limitations of cell lines, patient-derived xenografts, and three-dimensional culture of patient-derived cells. Cells 8(1):74. https://doi.org/10.3390/cells8010074

16. Williams J (2018) Using PDX for preclinical cancer drug discovery: the evolving field. J Clin Med 7(3):41. https://doi.org/10.3390/jcm7030041

17. Li G (2015) Patient-derived xenograft models for oncology drug discovery. J Cancer Metast Treat 1(1):8–15. https://doi.org/10.4103/2394-4722.152769

18. Khan FR, Henderson JM (2013) Deep brain stimulation surgical techniques. In: Handbook of clinical neurology, vol 116. Elsevier, Amsterdam, pp 27–37. https://doi.org/10.1016/B978-0-444-53497-2.00003-6

19. Wu M, Shu J (2018) Multimodal molecular imaging: current status and future directions. Contrast Media Mol Imaging 2018

20. Janib SM, Moses AS, MacKay JA (2010) Imaging and drug delivery using theranostic nanoparticles. Adv Drug Deliv Rev 62(11):1052–1063

21. Xia J, Wang LV (2013) Small-animal whole-body photoacoustic tomography: a review. IEEE Trans Biomed Eng 61(5):1380–1389

22. Moses WW (2011) Fundamental limits of spatial resolution in PET. Nucl Instrum Methods Phys Res A 648(supplement 1):S236–s240. https://doi.org/10.1016/j.nima.2010.11.092

23. Jost SC, Collins L, Travers S, Piwnica-Worms D, Garbow JR (2009) Measuring brain tumor growth: combined bioluminescence imaging-

magnetic resonance imaging strategy. Mol Imaging 8(5):245–253

24. Lenting K, Verhaak R, Ter Laan M, Wesseling P, Leenders W (2017) Glioma: experimental models and reality. Acta Neuropathol 133(2):263–282

25. Commission E (2010) Directive 2010/63/ EU of the European Parliament and of the council of 22 September 2010 on the protection of animals used for scientific purposes. Off J Eur Union 276:33–79

26. Costa A, Antunes L (2011) Handbook of laboratory animals: mice, rats and rabbits, vol 419. Série didáctica Ciências Aplicadas. Universidade de Trás-os-Montes e Alto Douro, Vila Real, Portugal

27. Paxinos G, Franklin KB (2001) Paxinos and Franklin's the mouse brain in stereotaxic coordinates, 2nd edn. Academic Press, Cambridge, Massachusetts

28. Frenster JD, Placantonakis DG (2018) Bioluminescent in vivo imaging of Orthotopic glioblastoma xenografts in mice. In: Placantonakis DG (ed) Glioblastoma: methods and protocols. Springer New York, New York, NY, pp 191–198. https://doi.org/10.1007/978-1-4939-7659-1_15

INDEX

Vivek Agrahari et al. (eds.), *Nanotherapy for Brain Tumor Drug Delivery*, Neuromethods, vol. 163,
https://doi.org/10.1007/978-1-0716-1052-7, © Springer Science+Business Media, LLC, part of Springer Nature 2021

Printed in the United States
by Baker & Taylor Publisher Services